MAKING THINGS INTERNATIONAL 1

MAKING THINGS INTERNATIONAL
Mark B. Salter, Editor

1. Circuits and Motion
2. Catalysts and Reactions

Making Things International 1

Circuits and Motion

Mark B. Salter, Editor

University of Minnesota Press
Minneapolis • London

"Symptoms" was previously published as John Law and Wen-yuan Lin, "Making Differently: On 'Modes of International,'" *CRESC Working Paper* 129 (Manchester and Milton Keynes: Centre for Research on Socio-Cultural Change, August 2013).

Copyright 2015 by the Regents of the University of Minnesota

All rights reserved. No part of this publication may be reproduced, stored in a retrieval system, or transmitted, in any form or by any means, electronic, mechanical, photocopying, recording, or otherwise, without the prior written permission of the publisher.

Published by the University of Minnesota Press
111 Third Avenue South, Suite 290
Minneapolis, MN 55401–2520
http://www.upress.umn.edu

Library of Congress Cataloging-in-Publication Data
Making things international 1. Circuits and motion / edited by Mark B. Salter.
Includes bibliographical references.
   ISBN 978-0-8166-9625-3 (hc)
   ISBN 978-0-8166-9626-0 (pb)
1. International relations—Philosophy. 2. Materialism—Philosophy.
3. World politics—21st century. I. Salter, Mark B. II. Title: Circuits and motion.
   JZ1319.M37   2015
   327.101—dc23

2014046648

Printed in the United States of America on acid-free paper

The University of Minnesota is an equal-opportunity educator and employer.

21 20 19 18 17 16 15   10 9 8 7 6 5 4 3 2 1

Contents

Introduction: Circuits and Motion   vii
   Mark B. Salter

Part I. World in Motion

Electronic Passports   3
   William Walters and Daniel Vanderlip

Passport Photos   18
   Mark B. Salter

The Traffic Light   36
   Katherine Reese

AVATAR   49
   Benjamin J. Muller

Containers   62
   Can E. Mutlu

Bicycle   72
   Oded Löwenheim

Boats   85
   Geneviève Piché

Ballast   98
   Charlie Hailey

Part II. Bodies in Motion

Symptoms   115
   John Law and Wen-yuan Lin

Corpses   129
   Jessica Auchter

Virus   141
   Melissa Autumn White

Microbes 156
Stefanie Fishel

Breathless 171
Peter Adey

Blood 184
Jairus Grove

Bodies 201
Lauren Wilcox

Tanks 212
Michael J. Shapiro

Drones 222
Joseph Pugliese

Part III. Things in Motion

MemeLife 243
Kathleen P. J. Brennan

Videos 255
Rune Saugmann Andersen

Garbage 266
Michele Acuto

Carbon 282
Chris Methmann and Benjamin Stephan

Currency 298
Emily Gilbert

Biometric MasterCard 311
Elizabeth Cobbett

Cocaine 328
Mike Bourne

Clocks 348
Yvgeny Yanovsky

Acknowledgments 364

Contributors 365

Introduction
Circuits and Motion

Mark B. Salter

The international, the globe, the world is made up of things, of stuff, of objects, and not simply of humans and their ideas. Our collection looks at how the international is assembled by the enrollment of objects, humans, and ideas. Things play a crucial role in the assemblage of the international. Borders are made with fences, maps, compasses, passports, guards, and gates. War is made with guns, cell phones, improvised explosive devices, helmets, depleted uranium, aircraft, satellites, electricity, meals ready to eat, and oil. Diplomacy is made by telegrams, the Internet, diplomatic pouches, chicken dinners, and cameras. The international economy depends on real and virtual currency, transatlantic cables, weather reports, insurance tables, commodities, ships, and trucks, which in turn depend on pipes, petroleum, steel, and aluminum. Each of these modes of organizing international life can exist only when a set of things, actors, and ideas circulate in particular patterns. How do we change our androcentric thinking to get at the emergent, complex set of circuits that produces our messy world? We put things first, and engage in a method of radical openness, letting human and nonhuman actants demonstrate their capacity for agency for making the international. By thinking through the international with things, we flatten our ontology, which has the effect of displacing the political subject as the star of the project of modernity. *Making Things International* aims to provoke readers and to act as a catalyst in this emerging conversation between new materialism and international relations.

Making Things International insists that the international has something to say back to new materialism, just as new materialism is our catalyst for thinking differently about international relations. In the extant new materialist literature, there is a renewed focus on differential scales: the atomic, the cellular, the bodily, the ecological, the global, the planetary, and so on. To this menu we add the international. Fresh work by William Walters, Jairus Grove, and Peer Schouten indicate the radical potential of this way of thinking, as do the recent forum in *International Political Sociology* and the special issue of *Millennium:*

Journal of International Studies on materialism and world politics.[1] By focusing on the material assemblage of political claims, Walters demonstrates how technics and scientific explanation connect to controversies: the very technical analysis of "missile fragments, rubble, human remains and witness reports" assembled by Human Rights Watch in the aftermath of the 2008–2009 Gaza military offensive made possible some very powerful political claims about responsibility and accountability.[2] Grove similarly charts an "ecology of global warfare" by tracing the different materials (abandoned ordinance), technologies (cheap cell phones), and techniques (know-how) of the improvised explosive device.[3] The emphasis on materiality, both he and Schouten argue, can counter the discourse- and speech-heavy analysis of much contemporary critical international relations.[4] For this emerging debate, the material enriches and challenges our current understandings of the international and of politics itself.

There is a real utility to flattening the ontology of the international sphere in all of its objects of interest: violence, economy, culture, environment, identity, the everyday. Environmental regimes cannot be understood without giving agency to the nonhuman actants that make up the biosphere. Global economic relations cannot be understood without reference to the independent agency of algorithms that act too quickly for human oversight or interference. The economy is not an external object but a set of assumptions, processes, and practices.[5] Security cannot be understood solely as a set of speech-acts but also requires guns, tanks, drones, tear gas, badges, and fences. In each of these areas, there are nonhuman actants that fundamentally alter the condition of human possibility in ways that are unpredictable and irreducible to their constituent elements.

The world is in motion; things circulate. In this volume, we highlight the circulation of things in ways that draw equal attention to human and nonhuman actants. In part, this focus on circulation and mobility speaks to an ongoing conversation in geography about mobility.[6] Peter Adey was instrumental in setting this agenda in my own work from his very early publication on the circulation of things and people at airports.[7] Michel Foucault highlights a deep connection between the evolution of sovereignty and the management of circulation. In *Security, Territory, Population* (1977–78), Foucault describes how sovereignty, discipline, and security as different modalities of control engage with space: "Baldly, at first sight and somewhat schematically, we could say that sovereignty is exercised within the borders of a territory, discipline is exercised on the bodies of individuals, and security exercised over a whole population."[8] He relates these modes to the organizations of the town, as a particular material instantiation of these forms of power: "In other words, [sovereignty] was a matter of organizing circulation, eliminating its dangerous elements, making a division between good and bad circulation, and maximizing the good circulation by diminishing the bad. It was therefore also a matter of planning access to the outside."[9] Whereas discipline attempts to structure the performance and behavior of individual bodies with the minimum application of direct power, security is about the management of the aleatory or the uncertain. Security is also concerned with circulation: not just the circulation of materials and bodies but also the circulation of risk and uncertainty. From this perspective, both security and sovereignty as modes of governance are intimately connected with the management of circulation, and in our case particularly international circulation. To understand sovereignty, security, and circulation, we must understand how things and people move, and how they are made to move in which particular circuits.

Interventions in *Making Things International 1: Circuits and Motion* will highlight how the international is created as things move—enabling and restricting different circuits and flows. For example, in Part I, "World in Motion," William Walters and Daniel Vanderlip, Mark B. Salter, Katherine Reese, and Benjamin J. Muller all look at how particular technologies and objects facilitate or structure the mobility of individuals, how borders filter passages, and how traffic flows are managed. Can E. Mutlu, Oded Löwenheim, Geneviève Piché, and Charlie Hailey analyze particular mobility platforms for the circulation of goods and persons, through containers, bicycles, and boats. Hailey's discussion of how ballast circulates, for example, links trade to urban planning and ecology, but ballast is also necessary for the way that ships move. In the second section, "Bodies in Motion," John Law and Wen-yuan Lin look at how the material facts of European maritime expansion can lead to thinking the international differently, using a counterfactual history of Chinese maritime expansion and Chinese ways of conceptualizing health. This focus on different ways of constituting and considering circulation in the body leads to Jessica Auchter, Melissa Autumn White, Stefanie Fishel, Peter Adey, Jairus Grove, and Lauren Wilcox, who discuss the body as a site of material circulation in itself. Breath, blood, microbes, and viruses all circulate within the live body, and Auchter looks at the circulation of corpses and the particular value attached to the dead body. Wilcox figures how the corporeal experience of targets is made by drones circulating above. Adey examines the constitution of the atmosphere through breath and being breathless. Fishel and White look at the international circulation of microbes and viruses. Grove traces a fascinating story about the literal circulation of blood globally. Michael J. Shapiro and Joseph Pugliese analyze two war machines that turn some bodies into corpses: tanks and drones. Shapiro figures the role of the tank in war films through the device of visuality, whereas Pugliese looks at the bodies that are imbricated in the global drone network. In the third section, "Things in Motion," Kathleen P. J. Brennan and Rune Saugmann Andersen look at how particular representations—memes and videos—travel through specific technological circuits. Michele Acuto focuses on the circulation of garbage as the necessary material surplus of our way of life, and Chris Methmann and Benjamin Stephan trace the circulation of carbon. Emily Gilbert and Elizabeth Cobbett connect the circulation of credit with its material traces, both currency and credit cards. Mike Bourne looks at the circulation of cocaine and its attendant assemblage of knowledge and the illicit. Yvgeny Yanovsky ends this section by demonstrating the way that clocks enable the technologization and standardization of time and its keeping. Each of these chapters engages differently with the questions of the material and the sovereign, but all together they provide a unique and provocative view of the international as constituted through the management and circulation of things.

Making Things International 2: Catalysts and Reactions will follow, demonstrating how material objects can serve as catalysts for understanding the entanglement of local, global, and planetary scales. That volume will examine cars, benches, jumpsuits, tents, phones, dirt, and other unexpected material foci for the exercise of sovereignty and the materialization of the international.

Inspirations

I want to point to some common touchstones by way of introduction, particularly given that many of our inspirational texts do not come from the discipline of international relations (IR) or speak precisely to the international. New materialist analyses often focus on very specific locations, and on specific assemblages of human and nonhuman actants. For example, one of the largest influences on actor-network theory (ANT) is *Laboratory Life: The Social Construction of Scientific Facts* (1979), by Bruno Latour and Steve Woolgar, which mapped out how knowledge claims were arrived at and transmitted at the Salk Institute for Biological Studies.[10] Displacing a divide between creative, active subjects and inert, inactive objects, Latour and Woolgar complicate the distinction between the "technical and social"[11] through an anthropological study of this very particular and precise setting. Things, they claim, are not "pregiven . . . merely await[ing] the timely revelation of their existence by scientists. . . . Rather, objects (in this case substances) are constituted through the artful creativity of scientists."[12] Working from this example, more studies of particular sites of the constitution of ideas and objects emerged: Michel Callon's study of scallops in Saint Brieuc Bay and the electric car, John Law's work on maritime expansion of the Portuguese and on the TSR2 aircraft, and Bruno Latour's history on Pasteur, culminating in this early stage with Latour's *Science in Action: How to Follow Scientists and Engineers through Society* (1987).[13] More unconventionally, these scholars have also examined sheep, pumps, and train accidents.[14] Annemarie Mol has examined the way the human body itself and, in particular, diseases like atherosclerosis are figured.[15] Two more recent texts, *Actor Network Theory and After* (1999), edited by John Law and John Hassard, and Bruno Latour's *Reassembling the Social: An Introduction to Actor-Network-Theory* (2005), refine the tools and concepts used by this group of scholars.[16] In general, these scholars share a sensibility that the investigator should take the world as it is, rather than try to draw conclusions through a prechosen theory about what the appropriate actors and relationships, structures and institutions, or even levels of analysis are. In particular, scholars must be open to the idea that objects or nonhuman actants might exercise as much agency as humans: "ANT has tried to render the social world as *flat* as possible in order to ensure that the establishing of any new link is clearly visible," that is, in order to open up the "black box" of its object.[17] Without engaging in the definitional or disciplinary skirmishes of ANT, we can identify key concepts that are frequently used.

Callon introduces the concept of "translation" to explore how "the identity of actors, the possibility of interaction and the margins of maneuver are negotiated and delimited."[18] Thus, any field is made up of those elements which identify themselves and are identified as interacting, often crossing assumed professional, local, legal, or international boundaries.[19] As these interactions evolve, different elements are deployed or arrayed to be included in the field by "interessements," or "those attempts to impose and stabilize the identity of the other actors"; and "enrollment" is that process by which "the group of multilateral negotiations, trials of strength and tricks that accompany the interessements and enable them to succeed" are included.[20] "Negotiations, trials of strength and tricks" are well within the competency of IR. This process, in Callon's view, yields a particular assemblage that comes to include different social and material actors. In his analysis of the scallop-conservation and fishing practices in Saint Brieuc Bay, he argues that the scallops themselves must be enrolled in conservation efforts. Although the scallops do not possess rationality or intentionality, the system of scallop conservation and farming cannot exist without the scallops' acting in relation to the tide, their starfish predators, the ocean currents, and the various human actors, such as fisherfolk and scientists. They do not have human agency, yet they do have a kind of scallop-agency to endorse or to disrupt the best-laid plans of scientists and fishers. Callon and John Law argue that "often in practice we bracket off nonhuman materials, assuming they have a status which differs from that of a human. So materials become resources or constraints; they are said to be passive; to be active only when they are mobilized by flesh and blood actors. But if the social is really materially heterogeneous then this asymmetry doesn't work very well. Yes, there are differences between conversations, texts, techniques and bodies. Of course. But why should we start out by assuming that some of these have no active role to play in social dynamics?"[21] Or, as Callon expresses the methodological doctrine of "free association: the observer must abandon all *a priori* distinctions between natural and social events. . . . [I]nstead of imposing a pre-established grid of analysis on these, the observer follows the actors to identify the manner in which these [divisions] define and associate the different elements by which they build and explain their world, whether it be social or natural."[22] The ANT sensibility is thus one of careful ethnography that attempts to be open to the agency of all kinds of human and nonhumans.

Clearly, there is a fundamentally new notion of agency and causation in this model. Whereas Callon talks about how both the fisher and the scallops are represented within this political forum, Jane Bennett uses a different metaphor—collaboration: "An actant never really acts alone. Its efficacy or agency always depends on the collaboration, cooperation, or interactive interface of many bodies and forces."[23] In acknowledging that the human is never independent from the material—and in fact that the human itself is material—we are inclined to see any social field or action as an assemblage of different forces. The term *assemblage* has found a large number of users, inspired by Gilles Deleuze and Félix Guattari and others. Bennett again: "[A]ssemblages are ad hoc groups of diverse elements, of vibrant materials of all sorts. Assemblages are living, throbbing confederations that are able to function despite the persistent presence of energy that confound them from within. . . . The effects generated by an assemblage are . . . emergent properties, emergent in that their ability to make something happen is distinct from the sum of the vital force of each materiality considered alone."[24] Using an empirical example of the 2003 Northeast power failure, Bennett demonstrates that the effect of a regional blackout affecting 50 million households was the result of an assemblage of regulatory changes, market incentives, and profit motives, individual and computerized decision models, electron flows, weather, and the capacities of the wires themselves. In painting this complex picture, she resists a reductionist view of causation or responsibility, which runs precisely at odds with the juridical model of assigning fault and liability. Traditional models of social science attempt to find necessary and sufficient conditions, which William E. Connolly labels a model of "efficient causality."[25] Instead, Bennett argues that "causality is more emergent than efficient, more fractal than linear.... [I]f efficient causality seeks to rank the actors involved, treating some as external causes and others as dependent effects, emergent causality places the emphasis on the process as itself an actant, as itself in possession of degrees of agentic capacity."[26] In each of these assemblages, agency is distributed among various material and human forces that operate independently but also have cooperative effects that are greater than the sum of their parts, in a way that is sensible but not predictable and is always driven from the open, active, engaged analysis of the forum or the problem without a bias toward a prior scale of analysis or assumption about the agentic capacity of humans or materials.

The International Challenge

In engaging with two rich and provocative edited collections, Bruno Latour and Peter Weibel's *Making Things Public: Atmospheres of Democracy* (2005) and Stephan Harrison, Steve Pile, and Nigel Thrift's *Patterned Ground: Entanglements of Nature and Culture* (2004), I noted a scale-of-analysis bias toward either the local or the global, without pause at the level of the international.[27] Latour's provocation, in *Reassembling the Social*, is that "objects—taken as so many issues—bind all of us in ways that map out a public space profoundly different from what is usually recognized under the label of 'the political.' It is this space, this hidden geography that we wish to explore."[28] In many of the contributions in these two collections, the modernist assumption is perhaps unconsciously repeated that politics is contained primarily within the sovereign state or at the level of humanity, rather than in the interaction between sovereign states or in the daily practices of sovereignty. One productive avenue might be to examine the things that are peculiarly international: treaties, border posts, aid packages, or the weapons claims made at the United Nations. In this collection, for example, Walters and Vanderlip, Muller, and Salter all look at some of the things involved in constructing and policing borders, and monitoring the flows across them. Piché and Mutlu look at boats and containers, respectively, to understand how things enable the mobility of other things, people, and governance structures. This lack of concern with the international as a level of analysis can be seen also in Latour.

Latour, for example, points to the abuse of facts in Colin Powell's presentation to the United Nations Security Council. International relations is really helpful here in providing the complex analysis of international politics and the bureaucratic politics of the Security Council. Powell's authority to make his claims as facts is not independent from international power structures: not all secretaries of state are equal at the United Nations. An exclusively object-oriented analysis might start with the state as analytically weak, just as an object-free analysis might focus exclusively on the state. This absence of the international is found in much ANT-inspired work. So, IR benefits from ANT, just as ANT benefits from IR.

Patterned Ground provided a model for the genre of *Making Things International*, inviting "authors to write short essays that discussed one element in the landscape that they [were] particularly passionate about [in order to unsettle] prevalent understandings of the relationship between nature and culture."[29] Inevitably, because geographic focus is the aim, Harrison, Pile, and Thrift's collection does not engage the question of the international systematically, although within the 108 chapters there are some provocative statements that engage the global. Anssi Paasi argues, in "Territory," that the forces of globalization lead to heterogeneous dynamics of deterritorialization and reterritorialization; Mark Gottdiener, in "Airports," discusses the airport as a unique space that connects mobile populations on the micro and macro scales. However, what is almost completely absent from *Patterned Ground* is consideration of the dynamic of sovereignty itself, which this collection seeks to remedy.

We argue here that sovereignty is an important and unique way of conceptualizing the limit of politics, not simply as the source of law and authority but also as the way of organizing multiple and diverse polities. The interaction of different communities may transcend "borders, boundaries and limits," but those international assemblages are fundamentally different from global assemblages or national assemblages because of that crossing.[30] International relations, and international political sociology in particular, is uniquely suited to plotting, tracing, and following these international assemblages.

The discipline of international relations has long been concerned with the way that material and geographic factors influence global politics and, indeed, constitute a unique social realm of the international. As a field, IR has been particularly concerned with the interaction between states, often through relations of violence, diplomacy, and trade. International political economy, as a subfield, takes resources seriously, and often foregrounds the international relations constituted by those material circuits, such as oil and other hydrocarbons, uranium and other fissile materials, cotton, sugar, and spices.[31] Strategic studies foregrounds the role and circulation of technologies, particularly weapons, such as nuclear missiles, land mines, biological weapons, and small arms.[32] The work of peace studies and strategic studies often involves the tracking and circulation of particular arms or weapons platforms, components, or precursors (nuclear materials, biological weapons, or particular technologies). Other disciplines investigate the migration of WMD toward nonstate actors, which has been identified as a major threat vector since 9/11. But these traditional approaches start from the assumption that the state should be the primary level of analysis. They articulate the influence of materials through the unit of the state: how states regulate or compete, how they interact and form global regimes, or how substate actors interact with states. This statist approach, even when it engages with the material elements of international politics, methodologically separates international politics from these global materials. This collection represents the value added of a materialist approach, which does not presume a separation between the technical and the material and the political.

We can point to some precursors in IR: Timothy Mitchell's *Carbon Democracy: Political Power in the Age of Oil* (2011), and Matthew Paterson's *Automobile Politics: Ecology and Cultural Political Economy* (2007).[33] Mitchell argues through his persuasive history of oil that the form of energy available has had a radical effect on the political culture. In effect, oil conditions the limits of possibility for our contemporary configuration of sociopolitical power. He argues that "democratic claims for a more egalitarian collective life [in the nineteenth century] were advanced through the flow and interruption of supplies of coal. In the second half of the twentieth century, governments sought to weaken this unusual power that workers had acquired by an equally simple engineer project: switching from using coal to using gas."[34] Mitchell focuses on the role of oil in disputes or controversies. For Mitchell and others working in this materialist tradition, such as Latour, there is no separation between the social and the material: "As the kinds of controversies we face [such as climate change and peak oil] show, our world is an entanglement of technical, natural and human elements. . . . In introducing technical innovations, or using energy in novel ways, or developing alternative sources of power, we are not subjecting 'society' to some new external influence, or conversely using social forces to alter an external reality called 'nature.' We are reorganizing socio-technical worlds, in which what we call social, natural and technical processes are present at every point."[35] Mitchell is not alone in giving oil a much more active role in the shaping of politics. Paterson's *Automobile Politics*, Paterson and Simon Dalby's "Empire's Ecological Tyreprints" (2006), and David Campbell's "The Biopolitics of Security" (2005) highlight the capacity of the carbon market or the circulation of oil or the culture of automobility to dramatically affect global relations, without reducing all global action to the level or interest of the state.[36] For example, Campbell argues that "the SUV has come to underpin U.S. dependence on imported oil. This dependence in turn underpins the U.S. strategic interest in global oil supply, especially in the Middle East, where the American military presence has generated such animus. As a result, the SUV symbolizes the need for the U.S. to maintain its global military reach."[37] Without directly invoking materialism, Paterson nevertheless invokes the power of things: "Embedded in globalising processes, and necessary for their reproduction, is a set of movements which are irreducibly physical and as such represent and enact flows of soil, water, food, minerals, energy, toxic waste and so on."[38] This critical perspective examines how materials, cultural forms, and political and commercial interests are enrolled in particular assemblages, without being directly in the materialist tradition.

Although we want to insist on the importance of the global and international scale, and take from IR a focus on the sovereign state as a crucial locus of controversies, authority, power, and law, the writers in this vein do not want to reduce all international politics to interstate politics or politics that are always articulated through the state. There is a rich tradition in IR, particularly international political economy and international political sociology, to which this goal is already familiar.

Flat Ontologies and Sovereignty

For the discipline of international relations, or any other similar disciplinary project, such as global studies or international political sociology, this new materialist sensibility toward open inquiry without pregiven scales of analysis poses a powerful and immediate intellectual challenge. International relations as an intellectual project already assumes that the international is an important—if not the paramount—"pre-established grid of analysis." This should cause tension with new materialist scholars. Rather than assume that any one scale of action is more important than any others, ANT scholars in particular attempt to unpack how differentiation or distinctions arise without any reference to pre-established grids of analysis. Thus, the international cannot be assumed to be important, but must be demonstrated to be so. This follows a number of critical trajectories: agential capacity is inherent not in the scale of action but rather in the assemblage of resources and the obedience of actors to those claims. Latour also makes significant broadsides at the key concepts of international relations, not only the scale of analysis. He advocates, "'Be sober with power.' In other words, abstain as much as possible from the notion of power in case it backfires and hits your explanations instead of the target you are aiming to destroy. No powerful explanations without checks and balances."[39] For a discipline that has been shaped—whether in a realist, liberal, feminist, or critical mode—by its fetish of power, this is a provocative challenge. Rather than assume an international—or indeed an organization, an issue, a field, a nation, a region, or a global—new materialist scholars must discover how different actants express their position in relation to one another.

And so, in preparation for this work, there were a number of challenges from colleagues and contributors: What's the difference between *the* international and *an* international, or between the international and the global? Aren't all materials or objects and their attendant assemblages sufficiently local to undermine any claim to the international? If by investigation we discover the international face of objects or material, all to the good, but doesn't a collection premised on this notion betray the promise of the openness of ANT and other forms of new or vital materialism? Why can't I just say the global and not the international? What does the international add to objects? In part, the multiple contributions to this volume are a polyvalent response to these questions, but I want to take up a couple of key points.

I want to claim that there is something distinctive about the international, distinct from the global, and that is the concern for the institution of sovereignty. Sovereignty is one of the foundations of the discipline of international relations, and it marks the organizing principle of a system of juridically equal states that have a right of self-determination and noninterference, as well as the separation between the domestic realm, where political theory is possible, and an international realm of repetition and anarchy, which establishes the principle of self-help and structures the daily practice of claim to authority. Whereas traditional analyses of sovereignty have focused on the international legal framework that structures interstate interactions, particularly through the organizing frame of anarchy, critical analyses take sovereignty to be a daily practice that has effects at multiple and diverse sites. Sovereignty is a regulative ideal that suggests that independence and noninterference are the ultimate political norms, as opposed to interdependence or mutual aid. Sovereignty places those norms of independence and noninterference above universal norms such as human rights or democracy. However, in addition to being a regulative ideal, sovereignty also instantiates an international system that is governed by these notions of independence and noninterference. Although the argument made by theorists of globalization is precisely on point—that networks, identities, communities, and authority structures all transcend boundaries in new and important ways—the death of sovereignty is greatly exaggerated. We would agree completely that borders, boundaries, and limits have fundamentally changed in an era of globalization, but those borders are still persistent. It would be as great an error to ignore the limits of sovereignty as it would be to retain a nineteenth-century idea of sovereignty. R. B. J. Walker's work in this area has been seminal. He argues that "anyone seeking to reimagine the possibilities of political life under contemporary conditions would be wise to resist ambitions expressed as a move from a politics of the international to a politics of the world, and to pay far greater attention to what goes on at the boundaries, borders and limits of a politics orchestrated within the international that simultaneously imagines the possibility and impossibility of a move across the boundaries, borders and limits distinguishing itself from some world beyond."[40] In his important recent work *After the Globe, before the World* (2010), Walker concludes that the international as a category retains a great critical purchase, more than the global, precisely because of its inherent contradictions, insisting on borders, boundaries, and limits that it constantly transcends. International relations, as we will demonstrate, is uniquely and powerfully suited to these analyses of international assemblages. It is our argument here that sovereignty as a central problematique of contemporary politics is undertheorized in contemporary new materialist studies. The new materialist turn presents a remedy to the dominance of the discursive turn in critical international relations theory that accepts the contingent nature of political formations but does not reduce all social or political action to discourse.

Our Project

A crucial aspect of thinking differently about the international is to precisely resist a big-picture grand narrative that defines key terms, identifies research agendas, or invokes the old philosophers and the new. Consequently, our collection is multiple and contradictory. There is a shared positive, constructive sensibility that from various directions and disciplines we have come to this juncture at which we want to include objects in our understanding of the world.

Building on the new materialist turn in critical and social thought, this broad collection of short interventions aims to provide a wide, diverse, and clear set of cases for how the international is evoked, enrolled, assembled, and deployed in a material world. *Making Things International* represents a multiplicity of methods and traditions in this field. Authors here write in their own voices on their own projects, and although there is an intellectual coherence to the volume, *Making Things International* is intended not as a programmatic or sovereign mapping of a research field but as an invitation to a party not quite in full swing. Each of the authors has made his or her best argument within a confined intervention, minimizing jargon and quotations, indicating his or her own relevant work and other important texts: our intention is to draw readers to a wider debate. Consequently, this collection is not a survey of all science and technology studies (STS), actor-network theory, new materialist philosophy, or critical international relations theory.

This project started as part of a workshop in critical research methods in security studies, hosted at the University of Ottawa with the support of the Social Sciences and Humanities Research Council of Canada, and of uOttawa's Faculty of Social Sciences and the Office of the Vice-President, Research. At that meeting, in 2010, it was clear that a number of graduate students had wandered afield into STS and other areas of social and political thought to engage with materialist theory. My colleague, friend, and coeditor Can E. Mutlu identified this as a productive area, and we included a section in *Research Methods in Critical Security Studies: An Introduction* (2013) on the materialist turn.[41] Once we became attuned to this growing community, it suddenly seemed visible everywhere. The *Millennium* annual conference in 2012, convened under the title "Materialism and World Politics," featured three days of papers and a special workshop afterward organized by Peer Schouten, Rocco Bellanova, Julian Jeansdeboz, and Christian Bueger. Inspired by *Making Things Public*, edited by Latour and Weibel, I was sure that a similar collection around the international would be exciting and useful—particularly to new scholars or scholars new to this area. One of my first contacts, my friend and colleague Peter Adey, directed me toward *Patterned Ground*, edited by Harrison, Pile, and Thrift, which looked at "entanglements of nature and culture," and I knew we had a successful model. It is a mark of the collegiality and generosity of this small, readerly community that all of these contributions to *Making Things International* are essentially positive and constructive, rather than negative or antagonistic.

When full yellow legal pads, the ghosts of pens long dead, innumerable sticky flags, and electronic files turn into another thing—proofs, galleys, and finally a book—I experience the distinct pleasure of the maker. No matter how good a lecture, seminar, debate, question, or turn of phrase is, I am still convinced that the feeling is ephemeral and will die. Despite having read "Plato's Pharmacy," by Derrida, when I was young and impressionable, my first instinct is to trust the tangible book rather than the fleeting ecstatic feeling of oral argument. But I am wrong, dead wrong. Without the reader, the book is an empty space, a hammer on the workshop floor. Like all books, our collection tries to create an emotional and intellectual disturbance in the reader; it wants to provoke, to enrage, to fascinate, to convince. Just as with the perfect seminar, in which the questions are incisive, the metaphors provocative, the very air magical and alive with electricity, that experience is more than the intent of its participants. An influential teacher of mine, Derek Gregory, once riffed on Heraclitus, saying that you cannot read the same book twice. This book will be mesmerizing and fascinating for some, shelf filler for others. And this book will die through the slow disintegration of its pages or as the changing of technology renders this format unreadable, like an eight-track tape or a stereographic photo, and so this thing too is part of multiple assemblages that will cohere, circulate, and then dissipate. Still, it will have a moment that exceeds any intention or design.

Notes

1. William Walters, "Drone Strikes, *Dingpolitik*, and Beyond: Furthering the Debate on Materiality and Security," *Security Dialogue* 45, no. 2 (2014): 101–18; Jairus V. Grove, *Thinking Like a Bomb: An Ecology of Global Warfare* (Minneapolis: University of Minnesota Press, forthcoming); Peer Schouten, "Security as Controversy: Reassembling Security at Amsterdam Airport," *Security Dialogue* 45, no. 1 (2014): 23–42; Jacqueline Best and William Walters, "'Actor-Network Theory' and International Relationality: Lost (and Found) in Translation," *International Political Sociology* 7, no. 3 (2013): 332–34; Nick Srnicek, Maria Fotou, and Edmund Arghand, "Materialism and World Politics," introduction to *Millennium: Journal of International Studies* 41, no. 3 (2013): 397.

2. Walters, "Drone Strikes, *Dingpolitik*, and Beyond," 103.

3. Grove, *Thinking Like a Bomb*.

4. Ibid.; Schouten, "Security as Controversy."

5. Michel Callon, "Actor-Network Theory—The Market Test," in *Actor Network Theory and After*, ed. John Law and John Hassard (Oxford: Blackwell / The Sociological Review, 1999), 181–95.

6. Tim Cresswell, *On the Move: Mobility in the Modern Western World* (London: Routledge, 2006); Mimi Sheller and John Urry, "The New Mobilities Paradigm," *Environment and Planning A* 38 (2006): 207–26; Mark B. Salter, "To Make Move and Let Stop: Mobility and the Assemblage of Circulation," *Mobilities* 8, no. 1 (2013): 362–65.

7. Peter Adey, "Secured and Sorted Mobilities: Examples from the Airport," *Surveillance and Society* 1, no. 4 (2004): 500–519.

8. Michel Foucault, *Security, Territory, Population: Lectures at the Collège de France, 1977–78*, trans. Graham Burchell (Basingstoke, Eng.: Palgrave Macmillan, 2007), 11.

9. Ibid., 18.

10. Bruno Latour and Steve Woolgar, *Laboratory Life: The Social Construction of Scientific Facts* (Beverly Hills, Calif.: Sage, 1979).

11. Foucault, *Security, Territory, Population*, 29.

12. Latour and Woolgar, *Laboratory Life*, 128–29.

13. Michel Callon, "Some Elements of a Sociology of Translation: Domestication of the Scallops and the Fishermen of St. Brieuc Bay," in *Power, Action, and Belief: A New So-*

ciology of Knowledge?, ed. John Law (New York: Routledge, 1986), 196–223; Michel Callon, "The Sociology of an Actor-Network: The Case of the Electric Vehicle," in *Mapping the Dynamics of Science and Technology: Sociology of Science in the Real World,* ed. Michel Callon, John Law, and Arie Rip (London: Macmillan, 1986), 19–34; John Law, "On the Social Explanation of Technical Change: The Case of the Portuguese Maritime Expansion," *Technology and Culture* 28 (1987): 227–52; John Law, *Aircraft Stories: Decentering the Object in Technoscience* (Durham, N.C.: Duke University Press, 2002); Bruno Latour, *The Pasteurization of France* (Cambridge, Mass.: Harvard University Press, 1993); Bruno Latour, *Science in Action* (Cambridge, Mass.: Harvard University Press, 1987).

14. John Law and Annemarie Mol, "The Actor Enacted: Cumbria Sheep in 2001," in *Material Agency: Towards a Non-anthropocentric Approach,* ed. Carl Knappett and Lambros Malafouris (New York: Springer, 2008), 57–77; Marianne de Laet and Annemarie Mol, "The Zimbabwe Bush Pump: Mechanics of a Fluid Technology," *Social Studies of Science* 30, no. 2 (2000): 225–63; John Law and Annemarie Mol, "Local Entanglements and Utopian Moves: An Inquiry into Train Accidents," *Sociological Review* 50, no. 2 (2002): 81–105.

15. Annemarie Mol, *The Body Multiple: Ontology in Medical Practice* (Durham, N.C.: Duke University Press, 2003).

16. Law and Hassard, *Actor Network Theory and After*; Bruno Latour, *Reassembling the Social: An Introduction to Actor-Network Theory* (Oxford: Oxford University Press, 2005).

17. Latour, *Reassembling the Social,* 19.

18. Callon, "Some Elements," 203.

19. This is strikingly similar to a Bourdieusian field analysis, which has played such an important part in the development of international political sociology.

20. Callon, "Some Elements," 211.

21. Michel Callon and John Law, "After the Individual in Society: Lessons on Collectivity from Science, Technology, and Society," *Canadian Journal of Sociology* 22, no. 2 (1997): 168.

22. Callon, "Some Elements," 196.

23. Jane Bennett, *Vibrant Matter: A Political Ecology of Things* (Durham, N.C.: Duke University Press, 2010), 21.

24. Ibid., 23–24.

25. William E. Connolly, "Method, Problem, Faith," in *Problems and Methods in the Study of Politics,* ed. Ian Shapiro, Rogers M. Smith, and Tarek E. Masoud (Cambridge: Cambridge University Press, 2004), 342–43.

26. Bennett, *Vibrant Matter,* 33.

27. Bruno Latour and Peter Weibel, eds., *Making Things Public: Atmospheres of Democracy* (Cambridge, Mass.: MIT Press; Karlsruhe, Ger.: ZKM/Center for Art and Media in Karlsruhe, 2005); Stephan Harrison, Steve Pile, and Nigel Thrift, eds., *Patterned Ground: Entanglements of Nature and Culture* (London: Reaktion, 2004).

28. Latour, *Reassembling the Social,* 15.

29. Harrison, Pile, and Thrift, *Patterned Ground,* 3.

30. R. B. J. Walker, *After the Globe, before the World* (New York: Routledge, 2010), 33.

31. Giorgio Riello, *Cotton: The Fabric That Made the Modern World* (Cambridge: Cambridge University Press, 2013); Gail M. Hollander, *Raising Cane in the 'Glades: The Global Sugar Trade and the Transformation of Florida* (Chicago: University of Chicago Press, 2008); Ben Richardson, *Sugar: Refined Power in a Global Regime* (Houndmills, Eng.: Palgrave Macmillan, 2009).

32. Graham T. Allison et al., *Avoiding Nuclear Anarchy: Containing the Threat of Loose Russian Nuclear Weapons and Fissile Material* (Cambridge, Mass.: MIT Press, 1996); C. J. Chivers, "Small Arms, Big Problems: The Fallout of the Global Gun Trade," *Foreign Affairs* 90 (2011): 110–21; Sonia Ben Ouagrham-Gormley, "Barriers to Bioweapons: Intangible Obstacles to Proliferation," *International Security* 36, no. 4 (2012): 80–114.

33. Timothy Mitchell, *Carbon Democracy: Political Power in the Age of Oil* (New York: Verso, 2011); Matthew Paterson, *Automobile Politics: Ecology and Cultural Political Economy* (Cambridge: Cambridge University Press, 2007).

34. Mitchell, *Carbon Democracy*, 236.

35. Ibid., 239.

36. Paterson, *Automobile Politics*; Matthew Paterson and Simon Dalby, "Empire's Ecological Tyreprints," *Environmental Politics* 15, no. 1 (2006): 1–22; David Campbell, "The Biopolitics of Security: Oil, Empire, and the Sports Utility Vehicle," *American Quarterly* 57, no. 3 (2005).

37. Campbell, "Biopolitics of Security," 944.

38. Paterson, *Automobile Politics*, 7.

39. Latour, *Reassembling the Social*, 39.

40. Walker, *After the Globe, before the World*, 3.

41. Mark B. Salter and Can E. Mutlu, eds., *Research Methods in Critical Security Studies: An Introduction* (New York: Routledge, 2013).

PART I
World in Motion

Electronic Passports

William Walters and Daniel Vanderlip

"An identity theft wet dream!" According to Barry Steinhardt, director of the Freedom and Technology Program of the American Civil Liberties Union (ACLU), this was the damning verdict on the new, technologically enhanced passport that the U.S. government was rolling out. Steinhardt was speaking on a panel at the fifteenth Computers, Freedom, and Privacy Conference in Seattle in 2005. His fellow panelists included the security technology expert Bruce Schneier and the then deputy assistant secretary of state for passport services, Frank Moss.[1]

The representative from the State Department had already spoken, championing the new electronic passport.[2] Among the major claims Moss made were that the passport would be even harder to forge, would offer a stronger assurance that the bearer was the person to whom the passport had been issued, and would enhance the facilitation of travel through busy airports and border crossings. Moss sought to assure the audience that the government had every confidence in the integrity of this new document and would not use Americans as a "test population." Not only did this new document incorporate advanced printing techniques, already in use in the later generations of U.S. currency, but it also featured biometric elements. The new passport would include a 64-kilobyte contactless chip in its rear cover. Information from the data page of the passport would be written onto this chip, accompanied by a digital image of the passport holder's face. This data would in turn be secured by public-key-infrastructure (PKI) technology. In particular, Moss sought to allay concerns that the chip made the passport vulnerable to unauthorized reading from a distance (skimming) and eavesdropping by unauthorized third parties when it was wirelessly communicating with its electronic reader at passport control. To this end, he stressed that the chip being used was consistent with standards determined by the International Organization for Standardization (ISO)—ISO 14443A or B—that prescribe a maximum reading distance of no more than ten centimeters.

Then the floor was passed to Steinhardt. The representative from the ACLU did not simply make counterarguments about the new passport; he offered a very practical, if not entirely smooth, demonstration of its potential vulnerability. Attaching a self-adhesive chip of the same kind specified by the ISO to his own passport, he waved the document at a radio-frequency identification (RFID) reader placed on the floor near the presenters' table. The reader was attached to a laptop and a data projector. Despite a few technical glitches, the experiment proved that the RFID reader was capable of reading Steinhardt's passport at a distance considerably greater than ten centimeters. "I want to show you what your future is going to look like. . . . I want you to think about whether you want to be walking around in the world with a document that can in fact be read from thirty feet," Steinhardt told the audience.

The ACLU's stunt was one of the first in a wave of demonstrations and attacks carried out by individuals and groups associated in media eyes with hacking. Some were staged before conference audiences, others before the watching eyes and lenses of journalists. For example, in the United Kingdom, a computer expert showed it was possible to break the security codes on the United Kingdom's biometric passport and download personal data.[3] In the Netherlands, another researcher claimed to be able not just to download data but also to upload new information onto the chip.[4] These and other similar demonstrations certainly caught the public's attention. Many were featured in headline stories and TV reports. If the electronic passport had been designed to be a black box, these performances would, if not fully open it up, then at least suggest that it was not entirely closed or immutable. If assorted manufacturers, standard setters, and governments set out a network of border posts, testing laboratories, and technical agencies in which this new travel document was to operate, these hacks and attacks would enroll it in a different network and connect it to other devices—machines capable of generating certain political effects. So, an object that was designed to engineer greater trust among the border authorities of the world's states, and to speed circulation while countering threats of forgery and deception, would now be shadowed by certain insecurities and doubts. For a while, the very ontology of the new passport seemed ambiguous.[5] Was it the hard enhancement of security and technology of speed that Moss and others claimed it to be? Or was it perhaps a kind of beacon, capable of making its bearer traceable as he or she passed through railway stations and airports, not unlike a UPS package?[6]

The electronic-passport project appeared momentarily unsettled. Bruno Latour has suggested that when controversies flare up around objects, those objects move from being "matters-of-fact" to "matters-of-concern."[7] In such moments, he argues, one sees in quite prominent ways how the properties, qualities, affordances, and material specificities of objects can move into the center of public attention and become mediators and stakes in political controversies. Most of Latour's examples of matters-of-concern involve highly prominent, technopolitical crises: the crash of the Challenger space shuttle, the allegation of Iraq's weapons of mass destruction, and so on. Such examples are what he calls "Dingpolitik," where *Ding-* refers both to the place of assembly (following the archaic Nordic word for parliament) and to the disputed thing. With Dingpolitik, there is fluidity. Not only are the constitution and effects of technological objects called into question, but so too is the makeup of the expert body that speaks for those things. In Dingpolitik, the lines and hierarchies distinguishing informed "insiders" and uninformed "outsiders" (a division Walter Lippmann held to be foundational to politics in technological societies)[8] are often unsettled as new voices struggle to represent existing and emergent objects. According to Jacques Rancière, politics happen when those with no recognized part speak up.[9] When those who are not formally recognized as insiders speak as though they are experts, we call this technological interruption.

We argue that not all Dingpolitik occurs with the same scale or intensity. The case of the electronic passport illustrates what we propose to call "minor Dingpolitik." What is minor Dingpolitik? As Latour describes it, Dingpolitik couples the energy of an assembled people, the demand for public accounting, and the practical knowledge of experts in order to open up black boxes. By contrast, minor Dingpolitik engages the black box in the absence of major public controversy. It is active rather than reactive. It does not wait for the accident, the disaster, or the scandal to reveal a problem. Minor Dingpolitik operates in what Gilles Deleuze and Félix Guatarri call "cramped space."[10] If cramped space calls for creativity,[11] then minor Dingpolitik has to be especially inventive to win public attention. There has been no major technological failure or political crisis concerning the electronic passport; but neither has its introduction passed without any incident or public concern. There have been minor skirmishes and microdisputes of the kind we have already documented. Privacy advocates and civil-liberties groups weighed in when the standards for the new passports were first announced, highlighting their implications for surveillance and misuse. These may or may not have shaped the developmental trajectory of this device. Nevertheless, we argue that attending to the minor Dingpolitik of the electronic passport is a worthwhile exercise. Moments of minor Dingpolitik offer us sites of critical understanding. They allow ordinary and otherwise self-evident features of an object to be grasped in a more critical light. For example, by demonstrating that electronic passports could be made to function like beacons, independent researchers and hackers brought into question the established relationship between passports and identification. Their instigation of minor Dingpolitik thus opens a kind of bridgehead space for other researchers to enter.

The technological interruptions staged by security researchers and hackers brought public scrutiny to several features of the electronic passport. There is one feature that interests us and that serves as the focus for the remainder of this chapter. We are interested in the fact that the passport is an object of standards and standardization. In the discussion of standards that follows, we argue that standards are key to understanding how certain objects shape, and even constitute, conditions of possibility for certain forms and experiences of international life. At the same time, we insist that standards are neither dull nor merely technical. Not only are political decisions and values encoded in standards, but sometimes standards can also serve as surfaces of political activity. In the following two sections, we discuss the standardization of the passport. In a final section, we examine the entanglement of the passport and its standards in politics.

Standards and Standardization

Practices of standardization are significant forms of governance and regulation in the contemporary context deserving of specific and concerted academic focus.[12] International standards are typically established through nongovernmental organizations on a voluntary basis, and have a significant impact in a broad range of issue areas, from business practices (such as auditing, accounting, and risk management) to terminological and definitional standards (such as disease typology and nomenclature) and technical standards governing the makeup of nonhuman objects (such as shipping-container dimensions and credit-card thickness).[13]

Standards work to reduce difference across space and time as a response to recurring problems of interoperability. In a globalized world, standards act as one way of establishing shared rules, practices, languages, design principles, and procedures to allow heterogeneous actor-networks to work together.[14] Standards are also often highly technical and specialized, resulting from the evidence brought forth by and debated among groups of experts in a particular field. The fact that standards often deal with highly specialized and scientific problems and issues lends a certain impermeability and durability to the processes of standardization in the contemporary context; expertise in this sense functions as a barrier to entry in the standardization process for the nonexpert.

By centering standards and practices of standardization as a key unit of analysis in the study of global politics, we argue for the need to take what Hans Krause Hansen terms an "extended view of authority,"[15] focusing on the ways in which standards are constitutive of and work to (re)produce understandings of national and international in specific contexts. We call attention to the need to focus on the ways in which standards are both productive of and are produced by the national and the international as techniques of government beyond the traditional nation-state. Practices of standardization require the continued existence of national and sovereign administrative systems to do the day-to-day work of governing. In this respect, standards work to effect a certain industrialization of trust, which functions to make national regimes of government interoperable within an international logic of sovereign control.

The Standardization of the Electronic Passport

The process to standardize machine-readable and electronic passports demonstrates the immense amount of work required to make an internationalized and interoperable object, as the result of significant bureaucratic and technical interventions and processes, negotiation, and contestation.[16] Established in 1944 as a multilateral agency to govern standards and procedures in civil aviation, the International Civil Aviation Organization (ICAO) is the international body responsible for overseeing and coordinating technical standards for passports and other travel documents. Far from being a purely technical and linear process, the standardization of the electronic passport under the auspices of the ICAO's Technical Advisory Group (TAG) on Machine-Readable Travel Documents (MRTD) reveals that the electronic passport is a somewhat open artifact, and requires consistent effort to be remade as an international object. Beyond technical specifications, the ICAO, through the TAG MRTD, has also established recommended compliance and security guidelines, information-sharing protocols, timelines, and milestones for ePassport implementation. Technical problems, interoperability issues, noncompliant technologies, geopolitical concerns, divergent national budgets, expertise, and bureaucratic systems, and conflicting timelines have all factored in to complicate the process to standardize electronic passports. The ICAO process to standardize ePassports is also voluntary, allowing states to opt in or abstain from participation. As such, the ICAO process is best characterized as an imperfect and nonseamless emergent interoperability subject to ongoing political work and contestation, rather than as a static fait accompli.

Approximately 104 states are currently issuing ePassports, with some four hundred million of these documents now in circulation.[17] The history of the ePassport is rooted in a long trajectory of technical reports, studies, and meetings under the ICAO.[18] For ICAO, the art of standards is a question of fostering confidence on the part of public authorities in the reliability of travel documents and in the inspection systems tasked with processing them.[19] The history of the standardization of the electronic passport and this engineering of an interbureaucratic and international state of confidence begins in an official sense with the ICAO Panel on Passport Cards. This body was established in 1968 to formulate recommendations for a machine-readable passport document that could facilitate processing of travelers at points of entry. The work of this panel culminated in the publication of the first edition of Document 9303, entitled "A Passport with Machine Readable Capability."[20] Sometimes called "the Bible" by regulators, this document has been continuously updated since its initial publication, and remains the touchstone publication outlining the standards and characteristics of ePassports issued by signatory nations.

Various editions of Document 9303 have set out specifications for a machine-readable passport, including the size and shape of documents, the layout of the data page, and the creation of a machine-readable zone, which is a set of lines of data on a passport that are printed in a standardized font and include information such as the passport holder's name and date of birth, as well as security codes. Machine-readable travel documents can be passed through standardized readers by border-inspection agents, reducing the possibility of errors in inputting data and streamlining the review process. However, although MRTDs enhanced the facilitation of travel, they did not improve the security of travel documents; consequently technical fixes were sought to better confirm "the validity of the traveler as the rightful 'owner' of the documents."[21]

In the late 1990s, the ICAO's New Technologies Working Group (NTWG), which was established as a subgroup of the TAG MRTD, began to study potential technical fixes to this identity problem. Through the combination of biometrics, computer chips, and cryptographic technologies, the NTWG sought to better fix the link between passports and passport holders, reducing the occurrence of fraudulent documents. As a result of advice provided by the NTWG, Document 9303 now contains detailed guidance regarding the technical specifications of the ePassport, defining it as a machine-readable passport that has a contactless integrated circuit chip, which stores data from the machine-readable page, a biometric measure to identify the rightful passport holder, and a cryptographic security technology to protect the data on the chip.

In 2001, the NTWG completed its first significant study in biometric capabilities for MRTDs. From this came the technical report *Selection of a Globally*

Interoperable Biometric for Machine-Assisted Identity Confirmation with MRTDs (2001).[22] This report sought to clarify which biometric measure was most appropriate for inclusion in an ePassport. Interestingly, this assessment process was not exclusively technical or performance-based. It was not simply the most accurate or best biometric that would be selected for inclusion in ePassports. In fact, the NTWG considered a range of factors in suggesting optimal biometric measures, including technical characteristics such as performance, durability, reliability, and backward compatibility, in addition to other, seemingly nontechnical factors, such as public perception, ease of use, privacy implications, legislative considerations, and potential health and safety considerations.[23] Ultimately, the 2001 technical report concluded that facial recognition best met the range of requirements and factors as a suitable biometric, followed closely by fingerprints and iris scans.[24] In June 2002, the NTWG officially endorsed facial recognition as the preferred biometric, noting that nations may choose to supplement ePassports with fingerprint- and iris-recognition technology as additional recognition technologies.[25] With this determination, the TAG MRTD subsequently issued a complex series of technical standards to govern ePassports, ranging from the lighting in and proportion of passport photos, the placement of data pages and chips in passports, storage capacities and compression protocols, and security regimes, and it continues to refine these specifications.[26] Although this process appears at first glance to be linear and technical at its core, it in fact reflects wider cultural considerations. For example, making the face, rather than fingerprints, the mandatory biometric speaks to the fact that fingerprinting, in many countries, carries strong associations with criminal suspicion.[27]

The NTWG is composed of a variety of passport experts. The core group of members includes bureaucratic representatives from signatory nations who are involved in the production of passports or are in border control. Technical expertise is drawn from experts accredited by the ISO, including physicists, engineers, chemists, information-technology experts, and lawyers, who are also supported by a number of observers, including security experts (such as Interpol), industry associations (such as the International Air Transport Association [IATA]), and representatives of airport authorities.[28] Despite the fact that this representation structure draws from a broad range of expertise, membership in the NTWG is closed. The NTWG is not a public forum in which to debate the merits or pitfalls of particular emerging ePassport technologies. Membership status in the group is restricted to member-government employees, accredited experts, and selected international organizations, and is controlled by the working-group chair. Membership thus functions as a key sorting function for the NTWG, empowering some with the ability to speak for and about passport technologies while rendering distant the perspectives of others. One could say that the NTWG helps to construct a hierarchy of insiders and outsiders when it comes to passport technology. As a result, although security hackers and privacy activists tend to engage with the NTWG on technological terms, challenging the professed capabilities of ePassports, their activity might also be read as challenging the boundaries of expertise associated with bodies like the NTWG, creating space for outsiders to voice perspectives and concerns.

In response to this challenge to its authority, the network of experts associated with the NTWG has taken further steps to police the field of technical expertise regarding ePassports and to shape who can speak authoritatively about this technology. For example, in 2009, Mike Ellis, a key expert from this network, published an article addressing what he described as "thirty-nine myths" about ePassports.[29] Ellis cast the work of hackers and activists as attempts to gain notoriety and attention, rather than serious technical and sociopolitical interventions that could help to shape the future of ePassports. He heralded the success of ICAO standardization processes and sought to debunk hackers and activists as uninformed and alarmist detractors: "Despite [its] success, some commentators have been critical of the e-passport. Most of this criticism is based on fiction, a misinterpretation of the facts, or a confusion of technologies. Some articles are written by hackers seeking recognition, others by security researchers working in pristine laboratories, a little divorced from reality."[30] In addition to challenges to the monopoly of expertise over the ePassport-standardization process, the ICAO also faces technological challenges to implementing a truly interoperable regime to regulate travel. Despite the fact that more than four hundred million ePassports have been issued by nation-states, with substantial political work having been committed to the establishment of extensive standards and protocols for ePassports, a significant number of ePassports issued do not, by the ICAO's own admission, conform to the standardized requirements for these documents.[31] This fact is of particular concern to the ICAO; if we recall that a key aim of adopting standardized travel documents is to regulate confidence regarding the veracity of particular documents, then noncompliant electronic passports represent a significant challenge to the standardization and internationalization of the passport. Noncompliant passports include documents that have poorly printed machine-readable zones (MRZs), incorrect radio-frequency identification chips, and nonconforming PKI technologies.[32]

PKI technology is essential in establishing a secure passport, as it provides a means of confirming the integrity of data stored on the document's chip.

The physical passport inspection, comparing traveler to travel document, represents the first stage of verification, and can be completed in absence of additional biometric data stored on an ePassport; the RFID chip contained within an ePassport allows for a second stage of verification, providing further information that can help to confirm that the bearer of the travel document is who he or she proclaims to be; PKI technologies provide a third stage of verification, signaling to border agents that the data contained on an ePassport has not been tampered with.[33] This third stage of verification is essential to the securitized and internationalized ePassport; without assurances that the data contained on an ePassport is accurate, the passport in effect becomes no more secure than a standard machine-readable travel document: "An ePassport . . . is only as secure as the biometric and biographic information contained in its chip. Information on the chip, in turn, is only useful if the data can be validated quickly and securely."[34] Countries each have a unique digital signature that is verified using public keys. This signature, relying on complex algorithms, allows border agents to determine whether particular ePassports are authentic or whether they have been tampered with. The ICAO has faced particular challenges in securely and quickly distributing certificates and associated information necessary for this verification process, and has been forced to rethink traditional bilateral exchanges of information. In response, the ICAO has recently established a public key directory for member states and has offered technical assistance in ePassport development, design, and implementation through a recent MRTD bulletin.

Far from being determined artifacts, ePassports are the subject of considerable political work, undertaken to internationalize the objects of travel facilitation and securitization. Security officials, technical experts, industry associations, government bureaucracies, and private-sector organizations have variously intervened in the standards-making process for biometric passports. Furthermore, technical aspects themselves have pushed back against the internationalization of the ePassport, as witnessed through the struggles that the ICAO and TAG MRTD have faced in implementing an interoperable PKI system.

Objects, Standards, and Politics

We have argued that extensive technical work goes into testing, making, adapting, and revising the standards that mediated the electronic passport. If the enhanced passport is to make good on the promises of greater security and enhanced movement of passengers that are made in its name, then it has to be equipped for interoperability. In a world in which governments jealously and proudly regard the passport as a potent signifier of national identity and sovereignty, in a world where much of the technology of personal identification and border security comes from a marketplace of rival security corporations, the only option seems to be that of getting diverse passports and their systems to cohere better. Technical standards are key in this respect. Standards are tasked with mediating between diverse products, materials, nations, travelers, and much more, and with ensuring that this mix will somehow hold together.

Returning to the theme of minor Dingpolitik, we argue that in its short history the electronic passport has been the setting of lively politics and, moreover, that these politics are closely connected to the very material properties and components of the new passport and to the standards that regulate those properties. In conclusion, we offer three observations about the relationship between objects, standards, and politics.

First, a point about standards and experts. Standard setters sometimes have to do a kind of politics themselves. The line between the political and the technical is sometimes blurry. Let us return for a moment to the incidents with which we began: the series of demonstrations, attacks, and hacks of the early versions of the electronic passport. These acts of technological interruption were quite widely publicized in the national press. If they did not catalyze a public outcry regarding the new passport, they did provoke an assertive and sometimes polemical response from official passport experts. In this response, experts published interviews and essays in which they sought to neutralize or discredit the claims and proofs of the hackers and researchers. We discussed the intervention of Mike Ellis already. Here we will mention another. Barry Kefauver, one of the experts involved in the setting of standards with the ICAO and the ISO, accused security researchers and hackers of seeing "these chip-based passports as a toy to be brought into the laboratory and made sport with on the basis of impractical and questionable scenarios."[35] This drew a response from Lukas Grunwald, one of Kefauver's principal protagonists, that he and his colleagues had done nothing other than adopt "the standard approach followed by well-established companies in the security industry for many years when auditing any system. A specific scenario may be impractical or questionable today, but it might be realistic tomorrow."[36]

We find this argument interesting not just because it illustrates how controversy, however minor, stalked the implementation of the standards for the passport but also because it demonstrates that the techniques of premediation, which imagine and enact disaster scenarios for the purpose of stress-testing security systems, are not the monopoly of the state and its experts.[37] In this case, we see them being mobilized by those positioned as outsiders. What Grunwald and his colleagues imagine and demonstrate are worst-case scenarios that exceed what the security agencies themselves have countenanced. Premediation, it seems, can be turned on the state from outside its apparatus.

Our second point concerns the way in which standards can act as mediators for politics of contestation, local politics with international ramifications. In some instances where experiments and demonstrations were performed—such as the ACLU case that we began with—it was not on the actual electronic passport, for that had yet to be issued to the public. Instead, the claim about the vulnerability of the chip to being read from a distance by unauthorized parties was built upon the manipulation of a chip that simply conformed to the technical standard.

In a different but not unrelated way, technical standards for the passport and its systems enables Grunwald's intervention. He claims to be able not just to read but also to clone the chip and its data and to load the data into a blank passport.[38] This, we are told, was based on careful research into the ICAO's published standards and on the use of the same passport-inspection reader technology as had been used by experts and industry to test the interoperability of their products. Grunwald demonstrated his attack on a new European Union German passport. But, thanks to standards, he is able to make a stronger claim: his method and the vulnerability it demonstrates should apply "to any country's e-passport, since all of them will be adhering to the same ICAO standard."[39] So here standards can be used to amplify and extend the scale of political claims.

One of the key features of technical standards is to allow for coherence in the midst of diversity. Standards are not about fostering a world of uniformity, whatever the claims of certain populist critics of regulation. Rather, standards allow nations and corporations to develop and deploy new products that will be compatible with existing (and future) systems and practices. Meet the requisite standards and you have the right to name your product an ePassport. It is as though the hackers benefit from this convention as well. As long as the networks of passports, scanners, and chips that they assemble and experiment with conform to the official standards, hackers can make technopolitical claims about not just those particular networks but also a wider class of objects. It doesn't always matter that it is not the actual or exact passport that they experiment with. Instead, there can be politics mobilized around *virtual ePassports*. The mere fact that one has accomplished something controversial or discrepant with an object that falls within the field of possible objects demarcated by standards is enough.

Finally, we offer an observation that is less about standards as such but is very much about the materials that are the subject matter of these standards. What do passports do? They materialize trust. Without passports, we would perhaps need a retinue of credible witnesses to follow us around, vouching that we are who we claim to be at every border crossing. As some impressive histories of the passport have shown, the passport is an invention that obviates such an unwieldy state of affairs.[40] In a more durable and tamper-proof way than the letter of reference, the passport materializes trust among strangers. It is a technology of freedom as well as a technology of control: not only does it afford a kind of freedom to travel internationally, but also it effects a kind of individualization.[41] Thanks to the passport, we can travel alone, as individuals, unaccompanied by all those witnesses and authorities who are now inscribed in its material form. They are its absent presence.

Like banknotes and coins, passports and their systems materialize trust. They employ specific materials and techniques, such as special papers, forms of printing, photographs, inks, certain metals, and magnetic strips, in order to do this. But once you materialize trust, you open up the possibility for replication or imitation. The passport becomes a research frontier: an ongoing experiment in the materialization of trust set against a "technological race" between issuers, regulators, and forgers.[42]

Seen from this perspective, the electronic passport is only the latest episode in the search for special materials—another cycle of the technological race. But from a perspective of political styles and forms, the turn to biometrics is very significant. It is significant, for one thing, because of the new forms of surveillance it opens up—a theme that has been explored at great length.[43] But it is significant in another sense as well. When the standard setters introduced the contactless chip into the passport, it is as though they engineered a new species, for they did not just connect up a very old technology—namely, the specially printed document—to some very new lines of technological development; at the same time, they exposed this old technology to some new political viruses. The point is simple: there hasn't always been a "hactivism" of passports. If hactivism represents an entire political genre, with its own codes, practices, styles, and cultures,[44] then making the passport increasingly digital has opened it up to these political currents. What we are seeing is perhaps a kind of political zoonosis. Hactivism has leapt over the boundary that previously confined it to the world of computing. As our various encounters reveal, it is now capable of striking at technical and social forms that were previously quite remote from it. Perhaps we can say that technical standards not only facilitate the international movement of passports and their bearers but also serve as vectors of political style.

To suppose that technical standards depoliticize the world, as some have assumed, is somewhat mistaken. Such standards may indeed give governance a technical and seemingly neutral face, but they also open up new possibilities for politics. The hacking of the first generations of the electronic passport did not effect a major political crisis; but these acts of minor Dingpolitik did show that standardized technological objects can become surfaces for creative political activity.

Notes

1. Computers, Freedom, and Privacy Conference, Seattle, Wash., April 13, 2005. The conference was blogged by Andrew Brandt ("Experts Debate Utility, Safety of E-Passports," *PC World*, April 19, 2005). For a sound recording and report, see http://web.archive.org/web/20050420030534/http://blogs.pcworld.com/staffblog/archives/000616.html. We are grateful to Christopher Alderson, who discovered this conference report. It is also discussed in Alderson and William Walters, "Making Technologies of Security Public" (paper presented at the International Studies Association Annual Meeting, San Diego, Calif., 2012). The sound recording and transcript of Moss's remarks are on file with the authors.

2. In this chapter, we use the term *electronic passports* to refer to the general field of biometrically enhanced passports. In contrast *ePassport* is a regulatory term. The International Civil Aviation Organization (ICAO)—the international body responsible for standardizing electronic passports—insists that those electronic passports which do not meet its technical specifications shall neither be called ePassports nor display the ePassport logo. See ICAO, *Machine Readable Travel Documents,* doc. 9303, pt. 1, *Machine Readable Passports,* vol. 2, 6th ed. (Montreal: ICAO, 2006), para. 2.3.

3. Sue Reid, "'Safest Ever' Passport Is Not Fit for Purpose," *Daily Mail* online, March 5, 2007, http://www.dailymail.co.uk/news/article-440069/Safest-passport-fit-purpose.html.

4. Steve Boggan, "'Fakeproof' E-Passport Is Cloned in Minutes," *Times* (London), August 6, 2008.

5. Daniel Neyland, "Mundane Terror and the Threat of Everyday Objects," in *Technologies of InSecurity: The Surveillance of Everyday Life,* ed. Katja Franko Aas, Helene Oppen Gundhus, and Heidi Mork Lomell (Abingdon, Eng.: Routledge-Cavendish, 2009).

6. Dario Carluccio et al., "E-Passport: The Global Traceability; or, How to Feel like a UPS Package," *Information Security Applications* 4298 (2007): 391–404.

7. Bruno Latour, "From Realpolitik to Dingpolitik; or, How to Make Things Public," in *Making Things Public: Atmospheres of Democracy*, ed. Bruno Latour and Peter Weibel (Cambridge, Mass.: MIT Press, 2005), 19.

8. Walter Lippmann, *The Phantom Public* (1927; repr., New Brunswick, N.J.: Transaction, 1993).

9. Jacques Rancière, *Disagreement: Politics and Philosophy* (Minneapolis: University of Minnesota Press, 1999).

10. Gilles Deleuze and Félix Guattari, *Kafka: Toward a Minor Literature* (Minneapolis: University of Minnesota Press, 1986), 17. On cramped space and minor politics, see Nicholas Thoburn, *Deleuze, Marx, and Politics* (London: Routledge, 2003).

11. Thoburn, *Deleuze, Marx, and Politics*, 19.

12. Stefan Timmermans and Steven Epstein, "A World of Standards but Not a Standard World: Toward a Sociology of Standards and Standardization," *Annual Review of Sociology* 36 (2010): 69–89.

13. Walter Mattli and Tim Büthe, "Setting International Standards: Technological Rationality or Primacy of Power?," *World Politics* 56, no.1 (2003): 1–42.

14. Timmermans and Epstein, "World of Standards."

15. Hans Krause Hansen, "Investigating the Disaggregation, Innovation, and Mediation of Authority in Global Politics," in *Critical Perspectives on Private Authority in Global Politics*, ed. Hans Krause Hansen and Dorte Salskov-Iversen (Basingstoke, Eng.: Palgrave Macmillan, 2008), 2.

16. Jeffrey Stanton, "ICAO and the Biometric RFID Passport: History and Analysis," in *Playing the Identity Card: Surveillance, Security, and Identification in Global Perspective*, ed. Colin Bennett and David Lyon (London: Routledge, 2008); William Walters, "Rezoning the Global: Technological Zones, Technological Work, and the (Un-)Making of Biometric Borders," in *The Contested Politics of Mobility: Borderzones and Irregularity*, ed. Vicki Squire (London: Routledge, 2011).

17. International Civil Aviation Organization, "ePassport Implementation and the ICAO PKD," *ICAO MRTD Report* 7, no. 3 (2012): 3.

18. This history is documented in International Civil Aviation Organization, *Machine Readable Travel Documents (MRTDs): History, Interoperability, and Implementation*, release 1, draft 1.4 (2007).

19. Ibid., 6.

20. Ibid., 7.

21. Ibid., 3.

22. International Civil Aviation Organization, *Machine Readable Travel Documents: Selection of a Globally Interoperable Biometric for Machine-Assisted Identity Confirmation with MRTDs*, ICAO TAG MRTD/NTWG Technical Report (2001).

23. ICAO, *Machine Readable Travel Documents* (2007), passim.

24. Ibid., 14–15.

25. Ibid., 15.

26. Ibid., passim.

27. Walters, "Rezoning the Global," 62.

28. Mike Ellis, "39 Myths about e-Passports: The Facts behind e-Passports and RFID Technology," *Keesing Journal of Documents and Identity*, no. 30 (2009); reprinted in *ICAO MRTD Report* 5, no. 1 (2010): 22–27.

29. Ibid. Note that in 2010, the article was republished in three parts (vol. 5, nos. 1–3) in ICAO's own publication *ICAO MRTD Report*.

30. Ibid., 14.

31. ICAO, "ePassport Implementation," 6–12.

32. Ibid.

33. Ibid.

34. Ibid., 7.

35. Barry Kefauver, "ePassports: The Secure Solution," *ICAO MRTD Report* 2, no. 2 (2007): 5.

36. Lukas Grunwald, "Response to Mr. Barry Kefauver on Security," *MRTD Analysis*, 2007, http://www.mrtdanalysis.org/.

37. Marieke de Goede, "Beyond Risk: Premediation and the Post-9/11 Security Imagination," *Security Dialogue* 39 (2008): 155–76.

38. Lukas Grunwald, cited in Kim Zetter, "Hackers Clone E-passports," *Wired*, August 3, 2006, http://www.wired.com/science/discoveries/news/2006/08/71521?currentPage=all.

39. Zetter, "Hackers Clone E-passports."

40. John Torpey, *The Invention of the Passport: Surveillance, Citizenship, and the State* (New York: Cambridge University Press, 1999); Mark Salter, *Rights of Passage: The Passport in International Relations* (Boulder, Colo.: Lynne Rienner, 2003).

41. Adam McKeown, *Melancholy Order: Asian Migration and the Globalization of Borders* (New York: Columbia University Press, 2008).

42. Andreas Fahrmeir, "Governments and Forgers: Passports in Nineteenth-Century Europe," in *Documenting Individual Identity: The Development of State Practices in the Modern World*, ed. Jane Caplan and John Torpey (Princeton, N.J.: Princeton University Press, 2001).

43. See, e.g., Aas, Gundhus, and Lomell, *Technologies of InSecurity*.

44. Douglas Thomas, *Hacker Culture* (Minneapolis: University of Minnesota Press, 2002).

Passport Photos

Mark B. Salter

Let others check our papers.
—Michel Foucault

The passport is a key artifact in the global mobility assemblage, understood as the *dispositif* of population in circulation. In addition to the more traditional story of a legal-sociopolitical regime that is based on domestic laws concerning citizenship, mobility, and immigration,[1] international standards about identity documentation and travel and frontier formalities, and a set of beliefs and behaviors about mobility—supplemented by a refugee regime that tidies up the loose ends of those excluded from the nation-state system[2]—this new materialist story includes the role of the physical infrastructure and objects of that regime. Following the work of Mika Aaltola, for example, on the hub-and-spoke pedagogical structure of global civil aviation illustrates the way that the actual physical infrastructure of air-travel routes conditions the kinds of travel possible, educates the traveler, and creates a particular imagination of the global.[3] The physical organization of the airport causes particular kinds of subject positions.[4] It is not simply that the passport makes this configuration of global mobility *possible* but also that the passport is a crucial physical part of the infrastructure that acts. If we decenter the human and stop viewing materials only in terms of how they serve the human, then we see the way that passports and data doubles relate to the material infrastructure and the human—how the passports themselves act.[5]

Standing in line to renew my Canadian passport, I overheard the following argument:

"Ma'am, I'm not putting down [on your passport] that your hair is
    blue."—Passport Canada official
"Is my hair blue?"—Blue-haired woman
"Yes."
"Well?"

"It's not naturally blue, is it? I'm not putting down that your hair is blue."

"Are you going to ask every woman in here what her natural hair color is?"

"I'm still not putting down blue."

This moment illustrates Engin Isin's "neurotic citizen" produced by the passport photo: the government official is anxious that the narrative description of the photo will undermine the authenticity of the document, even though the photograph will accurately present a blue-haired woman; the woman is anxious that the passport be consistent with itself and her body. In both cases, the anxiety is produced by the passport—and in the case of this woman, that anxiety will persist. The passport photograph—particularly in the age of biometric passports, in which individuals are instructed to have a passive face—continually speaks against the passport's authenticity, playing a crucial role in constructing the neurotic citizen.[6] The connection of the image of the face and the narration of the body to the government of identity and mobility puts new forces into relation with emergent results. Even in a predigital era, the presence of the photograph causes as many problems as it presumes to solve: first in standardization and then in interpretation. The questions of interpretation persist even in modern forms of measuring the body.[7]

This chapter builds on two literatures: the first encompasses science and technology studies and its interpreters in political and social theory; the second comprises critical literature on mobility from human geography and political science. The work of Bruno Latour, Michel Callon, John Law, and Jane Bennett has been engaged more seriously in political and social theory.[8] Latour argues, for example, that the production of scientific truths is dependent on an assemblage of human and nonhuman actors. Law makes a similar and powerful argument about causality: he argues that the actual empirical and ideational arrangement of things, people, and power is messy and that efficient causality can never capture the way that politics work.[9] My primary argument is that passports are such a technology that acts both toward its bearers and toward its interpreters.

The passport acts, in two ways. First, the passport is a physical object that circulates with other objects, and it conditions the possibility of different kinds of politics and subject positions. It is not simply a question of who has what kind of passport, but rather the fact that the object itself crucially represents a larger subject position, and indeed a prima facie grounding from which to claim certain rights. The passport "triggers new occasions to differ and dispute," such as the 2006 evacuation of Canadian citizens from Lebanon[10] or the case of Abousfian Abdelrazik, a Canadian citizen abandoned in the Sudanese embassy between 2008 and 2009.[11] In the first case, although dual Canadian Lebanese citizens were evacuated by Canada during the 2006 civil war in Lebanon, there was an immediate backlash within the government to question the validity of those citizenship claims. The legal basis to ask for assistance from the Canadian government was not disputed, only that these individuals did not live in Canada or have deep roots in Canadian society. These Canadian passports were seen as escape or insurance passports, to allow the Lebanese to escape their "proper" country. This is a similar discourse to the one surrounding the wave of Hong Kong Chinese arriving in Vancouver in the 1990s.[12] Abdelrazik was the only Canadian on the UN Security Council Terrorism 1267 blacklist. Despite his being cleared by the Sudanese and Canadian governments, Canada would initially not issue him travel documents but permitted him to live within the Canadian embassy.[13] In each case, the object of the passport acts. Second, the passport is a crucial object in the assemblage of global mobility. The passport, and in particular the passport photograph, acts; not only does it circulate in an open, complex system of other systems, including data flows, immigration controls, risk profiling, passenger-name data, tourism figures, and the like, but it also acts on the citizen-subject. The technology of the passport acts as a kind of mirror, insinuating a new anxiety that our identification documents do not resemble us, creating neurotic citizens who are constantly anxious that they may not be recognized. Thus, in addition to connecting the body to a number of open systems, making possible a set of politics, the passport also acts independently of governments, interpreters, and bearers.

Others have written on the passport and its central role in the governance of mobility.[14] The passport was formerly one tool that governments and individuals used to articulate the conditions of possibility for protection and claiming rights. I have argued (1) that the passport was a governmental tool that constituted, classified, and managed particular populations in relation to their political, economic, social, and biopolitical utility for the state and (2) that the status and identity represented by the documents and the claims that they represent were both made possible and fundamentally undermined by the discretion that is essential to the system.[15] Governments had used identification and authorization functions to police the legitimate means of violence, public health, and mobility. Security can be understood in relation to circulation.[16] Michel Foucault argues that the mode of security is an ever-expanding *dispositif* for the management of circulation—the circulation of ideas, of capital, of goods, of people. The contemporary global mobility regime is made possible and structured through the object of the passport and its attendant networks of visas, entry/exit systems, airline-booking systems, and automated risk evaluation. A passport makes certain kinds of identity interpretable and verifiable, rights claimable, border-control examinations possible, and certain circuits of mobility more or less easy. For example, if a passport can be linked to a body, dossier, or identity, then the document can be examined to make a decision about the bearer. The legal requirement for a passport for (the vast majority of) international travel is possible only if there is a large bureaucracy that can make, process, and validate passports: in the absence of the passport office, there can be no passport; equally, there can be no passport office without the passport. The passport also makes possible certain kinds of neurotic subjects: citizens who are anxious about their ability and authenticity of producing themselves as subjects that are readable as rights-claiming bodies.

Global Circulation Assemblage

Traditional scholarship on the global circulation of population is anthropocentric, focusing on the actions of humans as individuals, in groups, or in institutions. The mainstream literature in IR that describes or explains the global mobility regime is not just anthropocentric but also deeply liberal. Within this perspective, the global circulation of populations is a matter of individual intention and state policies. However, we can point to a mobilities turn in international relations and geography that is focused more on Foucauldian and Deleuzian models of power and is inspired by the Latour turn toward objects. Rather than take the state as a single unit of analysis, critical scholars look at the biopolitical governance of flows and sites. In this reading, the management of global population flows is not simply in the hands of states at the site of the frontier but is dispersed throughout different institutions, spaces, actors, and objects. There is an emergent research agenda in mobility studies that pays attention to the role of objects. Peter Adey and Alison Mountz, from a geographical perspective, often focus on the spatial and geopolitical dynamics of the management of mobility, both in its everyday and exceptional forms.[17] By focusing on the material emplacements of sites of mobility control (airports, islands, boats, camps), they both detail the way that architecture and state power go hand in hand and come to have very real effects. Adey specifically uses Latour to discuss the way that objects such as the passport, luggage, food, gifts, and airplanes themselves are assembled, surveilled, policed, and sorted at the airport.[18] Chris Rumford has used objects such as the pork pie and Stilton cheese as a locus for understanding "borderwork," in which local communities create and mobilize borders that function across levels of governance.[19] William Walters particularly engages with the work of Latour to talk about the assemblage of insurance, dogs, and checklists that are part of the governance of stowaways.[20] Walters also writes specifically on the development of the biometric passport, which "illustrates the emergence of new security spaces and practices that may be global in their aspiration but, in terms of their actual configuration, are better described in terms of the zone."[21] He emphasizes the "difficult and painstaking labour that goes into assembling, maintaining or extending . . . new spaces of security."[22] Can E. Multu examines the development of the ePassport and transport-container infrastructure to understand the material impacts of the European Neighbourhood Policy.[23] Connected to materialist-oriented writings in surveillance studies on the impact of particular technologies, there is a robust, if dispersed, literature on the ways that objects impact global mobility. This chapter aims to make the engagement with the new materialist literature more explicit and also to add another empirical case.

What does attributing agency to the passport do? The passport validates and structures claims to rights, and so it acts. What is the difference between the claim that the passport—or rather the modern standardized passport—acts and the more modest claim that the passport makes something possible? It is ontologically different to say that the passport is part of an emergent assemblage than to say that human actions are only possible because of a set of technical factors. In the case of the mobility assemblage, the passport not only makes things possible but also acts: we are able to make meaning from the passport, and really make meaning of ourselves, through the object of the passport in fundamentally different ways than we were able to when passports were simple letters. The passport photograph gives rise to a new kind of anxiety, a new neurosis for the citizen, that our face will be read against us, because the passport photo is always one that doesn't look like us. Lily Cho argues that the new instructions for photographs in new biometric passports encounter the presumed affective connection between identity and citizenship.[24]

The passport photograph creates a new ground for identification by the state: the presupposed isomorphism of the body to itself, the representation to the body, and the identity to both the body and the representation. It makes new kinds of authentication possible and creates the possibility that, for the first time, one cannot prove one's own identity, because one's body is not identical to its representations. The evolution of the modern biometric or machine-readable travel document demonstrates this point clearly. The technical standards that render identity data into machine-readable code open up a new space for political action: data doubles interact with security algorithms and risk profiles in entirely novel and unpredictable ways.[25] Automated systems that use passport data act, or more specifically interact, permitting or restricting mobility based on algorithms of risk and suspicion rather than the actual rights or identity of the travel.[26]

Gayatri Chakravorty Spivak details her experience of this system:

> I was supposed to take the airplane from Heathrow on Sunday. Air Canada says to me: "we can't accept you." I said: "why?" and she said: "You need a visa to go to Canada." I said: "look here, I am the same person, the same passport. . . . But you become different." When it is from London . . . I need a visa to travel from London to Canada on the same passport, but not from the United States. To cut a long story short, . . . I said to the woman finally before I left, in some bitterness:
> "Just let me tell you one small thing: Don't say 'we can't accept you' that sounds very bad from one human being to another; next time you should say: 'The regulations are against it'; then we are both victims."[27]

It is precisely the assemblage of the MRTD passport and the airline information system that prohibits Spivak's journey from England to Canada, even though the same assemblage permits that same Spivak to journey from the United States to Canada on the same airline with the same passport. This is not an error or bureaucratic mistake, yet the passport and Air Canada's boarding system prohibit Spivak from traveling, regardless of her actual rights status, and indeed the assemblage does not provide space for either the airline agent or Spivak to override the decision of the system. The automatic collection of passport data makes possible the automation of board/no-board decisions that are based on risk algorithms rather than rights or identity. Early passports and identity cards were understood to represent a body, but with "no evidence of character or loyalty . . . a physical means of attaching to a person a number through which particulars of that person can be traced in other records."[28] As Louise Amoore concludes, "[I]t is not strictly collected data that become an actionable security intervention, but a different kind of abstraction that is based precisely on an absence, on what is not known, on the very basis of uncertainty. . . . [T]he data derivative is not centred on who we are, nor even on what our data says about us, but on what can be imagined and inferred about who we might be, on our very proclivities and potentialities."[29]

Methodologically, is it possible to distinguish the agency of the things and the agency of the humans? The materialist argument would be that we can, but only when we have an emergent understanding of causality. The agency of the passport can be seen at a key moment in the evolution of that assemblage: the introduction of the photograph to the passport. This chapter examines that focusing event through archival work based on records at the British National Archives.

Passport Photos

During and in the immediate aftermath of the First World War, there was accelerated change in the assemblage of tools and agents for managing mobility. Important structural dynamics of that period include the breakup of large empires and the creation of new nation-states in Europe, the presence of large numbers of refugees who posed both a political and a public-health risk, and new international organizations that asserted authority (such as the Nansen Organization for Refugees and the League of Nations). With the proliferation of state borders and a growing concern over mobile populations, the passport became a more centralized tool of national and international governance, rather than a convenience or inconvenience for travel of the transnational elite and the increasingly upper-middle class. In this period, the passport became a standard part of the travel/mobility assemblage that helped concretize identity certification, affirmed state capacity for both identification and border policing, and expanded the domain of action in which the state could act. Crucially, moreover, in addition to being a way for states to define and manage their populations, the passport influenced the way national subjects understood themselves.

One of the most important moments in this construction of the postwar mobility assemblage was the collision of photography and the passport. Before 1915, portraits were not connected to identity documents, despite the availability of daguerreotypes in British metropolitan areas throughout the latter half of the nineteenth century, the invention of modern film in the late 1880s, and the wide availability of the Kodak Brownie camera at the turn of the century. The portrait photograph was available and standard before the passport. The integration of the photograph into the passport coincided with the invention of the standardized passport, which was both mandatory for international travel and made accessible through a more systematized bureaucratic application procedure. One moment of flux is a good focus point: the period between 1915 and 1920, as the UK passport was being developed but before the League of Nations Technical Subcommittee had formulated international norms for the passport. During this time, the passport photograph acted in the creation of a particular international mobility regime.

In 1914, passports were single folded sheets, sometimes with attached photographs of all styles.³⁰ The passport photograph was not yet trusted as a technology for identification, and so there were written descriptions of the body captured in the photograph. For example, descriptions included the following:

age
height
forehead (high, ordinary, oval, slightly receding)
eyes (colors—blue, green, brown, including gray)
nose (large, straight, Roman)
mouth (straight, firm, large, ordinary, medium, thick)
chin (round)
hair (brown)
complexion (fresh, pale, peachy, dark)
face (oval, thin)

By 1915–16, Form A included a requirement for a photograph, but there was no mandated format: some were seated formal portraits, one with a wooded background; others were more full-face; and in some cases individuals not named and not photographed (domestics or nannies) were covered by another person's passport.³¹ By the end of 1916, the standard was a fold-out passport of 5-by-2-inch panels with a blue binding, embossed with a gold coat of arms and the word *Passport* on the outer cover. Passport photos started to be regulated: full face, no hat or head covering (except in the case of "nuns, rabbis, doukhobor women, East Indians").³²

There is a lack of standardization of early photographs in the National Archives. The Passport Office initially did not require a particular kind of photograph and did not issue a standard document. Some of the key norms that the passport photograph made possible are absent: there was no imperative to name each individual, or to have a standard photo, or to associate one document with exclusively one person, or to make passports valid for more than one trip or longer than two years. Wives, children, and domestics were often included in a British national's application—indeed, adults could even have nonfamilial children placed on their passports for particular journeys with parental consent. Indeed, as late as the 1930s, one could be issued a Junior Collective Certificate, which acted as a group passport for "approved parties of students, boy scouts and girl guides."³³ For example, one passport was issued in 1915 to a woman, her unnamed children, and an accompanying anonymous nanny. The photograph included in the application (but not the document) is a family portrait without the nanny. Refusals were made on the bases of morality and politics but not identity: a woman of ill repute, for example, whom the officer suspected was traveling to engage in dance (understood as a euphemism for prostitution), or women who were traveling contrary to their parents' wishes. Over the space of a couple of years, the Passport Office came to integrate the photograph into the passport document itself, as well as the passport application form, but slowly and skeptically. The passport did not come in book form until 1920. The purpose of the passport is clearly indicated in the Foreign and Commonwealth Office's 1921 circular *Passports and Visas*: "[T]he primary object of the passport is to constitute official proof of the national status and identity of its holder, for purposes of either travel or of residence abroad."[34] Passports were valid only for two years and for a particular named set of countries.[35]

A photograph relies on both a technology and a set of reading skills—it is both a representation and a claim to representation.[36] But the turbulence of the collision of these technologies of identification came in the midst of the First World War.[37] The integration of the photograph into the Canadian passport application form demonstrates this turbulence: in the form of 1916, the photograph was given a space, but it was corroborated by accompanying physical descriptions of the photograph.[38] The passport application form used but also resisted the photograph; it incorporated the photograph as a representation/identification of the individual, but it also asked for a narrative corroboration. This narrative corroboration is also defined in Passport Office circulars, and directions for determining "visible distinguishing marks" are provided: "(a) only marks visible while wearing ordinary clothing are to be described, (b) only permanent marks are to be described, i.e. do not describe beards, moustaches, baldness, eyeglasses, tattoomarks, or warts . . . (e) use the word 'missing' rather than 'amputated' or 'cut off' . . . (g) use the word 'irregular' to describe deformities, e.g. a broken nose is described 'nose irregular.' A harelip is described as 'upper lip is irregular.' (h) use the word 'defective' to show an underdeveloped arm, leg, foot, hand or feature, e.g. 'right ear defective.'"[39] There lies a double move: the passport made the document and the body identifiable to itself, but there was also a self-limiting recognition that the photograph did not perfectly represent the body. Again, there was a kind of multiple anxiety engendered by the passport: anxiety of the bearer, the issuer, and the interpreter that the document was not faithful to the body or to the identity.

The application form was settled by 1921, and included "duplicate small unmounted photographs of the applicant (and wife if to be included) . . . one of which must be certified on the back by the recommender."[40] The two primary authenticators of identity were the declaration by a "responsible British subject possessing one of the following qualifications, vis: British official, resident

British merchant, banker, minister of religion, barrister, solicitor, physician or surgeon"; and a photograph, which was also signed by the guarantor.[41] The question of the guarantor is particularly interesting, because the Passport Office recognized in its own documentation that there was little room in the vetting process to actually verify that the applicant was known to the guarantor; it was important only that the guarantor existed in a professional register. Consequently, "only fools are caught and knaves escape"; those who made up the name of a judge or doctor were at risk, but those who faked the signature of an actual mayor or priest were very unlikely to be caught.[42]

The passport photo size was later standardized.[43] By 1948, the passport application Form A included specific instructions about the photograph: "a recent photograph . . . must be taken full face without hat, and the photographs must not be mounted. The size of the photographs must not be more than 2½ inches by 2 inches or less than 2 inches by 1½ inches. The photographs must be printed on normal thin photographic paper and must not be glazed on the reverse side."[44] There was a concern that the photographs not come from slot machines. At this time, then, the passport photo made possible and necessary a reading of the person and his or her image. Still, the key anxieties about identity verification and authentication remained unresolved: guarantors were still not verified, breeder identity documents (such as birth or baptismal certificates) still lacked biometric information, and passport photos still didn't look like the bearer. To examine this problem of isomorphism and misrecognition, we turn to the question of the neurotic citizen and his or her photograph.

Affect, Neurosis, and Identity

French psychoanalyst Jacques Lacan argues that one of the formative moments in the psychic development of a child is the mirror stage. When a child recognizes herself in the mirror, she sees herself, but a version of the self that is not anxious: the very possibility of an affirmative self that is not plagued by anxiety creates an anxious subject, a child who always knows that she is not unified, always insufficient.[45] The passport acts a mirror for the modern citizen-subject: it is the object in which we recognize a unified political subject whose claim to rights is not subject to discretion, a subject position that we cannot occupy. I argue elsewhere that the passport is a "technology of the international self,"[46] one of the technologies through which, as Foucault describes, "we have been led to recognize ourselves as a society, as part of a social entity, as part of a nation or of a state."[47] In the passport photo, we recognize an ideal citizen position that we cannot achieve. We are anxious now not that our photos do not look like us—to ourselves or to others—but rather that our photos look suspicious in their very composition.

Cho details these photos available for automated analysis: "They are natural only insofar as they are completely emotionally neutral. That is, they are not natural at all. They expose a citizen-subject caught and composed for identification purposes. This subject is neither angry, happy, sad, disgusted, nor even particularly present."[48] Cho is more concerned with the ways that artistic practice can trouble the attempt by the state to separate the affective register and citizenship. She argues that emotion is a vital aspect of the citizen that is denied in the state's interpellation in the photo. Lacan's mirror phase is a particularly apt frame: national subjects see themselves reflected through the neutral passport photo as being already too affective for the image of dispassionate citizen that the state desires. However, not only are the passport photo and the anxiety it induces vulnerable to artistic practice, but also anxiety circulates and produces a particular form of citizen-subject.

Isin's description of this position as the "neurotic citizen" uses a Freudian model of neurosis: "[T]he unattainable imago of wholeness and in its ultimately unfulfillable desire [the subject] could never attain a sufficient wholesomeness that is always posited as 'normalcy.'"[49] He argues that with the dominance of the paradigm of risk, governments are increasingly producing neurotic citizens: "The neurotic subject is one whose anxieties and insecurities are objects of government not in order to cure or eliminate such states but to manage them."[50] These neuroses are marshaled and managed at the border.[51] Travelers are required by law and obligated by the structure of the inquisition to disclose a narrative of self and belonging that is legible to the border guard's presuppositions about safe travelers.[52] Border interrogations thus produce anxiety and falsehoods, which fuel the anxiety of the border guards: the entire system works only because of the circulation of anxiety. The border guard is anxious not to be fooled, whereas the border crosser is anxious to be let across and to not be caught in a lie or in a crime. The system of border inspections works only because citizens and irregular migrants are conditioned to confess, and border guards are trained to ignore falsehoods. Psychologists make the argument that "deception is a major aspect of social interaction; people admit to using it in 14% of emails, 27% of face-to-face interactions, and 37% of phone calls."[53] Psychologist Stephen Porter argues that aggressive questioning by border guards precisely generates falsehoods: "In fact, psychological research shows that such techniques only serve to heighten anxiety in travellers and make them embarrassed and nervous. Travellers respond with anxiety, which then is mistakenly taken by border staff as a sign that the passengers have something to hide."[54] Even the technology of facial recognition requires human judgment. Automated systems, by contrast, have no way of detecting between the natural falsehoods that are part of any conversation, and dangerous falsehoods. The automated facial-recognition system that requires the neutral passport photograph, then, appears to both produce and manage the same neuroses: the technology itself produces photographs and unreliable readings that in turn produce anxiety in both the citizen and the border guard. The citizen must perform a certain degree of anxiety to demonstrate that he or she recognizes the authority of the border guard, but not so much as to suggest he or she has something significant to hide.

The effect of the passport photograph as Lacanian mirror, creating the neurotic citizen, is not reducible, however, to human intention or rationalities. The role of technology itself in the production of this Lacanian mirror is important. Because of the limits of biometric technology and computing capacity, facial-recognition programs simply cannot detect facial similarity with a high degree of accuracy (or even a low degree of accuracy) unless the primary image is neutral. Walters describes how "the work of standardization cannot confine itself to the specification of which images are to be captured or what kind of materials should be used in making a passport booklet. In the name of optimizing 'face image quality' it must extend itself into the fine details of how a person is to be photographed."[55] Facial-recognition technology itself is young and remains somewhat unreliable. Even the most advanced platforms fail to meet their own standards for reliability.[56] To match face to face in even a rudimentary way, photographs must control for as many variables as possible: shadows, background, expression, color, and so on. Technical limits with even the most up-to-date systems, just as with those early photographs in the post–WWI era, still render the photograph an imperfect mirror. The neutral photograph is also an artifact of the technology itself.

Lacan makes the point that when the child sees himself in the mirror, the reflection precisely seems whole and unified. The child might struggle to make a particular physical gesture, yet the mirror-child seems to make the same gesture effortlessly. Similarly, the passport photographic self purports to connect directly the body with the name with the information. However, as Shoshana Magnet writes, "[T]he use of biometrics to test identity is asked to perform the cultural work of stabilizing identity, conspiring in the myth that bodies are merely containers for unique, identifying information which may be seamlessly extracted."[57] However, the citizen knows how much effort (and dissimulation) it takes to present as a complete self for the judgment of the border guard and the automated system. Further, the failure of biometrics to successfully classify, and in particular to be efficient and effective at authenticating, vulnerable or marginalized populations is not an accidental outcome of the system but an inherent part of the system. The failure of biometric technologies is aleatory: it circulates in ways that reproduce other structures of power.

Maher Arar's Passport

Sociological analyses of the passport have analyzed the document exclusively as a tool of government or global governance, connecting individuals to their status as citizens. This chapter has made a strong claim that the passport as an object acts in the world, that it acts in concert with other systems within the global mobility assemblage not simply to make certain circuits viable and others impossible but also to generate action in the world that would not be predictable through a governmental analysis alone.

A final example will illustrate the power of the thing: Maher Arar, a Syrian-born Canadian citizen who was a subject of the American extraordinary-rendition program, was apprehended in the United States and tortured in Syria, then released and completely exonerated by an official commission of inquiry. Although the analysis of and conclusions found in this case focused on the errors and responsibilities of various government agencies and the ways that, in particular, the Royal Canadian Mounted Police (RCMP) failed to adhere to its own policies about information sharing, one could tell a powerful materialist story that places the passport at the center of an open assemblage of circulating bodies and data that had unintended and unexpected effects. For example, the report is careful to conclude that once the initial (unsubstantiated) intelligence assessment was shared with the American authorities, "there is no evidence that Canadian officials participated or acquiesced in the American authorities' decisions to detain Mr. Arar and remove him to Syria. It is very likely that, in making the decisions to detain and remove Mr. Arar, American authorities relied on information about Mr. Arar provided by the RCMP."[58] Arar himself was traveling on a Canadian passport and repeatedly asked for a Canadian consul; nevertheless, he was deported to Syria. The information that elicited such a decision is given as follows:

> detailed biographical material, various Canada Customs material, including material obtained from . . . secondary examinations at the Canadian border [that] were subject to explicit caveats, and shared with the American agencies without Canada Customs' consent . . . faxes, business materials, address books, phone lists, an agenda and

hard-drive data ... [including information that was] misleading: immigrant visa and record of landing, information that Mr. Arar applied for a gun permit in 1992 ... [and] an analysis of the names found on Mr. Arar's PDA, seized by Canada Customs on December 20, 2011.[59]

The information, extracted from both human and nonhuman sources, came to assemble a "threatprint" of Arar that made certain pathways and circuits possible and necessary.[60] In each moment of Arar's contact with border authorities, the passport was copied, photographed, and examined, and it came to connect that body with this dossier and a misleading risk assessment. In the words of the commission, "Swiping a passport or entering other identifying information generates a 'hit,' which informs the front-line officer that this individual must undergo a second, more thorough examination."[61] The passport plays a vital role in the assemblage that includes both human and nonhuman actors.

However, the passport does not simply facilitate government action and control; it also acts against the government. The Canadian government, in particular, was called to account by Arar's passport as much as by the individuals involved.[62] Just as the possession of Canadian passports by Lebanese dual citizens compelled Canadian action despite a great deal of reluctance, so too did Arar's passport compel an investigation. Once American authorities concluded that Arar was to be removed, Arar and others were concerned that he was going to be removed to Syria rather than Canada, even though Arar was traveling on a Canadian passport. Arar repeatedly asked to be deported and removed to Canada.[63] The fact that Arar was traveling on a Canadian passport compelled Canada to be involved, even when it did not wish to sour its relationship with the United States. At each stage in the report, when the RCMP and the Canadian Security Intelligence Service discussed Arar's fate with the FBI, there was a continual reference to notes: which e-mails and telephone calls were logged, and which were recorded in the officer's notebook. The object of the notes came to corroborate the memory or experience of the human. When Canada's Department of Foreign Affairs and International Trade first directed the consulate general in New York, where Arar was being held, to determine his situation, they noted that from his passport "there was no information on file concerning a problem with Mr. Arar's passport."[64] However, it was the data on the unproblematic passport that opened the connections to all of the misleading intelligence data. However, when Arar was finally repatriated to Canada, more than one year later, it was this unproblematic citizenship claim that provided the ground for a public demand for an inquiry and subsequent apology.

Bodies cannot be reduced to photographs, just as identities cannot be reduced to bodies. One of the primary insights of the materialist turn is to reject the subject/object dualism. Rather than reduce all agency and responsibility to the human, as the commission attempted to do in its report, we can see that information, data, and objects themselves come to assemble in ways that are unpredictable and emergent. The passport itself triggers the connections that make certain circuits of global mobility for both frequent travelers and for extraordinary rendition. This chapter has demonstrated the utility of a strong model of vital materialism by examining the passport and its attendant technologies of identification. Much more work in this new materialist vein remains.

Notes

Archival research for this project was conducted in 2002 with the support of the American University in Cairo and again in 2004 with the support of the University of Ottawa, and presented at the *Millennium* annual conference in 2012 with the support of the Social Sciences and Humanities Research Council of Canada. My thanks to Daniel Levine, Jarius Grove, Can E. Mutlu, and others for their critical comments.

1. Alexander Betts, ed., *Global Migration Governance* (Oxford: Oxford University Press, 2011); Rey Koslowski, ed., *Global Mobility Regimes* (New York: Palgrave Macmillan, 2011).

2. Vicki Squire, ed., *The Contested Politics of Mobility: Borderzones and Irregularity* (London: Routledge, 2011).

3. Mika Aaltola, "The International Airport: The Hub-and-Spoke Pedagogy of the American Empire," *Global Networks* 5, no. 3 (2002): 569–74. See also Peter Adey, "Airports, Mobility, and the Calculative Architecture of Affective Control," *Geoforum* 39, no. 1 (2008): 438–51; Aharon Kellerman, "International Airports: Passengers in an Environment of 'Authorities,'" *Mobilities* 3, no. 1 (2008): 161–78; Andrew Wood, "A Rhetoric of Ubiquity: Terminal Space as Omnitopia," *Communication Theory* 13 (2003): 324–44.

4. Gillian Fuller and Ross Harley, *Aviopolis: A Book about Airports* (London: Black Dog Books, 2004).

5. David Lyon, "Airports as Data Filters: Converging Surveillance Systems after September 11th," *Information, Communication, and Ethics in Society* 1, no. 1 (2003): 13–20.

6. Engin F. Isin, "The Neurotic Citizen," *Citizenship Studies* 8, no. 3 (2004): 231–32.

7. Shoshana A. Magnet, *When Biometrics Fail: Gender, Race, and the Technology of Identity* (Durham, N.C.: Duke University Press, 2011); Benjamin J. Muller, *Security, Risk, and the Biometric State: Governing Borders and Bodies* (New York: Routledge, 2010).

8. Michel Callon, "Some Elements of a Sociology of Translation: Domestication of the Scallops and the Fishermen of St. Brieuc Bay," in *Power, Action, and Belief: A New Sociology of Knowledge?*, ed. John Law (London: Routledge & Kegan Paul, 1986); Bruno

Latour, *Science in Action* (Cambridge, Mass.: Harvard University Press, 1987), Bruno Latour, *Reassembling the Social: An Introduction to Actor-Network-Theory* (New York: Oxford University Press, 2005); John Law, "Objects and Spaces," *Theory, Culture, and Society* 19, nos. 5–6 (2002): 91–105; Jane Bennett, "The Force of Things: Steps toward an Ecology of Matter," *Political Theory* 32, no. 3 (2004): 347–37; Jane Bennett, *Vibrant Matter: A Political Ecology of Things* (Durham, N.C.: Duke University Press, 2010).

9. John Law, *After Method: Mess in Social Science Method* (New York: Routledge, 2004).

10. Peter Nyers, "Dueling Designs: The Politics of Rescuing Dual Citizens," *Citizenship Studies* 14, no. 1 (2010): 47–60.

11. Reg Whitaker, "The Post-9/11 National Security Regime in Canada: Strengthening Security, Diminishing Accountability," *Review of Constitutional Studies* 139, no. 2 (2011–12): 139–58.

12. Shibao Guo and Don J. DeVoretz, "Chinese Immigrants in Vancouver: Quo Vadis?," *Journal of International Migration and Integration* 7, no. 4 (2006): 425–47.

13. Paul Koring, "Terror Claims Trap Canadian in Khartoum," *Globe and Mail*, April 28, 2008, http://www.webcitation.org/61w6Qg18N.

14. Martin Lloyd, *The Passport: The History of Man's Most Travelled Document* (Stroud, Eng.: Sutton, 2003); John Torpey, *The Invention of the Passport: Surveillance, Citizenship, and the State* (Cambridge: Cambridge University Press, 2000); Andreas Fahrmeir, "Governments and Forgers: Passports in Nineteenth-Century Europe," in *Documenting Individual Identity: The Development of State Practices in the Modern World*, ed. Jane Caplan and John Torpey (Princeton, N.J.: Princeton University Press, 2001), 218–34.

15. Mark B. Salter, *Rights of Passage: The Passport in International Relations* (Boulder, Colo.: Lynne Rienner, 2003); "Passports, Mobility, and Security: How Smart Can the Border Be?," *International Studies Perspectives* 5, no. 1 (2004): 71–91.

16. Michel Foucault, *Security, Territory, Population: Lectures at the Collège de France, 1977–78*, trans. Graham Burchell (Basingstoke, Eng.: Palgrave Macmillan, 2007), 18.

17. Peter Adey, *Mobility* (New York: Routledge, 2010); Alison Mountz, *Seeking Asylum: Human Smuggling and Bureaucracy at the Border* (Minneapolis: University of Minnesota Press, 2010).

18. Peter Adey, "Secured and Sorted Mobilities: Examples from the Airport," *Surveillance and Society* 1, no. 4 (2003): 500–519.

19. Chris Rumford, "Citizens and Borderwork in Europe," introduction to *Space and Polity* 12, no. 1 (2008): 1–12.

20. William Walters, "Bordering the Sea: Shipping Industries and the Policing of Stowaways," *Borderlands* 7, no. 3 (2008): 9–11, http://www.borderlands.net.au/vol7no3_2008/walters_bordering.pdf.

21. William Walters, "Rezoning the Global: Technological Zones, Technological Work, and the (Un-)Making of Biometric Borders," in Squire, *Contested Politics of Mobility*, 54.

22. Ibid.

23. Can E. Mutlu, "Material Cultures of Mobility: Security in the European Neighbourhood" (PhD diss., University of Ottawa, 2013).

24. Lily Cho, "Citizenship, Diaspora, and the Bonds of Affect: The Passport Photograph," *Photography and Culture* 2, no. 3 (2009): 275–88.

25. Louise Amoore, "Algorithmic War: Everyday Geographies of the War on Terror," *Antipode* 41, no 1 (2009): 49–69.

26. Louise Amoore, "Data Derivatives: On the Emergence of a Security Risk Calculus for Our Times," *Theory, Culture, and Society* 28, no 6 (2011): 24–43.

27. Gayatri Chakravorty Spivak, "Questions of Multi-culturalism," interview by Sneja Gunew, in *Interviews, Strategies, and Dialogues,* ed. Sarah Harasym (New York: Routledge, 1990), 65.

28. Great Britain Home Office, *Draft Booklet for Issuance to Police Consolidating All H.O. Circulars on National Registration Identity Documents* (1944), H.O. circular 700,600.17, PRO: HO 213 754.

29. Amoore, "Data Derivatives," 27–28.

30. Great Britain Foreign and Commonwealth Office, application for passport, no. 12037 (1914), PRO: FO 737/24.

31. Great Britain Foreign and Commonwealth Office, application for passport (1916), PRO: FO 737/24.

32. Canada Passport Office, revisions of Canadian passport regulations (1939), National Archives of Canada: RG 25 256 8. Doukhobors were an exiled group of Russian Orthodox Christians whose women wore head covering.

33. Canada Passport Office, Junior Collective Certificates 4662 (8799), June 1933, National Archives of Canada: 10-BK-40.

34. Great Britain Foreign and Commonwealth Office, *Passports and Visas* (1921), 21, PRO: FO 612/265.

35. Ibid., 14.

36. Michael J. Shapiro, *The Politics of Representation: Writing Practices in Biography, Photography, and Policy Analysis* (Madison: University of Wisconsin Press, 1988); Lene Hansen, "Theorizing the Image for Security Studies: Visual Securitization and the Muhammad Cartoon Crisis," *European Journal of International Relations* 17, no. 1 (2011): 453–63.

37. Paul Fussell, *Abroad: British Literary Traveling between the Wars* (Oxford: Oxford University Press, 1980), 24–31.

38. This practice was similar to that in other countries. My great-grandfather had a laissez-passer from the governor of Venice to go from Trieste to Belgrade in 1919, which included descriptions of build, hair, eyes, nose, mustache, beard, and color, as well as space for a photograph.

39. Canada Passport Office, *Passport Writing Sections* (1939), National Archives of Canada: RG 25 (8).

40. Ibid., 20.

41. Ibid., 14.

42. Great Britain Treasury Office, *Passport Office Reorganisation* 4933/54/505 (1938), PROF: FO 612/230.

43. Great Britain Foreign and Commonwealth Office, *Passport Office*, misc. 2031/7 (1940), PRO: FO 612/201.

44. Great Britain Foreign and Commonwealth Office, passport application Form A (1948), PRO: FO 612/267.

45. Jacques Lacan, "Some Reflections on the Ego," *International Journal of Psychoanalysis* 34 (1953): 14.

46. Mark B. Salter, "The Global Visa Regime and the Political Technologies of the Self: Borders, Bodies, Biopolitics," *Alternatives* 31, no. 2 (2006): 167–89.

47. Michel Foucault, "The Political Technology of Individuals," in *Technologies of the Self: A Seminar with Michel Foucault,* ed. Luther H. Martin, Huck Gutman, and Patrick Hutton (Amherst: University of Massachusetts Press, 1988), 146.

48. Cho, "Citizenship," 276.

49. Isin, "Neurotic Citizen," 223.

50. Ibid., 225.

51. Orvar Löfgren, "Crossing Borders: The Nationalization of Anxiety," *Ethnologia Scandinavica* 2 (1999): 5–27; Mark B. Salter, "When the Exception Becomes the Rule: Borders, Sovereignty, and Citizenship," *Citizenship Studies* 12, no. 4 (2008): 365–80, esp. n44.

52. Michael R. Curry, "The Profiler's Question and the Treacherous Traveler: Narratives of Belonging in Commercial Aviation," *Surveillance and Society* 1, no. 4 (2004): 476.

53. Stephen Porter and Leanne ten Brinke, "The Truth about Lies: What Works in Detecting High-Stakes Deception," *Legal and Criminal Psychology* 15, no. 1 (2010): 57.

54. Stephen Porter, quoted in Sarah Barmak, "Border Rudeness: Maybe the Jerk Method Doesn't Work," *Star* (Toronto), May 2, 2010, http://www.thestar.com/news/insight/article/803235--border-rudeness-maybe-the-jerk-method-doesn-t-work.

55. Walters, "Rezoning the Global," 64.

56. Benjamin J. Muller, "Borderworld: Biometrics, AVATAR, and Global Criminalization," in *Globalisation and the Challenge to Criminology,* ed. Francis Pakes (New York: Routledge, 2012), 129–45.

57. Magnet, *When Biometrics Fail,* 123.

58. Commission of Inquiry into the Actions of Canadian Officials in Relation to Maher Arar, *Report of the Events Relating to Maher Arar: Analysis and Recommendations* (Ottawa: Government of Canada, 2006), 14.

59. Commission of Inquiry into the Actions of Canadian Officials in Relation to Maher Arar, *Factual Background I* (Ottawa: Government of Canada, 2006), 95–96.

60. Alexandra Hall and Jonathan Mendel, "Threatprints, Threads, and Triggers," *Journal of Cultural Economy* 5, no. 1 (2012): 9–27.

61. Commission of Inquiry, *Factual Background I,* 58.

62. Monia Mazigh, *Hope and Despair: My Struggle to Free My Husband, Maher Arar* (Toronto: McClelland & Stewart, 2008).

63. Commission of Inquiry, *Factual Background I,* 168–70.

64. Ibid., 181.

The Traffic Light

Katherine Reese

From the first highways, which extended the reach of emperors and kings to the hinterlands, to contemporary national motorway systems, the road has long acted to order space into particular frameworks of authority. However, the twentieth century saw the traffic light subtly help to transform the road into a well-ordered and internationally legible space. In the everyday lives of the world's city dwellers, the traffic light has come to act as a powerful but sometimes unnoticed technology of control. It regulates automobility—both in the narrow sense of the word, as movement in automobiles, as well as more broadly, in the sense of autonomous mobility.[1] A green light appears, and one is expected to move forward in a given trajectory; a red light appears, and one is legally obligated to stop. By projecting onto the road a set of rules about proper movement, the traffic light renders the road a legible, well-ordered space. At the same time, the traffic light is thoroughly bound up in state authority. It translates legal rules of the road into a series of illuminated colors, yet its existence as a technology is conditional upon state funding, implementation, and maintenance. It is an object through which the state continually renews its promise of providing order through control in a chaotic world.

The twentieth century saw repeated and really quite successful efforts at the standardizing of road signs, including traffic lights, internationally. These efforts have understood the traffic light to have a latent universal quality, because it communicates through colors rather than words: it renders order visual in a way that transcends national languages. Traffic lights have made the road a universally legible space, a space that anyone can understand, regardless of national origin or linguistic training. In turn, these efforts understand order to be clearly linked to safety efficiency: the orderly road is the safe and convenient road. As such, the international standardization of the traffic light is part of a broader project to make the world an easily navigated space, a world safe for international trade and travel. The traffic light, then, has played a small but meaningful role in the incredible proliferation of international mobility—as both an actual and a normative force—in the twentieth and twenty-first centuries.

The Order-Producing Traffic Light

If "technologies are not merely aids to human activity, but also powerful forces acting to reshape that activity and its meaning,"[2] then the traffic light is a technology that reshapes human movement in quite clear ways. It projects its directives to the driver: Stop. Go. For this reason, traffic-light systems have been accused of attempting "to impose a strong social control over the most fundamental of human behaviors, whether to move or be still."[3] Although lights are occasionally enforced by means of cameras or police officers, more often it is the sheer knowledge of what the code of conduct is that keeps one motionless on red. The driver stopped at a red light, as a reader of Foucault would recognize, is a disciplined body: a dynamic of utility and docility work together to keep the driver's foot on the brake until she is permitted to move.

Like any form of governance, the traffic light neither universally succeeds in its intended purpose nor wholly determines behavior. Perhaps the reader has her or his own experience of the ways in which the traffic light's authority sometimes fails to govern one's behavior, depending on the nation, the city, the time of day, and the degree to which one is running late or otherwise distracted. If nothing else, the frequent failure of traffic lights' authority is illustrated easily enough by the impressive number of studies attempting to determine why drivers run red lights. Judging by the long and prolific history of this genre of study, proneness to red-light running (or RLR in the lingo) seems to be the unbanished bane of traffic engineering's existence. There are, for instance, studies trying to determine why drivers run red lights in Hong Kong,[4] why cyclists run red lights in Australia,[5] and why pedestrians seem to ignore lights altogether all over the world. The traffic-engineering literature seems to harbor a particular frustration toward pedestrians and cyclists, who are seen as frequent lawbreakers. This dates back at least to 1955, when a study of jaywalking was published in the *Journal of Abnormal and Social Psychology*.[6] All these studies have produced a number of findings—timing seems to matter, as does the geometric design of the intersection, and sometimes a broad category of so-called human factors is involved—but what they generally seem to be unable to capture, even with a nod to human factors, is the particular strain of sheer contrariness that the traffic light provokes precisely because it is designed to project law and order. This author's favorite expression of this is found in *Esquire* magazine's "Guide to Minor Transgressions" column, which has this to say about running a red light: "It's happened hundreds of times. There is no one around. It's nighttime. I'm alone. I'm at an intersection—a crossroads, as it were—and the light is red. I run the light and I'm saving seconds, tops. So it's not about time. (If you can't wait a couple minutes for anything in this world, you have bigger problems than the red light.) No, it's about freedom."[7] The traffic light as a technology of control, then, occasionally fails to control.

Furthermore, the traffic light's governance is not merely about individual behaviors. Theorizing the traffic light as a technology, like theorizing any political technology, "draws attention to the momentum of large-scale sociotechnical systems, to the response of modern societies to certain technological imperatives, and to the ways human ends are powerfully transformed as they are adapted to technical means."[8] The traffic light does not merely govern each individual's movement but also renders that movement part of a larger system. By reshaping each individual's movement into regulated stops and goes, it attempts to produce and maintain an efficient system of movement (at the level of the city, say, or the nation, or the globe). It is no coincidence that traffic engineering has its methodological roots in industrial engineering;[9] for much of its history, traffic engineering has seen the city as a machine whose parts and processes can be understood and managed to produce an optimally functioning whole. Individuals may occasionally experience the personalized technological governance of the traffic light as frustration, for example, or as oppression, or merely as unthinking habit, but they may value the systemic order it presumably produces: the traffic light means that it's not every driver for herself, fighting for space on the road to move. The traffic light, then, is a technical means to a systemic end of ordered movement.

Of course, neither technical means nor systemic ends are politically neutral. After all, "machines, structures, and systems of modern material culture can be accurately judged not only for their contributions to efficiency and productivity and their positive and negative environmental side effects, but also for the ways in which they can embody specific forms of power and authority."[10] For its part, the traffic light embodies a very particular, very modern form of state authority. This is a form of authority that seeks to make space "legible from without"[11]—that is, one does not need to have local knowledge to navigate the space. This legibility partly facilitates better governance: "[T]he city laid out according to a simple, repetitive logic will be easiest to administer and to police."[12] However, it also operates within a broader utopian logic of the high-modern state, in which control is exerted for the benefit of the public.[13] The state, wielding scientific expertise, hews order out of the chaos of the world. This order, in turn, ensures the general welfare, even the earthly perfection, of society. State interventions into the daily landscapes and mobilities of the people are merely means to an end: order, safety, and the presumable achievement of human happiness.

In the case of the traffic light, this means that, as rising motorization rates in mid-twentieth-century North America and Europe led to an increase in traffic collisions, the high-modern state had to ensure simultaneously both efficient mobility and safety. Where the state had earlier claimed disease as a public-health issue and therefore within its purview, traffic collisions now came to be claimed as a public-safety issue. As with public health, the way to ensure public safety was through order based on expertise. This era also saw the professionalization of traffic engineering. The adoption of automated traffic lights in the early 1920s required professional electricians to maintain the lights while specially trained engineers optimized them systemically. A technology that was once squarely in the domain of law and order now fell within the domain of traffic experts, who set to work studying human driving behaviors in order to govern traffic flows. Through the traffic light, the state promised order for the benefit of the people. As it diffused through North America and Europe in the early twentieth century, the traffic light was intended merely to supplement rather than to replace the police officers who directed traffic at intersections; no one believed that drivers would obey a traffic light without the physical presence of a living, breathing, ticket-writing representative of law enforcement.[14] However, as the invention of the automated traffic light in the early 1920s was followed by the phase-out of traffic lights attended by a person, drivers seemed on the whole to obey the automated, unattended lights. They tended to experience the traffic light as faster and more efficient than police officers at directing traffic, so they were happy enough to obey a technological form of state authority that allowed them to move through space more quickly.[15]

One implication of the traffic light is that, by instantiating scientific expertise and state authority, it assumes responsibility for the safety of the individual driver (who, absent a traffic light or well-known rules of the road, must rely upon her own instincts as to what constitutes safe behavior in a given situation). It is useful in thinking about this to consider an alternative. In the past few decades, a somewhat rogue school of thought in traffic engineering has arisen under the heading of "traffic calming." This school of thought, which emerged in the Netherlands with Hans Mondeman and is now gaining a degree of traction across Europe and North America, argues that uncertain roads are safer than efficient ones. Complexity in the road is seen to slow drivers down and make them more attentive to their surroundings. This school of thought prefers intersections that confuse, such as roundabouts, to intersections that command, such as those governed by traffic lights. Whereas the traffic light pits the individual driver against a technological form of state authority, the roundabout is a site where different movers—drivers and cyclists, rickshaws, pedestrians of all mobility levels and ages—must literally sort themselves out spatially through a social process of negotiation. Rather than following a bureaucratic rationality, where the state is the arbiter of who goes where and when, the roundabout invokes a sociality, where access to space and the sequencing of movement are worked out among whoever is moving through the space at that time. They are certainly not always worked out among equals (drivers, being equipped with massive chunks of steel with enormous potential momentum, have a certain power over others using the space), but they are worked out through processes of communication and negotiation rather than appeal to expert authority.

Fundamentally, traffic lights are designed as part of a system of ordered movement. Roads themselves are ordering spaces; they channel movement into given spaces, whereas lanes separate movement in one direction from movement in other directions. Traffic lights, as a part of the system of mobility, build order within those roads. They govern individual driving behaviors in a systematic way, such that if each individual obeys the commands of the traffic light, all movement within the road system will be well ordered. More importantly, though, in controlling movement, traffic lights render the movement of others predictable; and in conventional traffic engineering, the predictability of other drivers is effectively synonymous with safety. Thus, the traffic light becomes a technology that exchanges drivers' individual mobility for a promise of state-ensured safety.

The Universal Traffic Light

If the traffic light is a political technology, what characteristics are meaningful in its social functioning? First of all, the traffic light does not in any material way shape conditions for movement. In contrast to the physicality of roads, which quite literally structure movement, or speed bumps, which alter movement,[16] the traffic light is entirely symbolic. Yet the material characteristics of its symbolism—namely, the differentially colored lights—allow the traffic light to be imbued with a particular meaning: universality.

This is not to say that there is something innately, universally self-evident about the meaning of the colors red, yellow, and green. Though one might be tempted to impute a certain primal quality to the choice of the color red as meaning stop, the red/green code was in fact adopted for the traffic light from railroad signals, which in turn took the code from British admiralty signals, which were in turn adopted from lighthouse signals. Lighthouse signals were red because that was the most transparent color one could produce with the glass-staining technologies available in early nineteenth-century Britain.[17] Nevertheless, a simple code of three colors has two important characteristics. First of all, it is "supralingual"[18]—it associates a meaning to a sign without relying upon any particular spoken or written language. Second, a color code is visual. The visuality of the traffic light means that it is understood to be easily perceived by the human body, though there is an assumption underlying this that a normal body is a sighted body without difficulty distinguishing between red and green. The traffic light, then, has the potential to enforce the rules of proper movement across languages to anyone who can see red, yellow, and green.

Being thus understood as potentially universal, the three-color traffic light played a significant role in the international expansion of automobility throughout the last century. Beginning in the late nineteenth century and continuing into the 2000s, there has been a continuing international effort to make traffic signals and signs uniform across borders. The first attempt at standardization was a League of Tourist Associations meeting in the late 1890s, and in 1908 the Permanent International Association of Road Congresses agreed to standardize internationally signs indicating crossroads, railway crossings, steep gutters, and sharp curves.[19] The 1926 International Conventions Relative to Motor Traffic and Road Traffic followed, as did the 1931 League of Nations Geneva Convention concerning the Unification of Road Signals. The traffic light itself was first introduced into this genre in the 1949 United Nations Geneva Protocol of Road Signs and Signals, which was ultimately superseded by the 1968 Convention on Road Signs and Signals (which was then amended in 1995 and 2006, and further elaborated upon in the 2010 Consolidated Resolution on Road Signs and Signals). Altogether, the past century has seen a lot of work going into making a universal language for the rules of the road.

All these international conventions rely on the idea, stated most explicitly in the 1968 convention, that "[i]n order to facilitate international understanding of signs, the system of signs and signals prescribed in this Convention is based on the use of shapes, and colours characteristic of each class of sign and, wherever possible, on the use of graphic symbols rather than inscriptions."[20] The simple visual code of red, yellow, and green is a characteristic that is seen here to "facilitate international understanding." The 2010 restatement of international traffic standards goes so far as to use the term "'self-explanatory' roads," which it defines as "an environment that can be easily understood and safely operated by all its users."[21] The definition is remarkably similar to

James C. Scott's concept of an externally legible space, where one does not have to be an insider or a local to be able to read or navigate the space. To make a road self-explanatory, then, is to open it up to outsiders.

From the beginning, the international conventions on traffic signals have shared this understanding. They have sought "to increase the safety of road traffic and to facilitate international road traffic by a uniform system of road signalling."[22] Indeed, the language throughout the conventions is remarkably similar: the 1931 convention, the 1949 protocol, and the 1968 convention use virtually the exact same terms. The standardization of the traffic light, then, formed part of a series of conventions whose purpose was to make the road safer precisely by making it internationally legible and thereby (presumably) universally well ordered. One interesting illustration of this comes from a 1970 *New York Times* report on changes in U.S. road signs to pictures rather than words after the 1968 convention. The report invokes this image: "For the foreign motorist, faced with the sudden warning, 'Danger, Slippery When Wet,' or 'Two-Way Traffic Ahead,' the change could save his life."[23] This hypothetical foreign motorist survives on the newly universal road because, in the words of a U.S. Federal Highway Administration officer, "It's more automatic. The response comes quicker than with a worded sign."[24] The supralingual character of pictorial signs and colored lights, by triggering seemingly more automatic reactions, is seen as the most useful tool for controlling the actions of foreign drivers—thereby saving their lives.

This unification of signs and signals forms part of a broader attempt to regulate the safety of cars and roads more generally. To quote a different report from the *New York Times,* the 1968 convention is part of a "campaign in the United States for safer cars and highways" that is "spreading to Europe and to a number of countries in Latin America and Asia. . . . Committees and commissions on almost all continents are seeking ways to build better cars and highways and are working out international safety standards and laws." The report goes on to say that "[i]t is hoped" that the convention will be "one that will save a significant number of lives."[25] The logic clearly links technical uniformity to lives saved.

This project did not unfold merely in the domain of international legal conventions. Over the course of the twentieth century, the international laws on traffic signs and signals developed in tandem with a growing and increasingly professionalized international network of traffic engineers. Traffic engineering was an internationally minded endeavor from the beginning; like architects and urban planners, traffic engineers would often go on trips to other nations to learn from other cities' experiences. For example, in the early twentieth century, U.S. civil engineers drew from the work of French and British engineers to build roads designed to handle heavier loads of traffic.[26] There was also international communication among traffic engineers through journals: traffic engineers in Stockholm and Tokyo wrote to a U.S. traffic-engineering journal in 1931 to report that they had "installed their first traffic lights. Tokyo's engineer, Taiji Hirayama, even thanked Americans for developing the technology."[27] In the postwar era, traffic engineering began using standard road-building manuals internationally, in particular the U.S. *Highway Capacity Manual* (first published in 1950) and the American Association of State Highway Officials' *Policy on Geometric Design of Highways and Streets,* also known as the "Green Book" on roadway design (first published in 1956). There was a dynamic of international standardization occurring in the field of engineering expertise; the international legal road-sign conventions inscribed that engineering expertise into the domain of international law.

For anyone who has the traffic light woven into the texture of her or his daily life, it is a little odd to read the texts of these international conventions. The weighty language of international law seems incongruous, as if there were a series of conventions regulating the alarm clock as a technology, or the spoon. Take article 53, paragraph 1, of the 1949 Geneva Protocol, which establishes the traffic light as a standard road signal: "The lights of the traffic light signals shall be given the following meaning : (a) In a three-coloured system: Red indicates that vehicular traffic must not pass the signal; Green indicates that vehicular traffic may pass the signal; When amber is used after the green signal, it shall be taken as prohibiting vehicular traffic from proceeding beyond the signal unless the vehicle is so close to the signal, when the amber signal first appears that it cannot safely be stopped before passing the signal." Note that, contrary to what most of us think, officially the light is amber and not yellow. For another example, the 1968 Vienna Convention stipulates the arrangement of the lights: "Where the lights are arranged vertically, the red light shall be placed uppermost; where the lights are arranged horizontally, the red light shall be placed on the side opposite to that appropriate to the direction of traffic."[28] Experiencing that formality of language as incongruity—reading it as an official decree of the incredibly obvious—points to one of the most remarkable things about the traffic light. It is thoroughly, habitually, part of everyday life for people all over the globe. Though it has intentionally been made identical across borders, it is rarely experienced as something that is particularly international in character. Most of the time, it seems as local as the major intersection nearest to one's home, whose changing lights hold one up or usher one forward several times a day. Yet the traffic light is enshrined in international law: its technical form is fully specified, authorized, and standardized at the international level.

To be sure, the traffic light is certainly not experienced identically all over the globe. Cultures of driving remain remarkably distinct from nation to nation, city to city, neighborhood to neighborhood. The experience of the traffic light as an integral part of those driving cultures varies widely. The vagaries of local language or infrastructure construction can mean global difference. In South Africa, traffic lights are known as robots, invoking a sense of the high-tech, whereas the lazy-looking lights strung from wires in dusty West Texas towns seem more lonely and bored than anything else. Even the experience of the same traffic light at the same intersection can vary widely, depending on whether one is driving a new, fast car, tapping on the wheel in time to the radio, ready to roll; or whether one is crossing on foot in front of the shiny new car, giving a death-stare to the driver, knowing that the red light has only a tenuous hold on that car's impossibly fast trajectory. Despite its official standardization, the traffic light does not necessarily flatten out differences.

Indeed, sometimes the traffic light is actively deployed to express opposition. The account of one Palestinian writer powerfully contrasts the rigorous order created by traffic lights in Israel, "where everyone obeys the semiotics of their colours" and where they act viscerally as "symbols of modernity, order, prosperity and superiority," with the active resistance to and destruction of traffic lights in Ramallah and al-Bireh, where they are seen as all too obviously political "tools of governing" and "restrictions of freedom."[29] In a slightly different vein, at one point China's Red Guard reportedly wanted to reverse the meaning of the traffic light so that red meant go, as a grand gesture inverting Western norms.[30] The fact that the traffic light has become a part of everyday life internationally does not mean that it shapes lives in identical ways.

However, the standardization of the traffic light plays a significant role in a universalizing project. It is well established and thoroughly theorized that the experience of time and space changed over the course of the last two centuries. What the international efforts to standardize traffic lights in the name of safety indicate is a smoothing of the road for international mobility. The justification so often given for the standardizing of traffic controls is that "international uniformity of road signs, signals and symbols and of road markings is necessary in order to facilitate international road traffic and to increase road safety."[31] While transportation technologies and the increasing motorization of the globe in the years after the Second World War brought people more easily to different landscapes, the standardization of the traffic light, as part of a set of other landscape-homogenizing efforts, attempted to make the transition between national landscapes less jarring, less different, less collision-prone. They were a conscious effort to encourage international travel and tourism, but to channel it in very particular ways.

Consider the way we can now look back to medieval travel as "almost always full of 'dangers and difficulties,' especially since there were relatively few 'expert systems' that would mitigate the risks of such physical travel."[32] Certainly, twenty-first-century travel is not entirely painless and difficulty-free. But the premise of international tourism is that difference can be stimulating without being uncomfortable, new without being unintelligible. It is certainly no coincidence that the first efforts to implement road-sign standards stemmed from the nascent tourism industry in the late nineteenth century. Traveling in another nation no longer means encountering confusions and misinterpretations; as long as one stays on paved and signed roads, the symbolic and physical collisions produced by differences in language and practice are prevented by international law wedded to technical expertise.

If we acknowledge that road systems form part of the "infrastructures of social life"[33] and that the international road system has been unified and made symbolically identical, then what that presents us with is a particularly international infrastructure of social life. Though it is clear that "of course most people in the world can only dream of voluntarily sampling" the joys and choices of international travel, tourism now constitutes the largest industry in the world.[34] The internationalization of mobility means that international forms of social life are possible through tourism, business travel, adventure travel, discovery travel, or international trade. What this also means is that the traffic light helps to produce an infrastructure of social life that is hierarchical. There are internationally legible roads, on which businesspeople, tourists, and wide-eyed students can circulate without too much discomfort; and then there are spaces that remain out of sight of traffic lights. Despite its claim to universality, the traffic light requires electricity to function; areas of the world where the traffic light cannot be illuminated thus remain beyond the self-explanatory conduits of international travel.

The standardization of traffic lights unified the various national roadways into an international system of automobility. Mobility systems "make possible movement: they provide 'spaces of anticipation' that the journey can be made, that the message will get through, that the parcel will arrive. Systems permit predictable and relatively risk-free repetition of the movement in question."[35] If the traffic light sublimates individual movement to produce order at the level of the system, then the universal traffic light expands the boundaries of that system to a potentially planetary scale. At its beginnings, the traffic light was intended to govern particular intersections to improve a city-level system of movement; but to standardize the traffic light internationally is to expand the space of anticipation to cover the world.

The Traffic Light in a Mobile World

"As they become woven into the texture of everyday existence, the devices, techniques, and systems we adopt shed their tool-like qualities to become part of our very humanity."[36] One can take this sentiment in two ways. First, one can see that the traffic light has become part of what it means to be human in the late-modern world: to be mobile and to have that mobility constituted at least in part by illuminated green-tinted glass. Mobility may often feel like freedom; but in our times, it is made possible by a variety of interrelated engineered systems of lights, surfaces, fuels, and signs authorized by state expertise and sanctioned by international law. Second, one can take "our very humanity" not merely to mean the human quality of ourselves but the extent of lived human experience globally. Though it is quite a claim to assert anything about humanity as a whole, nevertheless the traffic light has given humanity an intriguingly common touchstone object. Whether it is a thing that commands respect, provokes resistance, creates a safe space for pedestrians to cross a road, is guiltily or gleefully disobeyed, or simply does not function, the traffic light does in fact universally project a sense of authority, of what the rules of the road are supposed to be. International efforts to standardize the traffic light have succeeded in this sense in producing an object universally imbued with a logic of order, regardless of the many ways in which that order might be reproduced, or transgressed, or ignored.

Ultimately, "the important question about technology becomes, As we 'make things work,' what kind of world are we making?"[37] The traffic light is part of an assemblage of international law, colored glass, transnational engineers, asphalt, and drivers: a panoply of things, trajectories of expertise, and forms of authority and political impulses that together order movement in the late-modern era. This assemblage produces a mobile world, a world of mobile subjects, while promising a world that is easily navigated. It is a world of presumably well-ordered movement, a collisionless world. That the traffic light operates within a universalizing logic does not mean that it succeeds in universally ordering human movement. Yet the traffic light, as part of a broader assemblage of international automobility, does expand automobility's space of expectation to planetary dimensions.

Notes

1. John Urry, *Mobilities* (Cambridge: Polity Press, 2007), chap. 6.
2. Langdon Winner, *The Whale and the Reactor: A Search for Limits in an Age of High Technology* (Chicago: University of Chicago Press, 1986), 6.
3. Clay McShane, "The Origins and Globalization of Traffic Control Signals," *Journal of Urban History* 25, no. 3 (1999): 379.
4. N. N. Sze et al., "Is a Combined Enforcement and Penalty Strategy Effective in Combating Red Light Violations? An Aggregate Model of Violation Behavior in Hong Kong," *Accident Analysis and Prevention* 43, no. 1 (2011): 265–71.
5. Marilyn Johnson et al., "Why Do Cyclists Infringe at Red Lights? An Investigation of Australian Cyclists' Reasons for Red Light Infringement," *Accident Analysis and Prevention* 50, no. 1 (2013): 840–47.
6. Monroe Lefkowitz, Robert R. Blake, and Jane Srygley Mouton, "Status Factors in Pedestrian Violation of Traffic Signals," *Journal of Abnormal and Social Psychology* 51, no. 3 (1955): 704–6.
7. See Ross McCammon, "Why Running a Red Light When No One's Looking Is (Probably) Worth It," Guide to Minor Transgressions, *Esquire,* March 26, 2009, http://www.esquire.com/features/Dilemmas/running-red-lights-0409.
8. Winner, *Whale and the Reactor,* 21.
9. McShane, "Origins and Globalization."
10. Winner, *Whale and the Reactor,* 19.
11. James C. Scott, *Seeing like a State: How Certain Schemes to Improve the Human Condition Have Failed* (New Haven, Conn.: Yale University Press, 1998), 55.
12. Ibid.
13. Ibid.
14. McShane, "Origins and Globalization," 383.
15. Ibid., 387.
16. Bruno Latour, *Pandora's Hope: Essays on the Reality of Science Studies* (Cambridge, Mass.: Harvard University Press, 1999).
17. McShane, "Origins and Globalization," 382n21.
18. Martin Krampen, *Icons of the Road* (Amsterdam: Mouton, 1983), 32.
19. M. G. Lay, *Ways of the World: A History of the World's Roads and of the Vehicles That Used Them* (New Brunswick, N.J.: Rutgers University Press, 1992), 190.
20. United Nations Economic Commission for Europe, *Vienna Convention on Road Signs and Signals,* November 8, 1968, art. 8, para. 1.
21. United Nations Economic Commission for Europe, *Consolidated Resolution on Road Signs and Signals*, ECE/TRANS/WP.1/119/Rev.2, May 31, 2010, iii.
22. League of Nations, *Convention concerning the Unification of Road Signals*, March 30, 1931, 249.
23. "Symbols to Replace Words on U.S. Traffic Signs," *New York Times*, May 31, 1970.
24. "U.N. Parley Seeks to Cut Road Deaths," *New York Times*, December 1, 1968.
25. B. Drummond Ayres Jr., "U.S. Safety Drive for Cars and Roads Spreading Abroad," *New York Times*, June 25, 1967.

26. Stephen B. Goddard, *Getting There: The Epic Struggle between Road and Rail in the American Century* (New York: Basic Books, 1994), 95.

27. McShane, "Origins and Globalization," 395.

28. *Vienna Convention*, art. 23, para. 5.

29. Yazan Al-Khalili, "(R&B) Rhythm and Blues: Post-Traffic Lights in Ramallah and Al-Bireh City," *Race and Class* 52 (2011): 47, 48.

30. McShane, "Origins and Globalization"; Tom Vanderbilt, *Traffic: Why We Drive the Way We Do (and What It Says about Us)* (New York: Knopf, 2008).

31. UN Economic Commission for Europe, *Vienna Convention*, art. 2.

32. Urry, *Mobilities*, 21, quoting Georg Simmel, *Simmel on Culture: Selected Writings*, ed. David Frisby and Mike Featherstone (London: Sage, 1997), 167.

33. Urry, Mobilities, 12.

34. Ibid., 4.

35. Ibid., 13.

36. Winner, *Whale and the Reactor*, 12.

37. Ibid., 17.

AVATAR

Benjamin J. Muller

A time will not come, but is in fact already here, when travelers will confront a machine and not a human when crossing the border. Questions about one's luggage, destination, and reasons for travel are delivered by a relatively sleek but nonetheless inhuman machine. An artificial-intelligence kiosk developed, as its developers euphemistically suggest, to provide "a noninvasive credibility assessment," the Automated Virtual Agent for Truth Assessments in Real Time, or AVATAR, is already undergoing field testing on two continents. As a result of both the complex global interconnections necessary to this technology's material evolution, the policy and alleged practical reasons provided to rationalize its use, and the extent to which it demonstrates an escalating commitment to identification and surveillance technology in the exercise of discretionary sovereignty power, AVATAR is international. The public and private connections and tensions, a global network of universities, research institutes, communities of experts and technocrats, governments, and various other organizations are instrumental in the development of this particular technology, destined and in some cases already used for applications in border security. Similarly, the specific American-led articulation of the events of 9/11 and its aftermath, and the alleged technological solutions to the conceived problems (insecure borders, slack migration policies, and so on), are excellent examples of emerging global norms of border-security management and of the mediation of the primary sovereign function of inclusion and exclusion by an algorithm across a range of jurisdictions. This trend has allowed for the rapid adoption of AVATAR in contemporary border-security systems, a story that is both local and global. AVATAR is international in the development of the material technology itself, and the local/global border-security norms provide the necessary receptivity to adopt this technology. Moreover, technologies such as AVATAR play a role in the mutual constitution of particular global norms of border-security management and an emerging model of global politics saturated with assumptions of danger, risk, otherness, and insecurity, where the discretionary powers of sovereignty are deemed to be more effectively exercised by the alleged less fallible and more precise algorithm.

Looking similar to an ATM machine, AVATAR relies on artificial intelligence, algorithms, and a range of biometric technologies to "detect deception" in allegedly more reliable ways than humans are capable of. Biometric technologies that involve the digitization of human characteristics and traits predate 9/11 but have emerged in the post-9/11 context as something of a panacea to the alleged insecurities of borders and the bodies that cross them.[1] AVATAR is the product of international scientific collaboration and technical development. It is also a material technology through which we can understand the role of U.S. hegemony in benchmarking international border-security norms, the global transformation of border management, and the extent to which local borders are nearly always global sites of differential socioeconomic, political, and cultural exchanges and essential sites for discretionary sovereign power. This intervention focuses on how AVATAR is international by virtue of the proliferation of specific norms in border management and security, local connections and global trends in global political economy, and an almost insatiable hunger for limitless securitization and exceptionalism vis-à-vis biometric and surveillance technologies, in this case people's giving over the primary sovereign function of determining inclusion and exclusion to the algorithm and the artificial intelligence of AVATAR. Joseph Pugliese (this volume) engages another important story, about the drone, from the perspective of science and technology studies about the evolution and development of AVATAR as a machine.

The local political, cultural, and socioeconomic context of the Arizona–Sonoran borderlands is essential to the story of AVATAR. The integral role of the border towns of Nogales, Sonora, and Nogales, Arizona; the University of Arizona; and the Arizona state legislature are critical nodes in the evolution, application, and implication of AVATAR. The evolution and development of AVATAR extol the virtues of global/local connectivity and a vision of global politics stained by an obsession with the politics of fear, threat, danger, and unpredictability, and a corresponding faith and fetishization of technology, alleging that ceding some sovereign functions to identification and surveillance technology will resolve this uncertainty. In this sense, the story of AVATAR, although in many ways local, provides an account of contemporary sovereignty and (in)security in the international. Unpacking the story of AVATAR allows for a fuller comprehension of the norms, trends, and benchmarks of border security and the generally uncritical embrace of technology for the management of the mobile bodies intending to cross borders, which, although bound up in various local concerns, are essential parts of global politics as sovereignty comes to be mediated, regulated, and even ruled by the algorithm.

What are some of the wider trends in contemporary border security, and how has the management of mobility and borders changed over the past decade? Existing trends, the dreams of policy makers and technology developers, and the general intensified securitization of migration and mobile bodies predated the events of 9/11 both locally and globally; however, specific articulations of 9/11 provided the necessary conditions of possibility for the specific policy and sociocultural window of opportunity necessary for the aggressive implementation of these technologies and strategies in contemporary border security. A discussion of the development and application of AVATAR both helps to unpack the local/global dimensions of this technology and moves toward an analysis of how AVATAR is made international. It is not simply a local solution for a relatively small border town in the Sonoran borderlands but a specific story of a broader trend to cede functions of sovereign power to technology.

Methodologically, such analyses are complicated. Those responsible for developing technologies such as AVATAR and the government officials and agencies involved in testing are often among the most reticent. Depending on case and context, arguments relating to the sensitive nature of the information discussed, whether they are due to insecurity concerns or to patent law, and the extent to which the lines between the arts, humanities, social sciences, and sciences require active maintenance, provide effective cover for those involved in the story of AVATAR. For the researcher, such impediments require creative innovation, and as a result, the bulk of research on AVATAR has thus far relied on public media releases, scientific articles, and public demonstrations of AVATAR. Making connections with those in the humanities and social sciences at the University of Arizona, where the majority of the development of AVATAR has taken place, has provided the possibility for a conversation. Moreover, the feeling of obligation, however slim, among the developers of AVATAR, due to the reliance on public funds, has helped to nibble away at the almost entitled reserve of such actors.

Borders, Bodies, Biometrics

Borders, as a focus of government policy and site of state "control" of some sort, are not at all new. Throughout history, borders have been treated as liminal sites of sovereign power, productive spaces of sociocultural, political, and economic exchange, and narrow bands of discretionary state power, as exertions of what

Victoria M. Basham and Nick Vaughan-Williams call "muscular liberalism."[2] As spaces of examination and exclusion, the borders and borderlands have also been regularly treated as spaces where deception was both common and, in some instances, as consumers tried to escape excise taxes for their cross-border purchases, tolerable. The state of borders did not altogether change after 9/11, but existing trends and practices intensified. The increasing use of technology moved toward full-scale reliance and the transformation that accompanied the uncritical embrace of biometrics and surveillance technologies. Furthermore, with the aid of institutional transformations—such as the creation of the Immigration and Customs Enforcement and the Customs and Border Protection sections of the Department of Homeland Security in the United States, and the coterminous creation of the Canada Border Services Agency as a part of Canada's Ministry of Public Safety—there was a marked shift in border management and security from a regime of visa examination toward surveillance and risk assessment.[3] Rather than remain a site of culture and socioeconomic exchange or simple revenue collection, the reconfiguration of the institutional architecture and reliance on various biometric and surveillance technologies allowed the border and borderlands to be rearticulated as spaces of unpredictability, incalculability, threat, and even violence, so much so that mobility itself became framed within calculations of risk and danger. Moreover, the technology and the algorithms and assumptions upon which the technology sat were increasingly given sovereign discretionary powers of inclusion and exclusion.[4]

The possibilities and potentialities of technology in border-security management are multiple. For the most part, the focus by border-security officials, policy makers, and governments is on the extent to which the introduction of surveillance and identification technologies in contemporary border security might expedite the processing of persons and cargo at the border; increase the reliability of the identity of those crossing; more reliably detect deception, and so on. These alleged objectives are all premised on particular assumptions about what the role of the border ought to be—a space for the discretionary sovereignty power to be exercised vis-à-vis technology—thus reinforcing preconceptions about the mobile bodies and goods destined to cross borders. Increasingly, a global norm in border-security management is emerging that is reliant on technology and on a particular belief in the reliability and possibilities of technology, namely, algorithms. In particular, the notion is spreading that those who wish to cross borders are often attempting to deceive, whether about their reasons for crossing, about what they may or may not be carrying with them, or even about their own identity. What might be characterized as the suspicion of mobility itself has also increased since 9/11, and such assumptions and suspicions have powerfully underwritten the alleged need for increased technology to facilitate the intensification of the examination and sovereign differentiation of those attempting to cross borders. These assumptions and suspicions, to a great extent, have been the result of specific American experiences and responses to various events since 9/11 that have come to set global norms and benchmarks of what constitutes effective border-security management.

A particularly cogent example of the manner in which U.S. norms in border security have become global is the case of AVATAR's development and rollout. First, AVATAR was field-tested by the internal European Union border-security agency Frontex. Embedded within such a test was the implicit assumption that all borders share enough characteristics that such a test would be viable. In other words, although AVATAR was clearly developed both in and for the Mexico-U.S. border, the field testing in Poland implicitly suggested that all borders are basically the same. Moreover, the ready acceptance and alleged success of AVATAR during these tests affirmed the general belief in the use of technology in and at borders. The challenges faced at all borders across the globe are effectively the same, and the rational solution to these challenges and problems is thought to be technological. As such, the use of AVATAR, developed predominantly in, by, and for the United States, nevertheless helps to foster a "borderworld"—a global reconfiguration constituted by an emphasis on "borders, borderlands, and technological rearticulations of these spaces and the bodies that traverse them."[5] Body scanners, radio-frequency identification (RFID), biometric-enhanced identity cards, and the like not only provide sound cases for the globalization or internationalization of particular strategies in border security but also provide compelling accounts of the mutually constitutive relationship these technologies have with institutional change, infrastructural innovations, and the functions of discretionary sovereign power in the borderlands vis-à-vis the algorithms. As such, the broad testing (and, one can safely assume, the international marketing) of AVATAR relies on a shared set of global norms among policy makers, government officials, and law enforcement agencies that accept the ongoing amplification of technology in the management of border security and the associated changes to borders, bodies, and the mediation of the sovereign function to differentiate between the inside and outside, the included and excluded. Borders and borderlands throughout the world do not reflect the homogeneity that the wide-scale reliance on such technologies implies—such as various levels of abuse among trusted traveler programs in Canada, the United States, and Mexico—and yet in moving forward on the assumption that technology provides the "one size fits all" solution for border security, the introduction of technologies such as

AVATAR helps to further erode the distinctiveness of borderlands as well as the ability of the stakeholders that occupy these spaces to have a hand in their management.

As I noted earlier, there is an extent to which biometrics are international but not global, insofar as their development, proliferation, and amplification are fostered and relied upon by the states of the global North. Although the application and implications of these technologies have proliferated and emerged as a significant global phenomenon even in the global South,[6] these technologies also operate within restrictive regimes of truth and are, as Pugliese notes, operations of infrastructural whiteness. Biometric technologies "operate as systems for the discrimination of non-normative subjects, including people of colour, refugees and asylum seekers, transgender subjects, labourers and people with disabilities."[7] Moreover, the reliance on technology is both materially and ideologically connected with development and notions of modernity. The failure to embrace and participate in what is presented and packaged as a reconceptualization of border security and border management is often maligned as not only marking inferiority but also turning one's back on modernity itself, particularly by the savvy marketers of such technologies and the politicians and law enforcement agents who are often beholden to these actors, if not nefariously interrelated in what Didier Bigo calls the "governmentality of unease."[8] Such assumptions go hand in glove with the generally uncritical embrace of technology by policy makers in Western governments, who regularly present such technologies as the panacea to past, present, and future challenges of mobility and border management, as certain sovereign functions are ceded to technology. Add the issue of cost, for AVATAR or something like it, as well as the technical, bureaucratic, and cultural infrastructure necessary for such systems, and this transformation of borders and borderlands into a complex technoscape is far less possible for developing and less-developed states.

AVATAR: The Global/Local Story

What is the global and local story of AVATAR's evolution, application, and implications? The local geopolitical context from which AVATAR emerges is relevant and is related to its global dimensions and to the manner in which it is international. Draconian, controversial legislation such as SB 1070 (about which I will have more to say later) and the perfunctory support for the $1 billion failure of the Secure Border Initiative conceived in 2005, as well as the complex politics of the Sonoran borderland, are essential local dimensions in the global story of AVATAR's emergence. The proliferation of border technologies, whether surveillance towers, virtual fences, ePassports, or RFID technologies, and the general tendency toward a so-called multilane strategy at borders where political, ethnic, racial, and socioeconomic sorting of all types preassesses the risk of mobile bodies, have fostered the global receptivity for technologies such as AVATAR. In spite of the long list of shortcomings and failures associated with the reliance on these technologies in border security, such as misidentifications, false positives, and racial and gendered bias, the vulnerable status of the mobile bodies wishing to cross borders, particularly when facing the enthusiastic exercise of the discretionary power of the state, poses a challenge to mounting a more significant and effective critique of these technologies. In other words, those subjected to these technologies may be among the most vulnerable in society and, if not, are nonetheless in one of the most vulnerable spaces in even liberal societies, at the border where the discretionary power of the state is most intense. In such a context, policy makers and law enforcement officers are quick to engage in zero-sum discourse, where failure to advocate such technologies and to uncritically embrace their application is equated with advocating crime, human trafficking, and even terrorism. Moreover, it is in such conditions that the aspects of the discretionary power of sovereignty are handed over to technology and the algorithm.

According to its makers, AVATAR is responding to the need to "rapidly and accurately determine credibility" in a border-crossing scenario. As such, AVATAR is designed to "flag suspicious or anomalous behavior that should be investigated more closely by a trained human agent in the field."[9] Developed through the BORDERS project at the University of Arizona with significant federal-government funding, AVATAR has already been field-tested by the European Union's border agency Frontex.

The Mexico-U.S. border has enjoyed primacy on the U.S. political agenda for some time, particular among the states bordering Mexico, but as Anne McNevin persuasively demonstrates, that preoccupation is not politically homogenous over time, nor is the particular penchant for walls and fences, both virtual and barbed.[10] In the 1990s, the history of successive U.S. government administrations dealing through the metrics of war with issues such as drugs, education, cancer, and so on came to the Mexico-U.S. border.[11] A corresponding rise in public and political concern over drugs and migration became associated with a particular articulation of the insecurity of the border. Particularly in times of economic recession, such as during the 1990s and after 2008, technology has been perceived to have superior capabilities for the refined task of classifying mobility in order to manage it, impeding or expediting accordingly.

In considering the context of the Sonoran borderlands in which AVATAR emerged, I will focus next on a series of local/global dimensions that are essential aspects to the emerging "borderworld": labor migration; discretionary state power, or exceptionalism; the uncritical embrace of technology's taking on this discretionary sovereign power vis-à-vis the algorithm; and what we might call the rebordering of borderlands.

Labor Migration

The increased demand for cheaper, more flexible labor in the global political economy is long-standing and well accounted for. The regularization of undocumented migration by state policies to provide a cheap labor pool for a range of industries, most notably agriculture, is a well-known aspect of the neoliberal global political economy. However, for various reasons related to identity, assumptions about migrants' intentions to change residence, and so on, these migration movements are often compartmentalized away from typical migration. Thus, although borders thickened and proliferated after 9/11, the economic necessity of labor mobility was believed to be worth protecting by the captains of industry and those beholden to the promises of neoliberalism and free trade, and it was central to many so-called multilane strategies in border security. According to law enforcement officials, border-security agents, policy makers, and the oft-quoted "security experts," schemes such as AVATAR and various trusted or registered traveler schemes that sort and create taxonomies of mobile bodies are allegedly facilitated more easily and effectively by technology: the sovereign capacity of inclusion and exclusion is outsourced to the algorithm. The complexities associated with various categories of migration, not least being the movement of cheap labor, saturate the political economy of the Sonoran borderlands. Manufacturing operations in a free-trade zone called maquiladoras dot the landscape of Nogales, Sonora, the Mexican town on the border with Arizona, where cheap labor is exploited under the conditions of NAFTA, often run by a managerial class for whom cross-border mobility is essential. In southern Arizona, defense contractors such as Raytheon, maker of the Sidewinder missile, are major employers and attract many educated Latinos, who often reside or have strong familial ties in Sonora, Mexico. These local dimensions of labor mobility that for government and law enforcement officials require a certain form of border management are also exemplars of global trends in political economy. Moreover, the close relationship between universities and research and defense contractors in these borderlands fosters both the alleged necessity to create differential mobility at the border, and the synergy necessary to produce a technology such as AVATAR. Under these complex conditions, the algorithm is perceived to be a more effective arbiter in the application of the sovereign function of inclusion and exclusion.

Discretionary Sovereign Power

The space of the border has become one where even in liberal democracy the discretionary power of the state is most evident. Particularly since 9/11, the representation of liberty and security as opposite ends of a continuum has provided a persuasive discourse to enable the expansion of the discretionary or exceptional power of sovereignty. Thickening and proliferating the border, through virtual borders and preassessment techniques, and defining various spaces and places as critical infrastructure enable the expansion of exceptional sovereign power vis-à-vis the algorithm. As such, the exercise of discretionary sovereign power that was once tolerated only at the border becomes normalized in a variety of other contexts and by other identification and surveillance technologies, whether it is the securitization of national monuments or wiretapping, dataveillance, and other exercises of state power that undermine citizens' liberty. In this way, the vulnerability of the border is expanded and multiplied, leaving more at-risk populations, such as undocumented migrants, asylum seekers, and visible minorities, in an almost permanent state of vulnerability, sanctioned through state policy and the outsourcing of sovereign functions of exclusion and inclusion to technology. Here, the Sonoran borderlands provide a particularly frightening exemplar.

On April 23, 2010, Arizona governor Jan Brewer passed Senate Bill 1070 in the Arizona legislature. SB 1070 is a worthy exemplar of the global/local context of AVATAR's emergence not only because of its expansion of discretionary state power but also because of the preoccupation with verifying identity, which mirrors the alleged primary benefit of adopting biometric technologies.[12] Expanding police powers that would be assumed at borders but nowhere else, SB 1070 indicates that "during any stop, detention or arrest, a police officer must try to determine a person's immigration status if the officer has reason to suspect that the person is here illegally."[13] Enhancing both the discretionary power of the state and the discretionary power of the law enforcement officer as a petty sovereign, SB 1070 compels officers to inquire about immigration regardless of the nature of the lawful stop, detention, or arrest, in much the same way that suspicion is cast on the mobile body crossing a border regardless of the legitimacy and legality of the crossing. Under such conditions, the introduction of AVATAR as a petty technosovereign is troubling but not unsurprising.

Technology as Panacea

The tendency toward sorting and categorizing populations as part of "risk management" or "governing through risk"[14] tends to rely on familiar taxonomies, reinforcing gendered, racial, and socioeconomic divisions. Ignoring racial and gendered genealogies of identification technologies,[15] biometrics and other technologies are regularly represented as avoiding or neutralizing the problem of racial profiling and thus depoliticizing while securitizing the sovereign function of exclusion and inclusion. It is in this milieu that AVATAR emerged.

As described in promotional literature, AVATAR's role is perceived to be nuanced and noninvasive, barely present in a simple border transaction in which deception is too easily unnoticed by fallible human border guards: "[AVATAR] is an interactive screening technology designed to be on the frontlines of the border crossings and airports. Individuals would approach the AVATAR kiosk, scan their identification, answer a few simple questions, and then move on. Meanwhile, the AVATAR kiosk has used non-invasive artificial intelligence and sensor technologies to gauge suspicious behaviour."[16] As one of the developers of AVATAR, Derrick, notes, "We're not interested in replacing human screeners, but the goal is to make an objective rapid assessment. . . . [T]hat's one of the advantages[:] it doesn't have any biases." However, when asked about the timeline of fully instituting AVATAR along the Mexico-U.S. border, Derrick claims, "[We'll] see how it responds to the kind of dirty, grimy environment, an uncontrolled environment, with people who aren't familiar with technology."[17] Such assumptions about the nature of the border and stereotypes regarding those intending to cross the border, voiced by a developer of AVATAR, come as no shock, considering Pugliese's points raised earlier regarding the discriminatory operations of biometric technologies. Questions about the role of these assumptions, the perceived lack of bias in such technologies, the technologies' increased effectiveness in the exercise of sovereign power, and the reliability of such technologies are sheltered from the public political debate. Instead, sharp contrasts are drawn between subjective and fallible overworked border guards and even escalating levels of corruption, on the one hand, and unreflexive valorizations of technology such as AVATAR, on the other.[18]

Rebordering the Borderlands

Throughout the 1990s, as arguments about an alleged borderless world emerged, often originating in liberal triumphalist quarters,[19] literature on the importance of borders and borderlands also increased, particular among anthropologists.[20] This literature emphasized the complexity, richness, and cultural, political, and socioeconomic exchanges in the global political economy associated with an alleged atrophy of borders and the sovereign politics that borders connote. After 9/11, a different set of global trends and institutional reconfigurations accelerated, often resulting in the centralization of authority[21] and what was referred to as "rebordering," and often challenging preexisting arrangements for trade and mobility[22] and enhancing exclusionary practices of sovereign power. This rebordering formally and informally challenged preexisting interconnections across borders, fostering the erasure of historical, cultural, and socioeconomic bonds, and often enhancing the exclusionary functions of sovereign power or exceptionalism, managed and conducted by allegedly more reliable identification and surveillance technologies.

The introduction of AVATAR into the Sonoran borderlands can be understood in part as an element of rebordering, strengthening sovereign power but subcontracting its functions to the algorithm. Instituting technology as a way of reinforcing the rigidity of the border and allowing the algorithm to exercise inclusion and exclusion place identification technology in the heart of discourses of otherness and difference. In the case of the Sonoran borderland, the Mission San Xavier del Bac, about 15 km south of Tucson, Arizona, is a powerful example of the complexity of the political community alongside the limit of sovereign power. The crisscrossing of borders, identities, and claims in this space—Native American, Spanish, Mexican, and American—exemplifies the diverse heterogeneity of the Sonoran borderlands, not unlike many heterogeneous populations globally. What emerges is AVATAR as a microcosm of a larger story of the reconfiguration of the globe and sovereign power, constituted by a heavy emphasis on borders, less so on borderlands, and almost always on the amplification and intensification of identification technologies, outsourcing the sovereign function of inclusion and exclusion to the algorithm. The story of AVATAR exhibits this set of emerging norms that have a mutually constitutive relationship with the international. Moreover, it highlights the primacy of technology in global politics and the way it takes a central role in particular exercises of discretionary sovereign power, as well as the broad and tacit acceptance for the so-called managers of unease. In helping to unpack the nouns—persons, places, things—of the international, the story of AVATAR also sheds light on how sovereign functions of exclusion and inclusion are outsourced to technologies and the algorithms that underlie them.

Notes

I thank Mark Salter and the anonymous reviewers for their engaging comments on an earlier draft of this chapter. I also thank Javier Duran for affording me the opportunity to engage more directly with AVATAR, its developers at the University of Arizona, and the borderland community in Sonora.

1. Benjamin J. Muller, *Security, Risk, and the Biometric State: Governing Borders and Bodies* (New York: Routledge, 2010).
2. Victoria M. Basham and Nick Vaughan-Williams, "Gender, Race, and Border Security Practices: A Profane Reading of 'Muscular Liberalism,'" *British Journal of Politics and International Relations* (2012).
3. See Muller, *Security, Risk, and the Biometric State*; and Mark B. Salter, "Imaginary Numbers: Risk, Quantifications, and Aviation Security," *Security Dialogue* 39, nos. 2–3 (2008): 243–66.
4. See Joseph Pugliese, *Biometrics: Bodies, Technologies, Biopolitics* (New York: Routledge, 2010).
5. Benjamin J. Muller, "Borderworld: Biometrics, AVATAR, and Global Criminalization," in *Globalisation and the Challenge to Criminology*, ed. Francis Pakes (New York: Routledge, 2012), 129–45.
6. Philippe M. Frowd, "The Field of Border Security in Mauritania," *Security Dialogue* 45, no. 2 (2014): 1–16.
7. Pugliese, *Biometrics*, 6.
8. Didier Bigo, "Security and Immigration: Toward a Critique of the Governmentality of Unease," *Alternatives: Global, Local, Political* 27, no. 1 suppl. (2002): 63–92.
9. "AVATAR: Automated Virtual Agent for Truth Assessments in Real-Time," BORDERS: National Center for Border Security and Immigration, University of Arizona, http://www.borders.arizona.edu/cms/projects/avatar-automated-virtual-agent-truth-assessments-real-time.
10. Anne McNevin, *Contesting Citizenship: Irregular Migrants and the New Frontiers of the Political* (New York: Columbia University Press, 2011).
11. J. A. Dowling and Jonathan X. Inda, eds., *Governing Immigration through Crime: A Reader* (Stanford, Calif.: Stanford University Press, 2013); Jonathan Simon, *Governing through Crime: How the War on Crime Transformed American Democracy and Created a Culture of Fear* (New York: Oxford University Press, 2007).
12. Simon, *Governing through Crime*, 131.
13. Brady McCombs, "Experts Go over SB 1070's Key Points," *Arizona Daily Star*, May 2, 2010.
14. Claudia Aradau and Rens Van Munster, "Governing Terrorism through Risk: Taking Precautions, (Un)Knowing the Future," *European Journal of International Relations* 13, no. 1 (2007): 89–115.
15. Simon A. Cole, *Suspect Identities: A History of Fingerprinting and Criminal Identification* (Cambridge, Mass.: Harvard University Press, 2007); Pugliese, *Biometrics*.
16. Liz Warren-Pederson, "AVATAR Kiosk Aims to Automate, Augment Border En-

forcement," *eller buzz,* Eller College of Management, University of Arizona, February 2011, http://www.eller.arizona.edu/buzz/2011/feb/avatar.asp.

17. Derrick, quoted in H. Stephenson, "Are We Telling the Truth? Ask AVATAR!," *Nogales (AZ) International,* February 11, 2011.

18. Muller, "Borderworld."

19. Thomas Friedman, *The Lexus and the Olive Tree: Understanding Globalization* (New York: Anchor, 1999); Kenichi Ohmae, *The Borderless World: Power and Strategy in the Interlinked Economy* (New York: HarperBusiness, 1990).

20. Hastings Donnan and Thomas Wilson, *Borders: Frontiers of Identity, Nation, and State* (New York: Bloomsbury Academic, 1999).

21. Muller, *Security, Risk, and the Biometric State.*

22. Peter Andreas and Thomas Biersteker, *The Rebordering of North America: Integration and Exclusion in a New Security Context* (New York: Routledge, 2003).

Containers

Can E. Mutlu

Containers are everywhere. They are stacked on top of each other in ports around the world, used as (temporary) accommodation in American military bases in Afghanistan and Iraq, and recycled to make up a shed in my university's community-gardening center. The intermodal shipping container, or simply container, which is how I will refer to it in this chapter, is a standardized, reusable steel box that ensures safe transportation of cargo freight. The container is designed for efficient and safe shipment of freight across multiple modes of transportation: trucks, trains, and ships. Containers come in different sizes, ranging from the most common twenty-foot-equivalent units (TEU) to units that measure fifty-six feet, depending on the needs of the region and the transportation network. Certain characteristics of the intermodal container, such as stackability and refrigeration, allow for a diversity of goods to be transported internationally at significantly higher volumes than before.

The versatility and the durability of the container are both important for its proliferation, but what makes the container truly an "international object" is its standardization. The standardization of the container is not fully complete; there are regional variations in size, and this undermines the full potential of the container to make international trade more efficient as a whole. But the primary purpose of the standardization process, as dictated by the International Organization for Standardization (ISO), is not to make things uniform in their appearance but rather to standardize their functioning. In other words, what makes a thing international, according to the ISO, is not the size of an object but its function.

The container, in addition to being a standardized object, is also a standard object used to order international trade. When goods are shipped internationally, quotes are issued for TEUs, which in turn are determined by a global container index based in China. The Shanghai Containerized Freight Index determines the price of a TEU based on global supply and demand; the going price for a TEU shipped from Asia to Europe was approximately US$1,500 in April 2014.[1] In international statistics provided by the World Bank, the amount of goods traded internationally is measured in TEUs[2]. In 2010, at the height of the global economic crisis, a total of 103,590,000 containers were shipped between China, the United States, and the European Union.[3] Today, it is estimated that roughly 90 percent of the global trade of nonbulk goods is transported in shipping containers.[4] Global supply chains and the internationalization of production have significantly benefited from the containerization of the global transportation industry. Today, megaships are commissioned by global transportation conglomerates to carry 11,000 TEUs in a single journey; ports and major transportation routes are being changed or further dredged to accommodate such ships.[5] Likewise, industry leaders have started to rethink cargo-port infrastructure and architecture to accommodate such ships and their vast cargo. To that end, port administrators have begun to rely on relevant port (security) technologies designed specifically to process containers quickly in order to avoid building vast storage areas for long-term storage of containers waiting to be cleared through customs.

One term that tries to capture this transformation is *containerization*. Containerization refers to the proliferation of containers as an international standard for global trade, moving away from specialized cargo ships or break-bulk shipping,[6] and the subsequent redevelopment of the transportation grid that includes various modes of vehicles, ports, protocols, practices, and technologies to facilitate the movement of individualized containers. The concept of containerization tries to capture the fact that international trade has increasingly been constructed around the container as a standardized object and to examine what that means for existing, precontainerization structures. The container is used as an object that enables other practices and technologies of international trade and transportation. The standardized container, in that sense, not only "makes things international" through enabling the conditions of possibility for international trade but also enables international trade as a whole, which in turn constitutes a major component of the international in the contemporary meaning of the term used in this volume.

Building on this brief introduction, the next section of this chapter looks at processes of containerization, starting with a reflection on the intermodal shipping container's history and function, and focusing on the significance of standardization on containerization of international trade. The chapter then discusses the intermodal container's role in enabling and transforming international trade by looking at its effect on the international architecture of trade. The final section focuses on the challenges faced by the containerization process: the question of empty containers that result from structural trade deficits between countries such as China and the United States, geological limitations to increased freight volume, and opportunities that stem from the afterlife of containers.

Containerizing the International

Trade is an essential part of the contemporary international. The stability of international trade is dependent on things, standards, and practices that facilitate the movement of goods and services across international borders. The prominent tension within this process is the one between speed and security. The question that occupies border-management professionals is how to make trade simultaneously efficient and secure. The container, as a closed and surveillance-friendly box, has emerged as a possible solution to this problem; given the technologies available to us today, we can easily track a container as it travels across the ocean and monitor its internal air pressure to ensure that its seal remains intact, all while surfing the web. However, the safety of the container is not the only reason we can speak of containerization of the international.

The international standardization of intermodal containers, palettes, railroad gauges, and ships, among other things, creates some of the material conditions for international trade to function. The disposition of these things is central to the ways we move things around and to the possibility of a global exchange economy. At face value, there is nothing appealing about the container; it is just a metal box. The idea behind the intermodal container, however, is an inspirational one: the ability to move freight uninterrupted across various modes of transportation and the establishment of a global standard that is not only efficient but also secure have profoundly changed global transportation practices and have helped shape international trade as we know it today.

Economic globalization refers to a set of complex and interrelated developments in information and communications technologies, relaxed tariff regimes, and interdependent global logistics networks, permitting just-in-time cargo-delivery techniques, producing global production chains. The introduction of intermodal containers made transnational production a profitable option for corporations. Growing cotton in Egypt and producing shirts in Italy, for example, was made much more profitable by reducing transportation costs.

Economic globalization, as a set of practices of interconnectedness and interdependence, requires a functioning global transportation infrastructure. In the last five decades, the container has emerged as a key object within the global transportation sector and has significantly contributed to economic globalization by making trade cheaper, more efficient, and more secure. The container has changed the physical infrastructure of logistics, and the containerization of global trade has forced border-security and customs officials, governments, port authorities, and shipping companies to adapt to this new normal. Containerization has both driven and contributed to the processes of technological innovation.

Containerization, as a process driven by intermodality and standardization, has led to the emergence of the intermodal container as a central object in international trade. I focus on three related issues: debates surrounding the emergence of global standards for intermodal containers, the concept of intermodality and its impact on the way we move things around the world, and the containerization of global transportation practices.

The idea behind transporting goods inside a closed container, rather than in break-bulk, goes back more than two centuries: "The British and French railways tried wooden containers to move household furniture in the late nineteenth century using cranes to transfer the boxes from flat railcars to horse carts."[7] The invention of the modern intermodal container is generally attributed to Malcolm Purcell McLean. McLean was an American trucker-turned-entrepreneur who pursued the idea of intermodal transportation by successfully combining maritime and road transportation for the first time in the 1950s by buying a ship called *Ideal-X*.[8] By using the intermodal container, McLean was able to bridge the administrative and practical differences between maritime and road transportation aboard his ship. This new intermodal route cut shipping costs significantly: "Loading loose cargo [break-bulk] on a medium size ship cost 5.83$ per ton in 1956, [whereas] McLean's experts pegged the cost of loading the *Ideal-X* at 15.8 cents per ton."[9] Although the *Ideal-X*'s journey did not save much on fuel costs, it saved a significant amount of time and money by cutting down costs associated with loading and unloading practices at ports.

Those early days of containerization, however, were plagued by the kind of format war familiar to historians of science: differences in widths, interlocking methods, and internal as well as external specifications generated turbulence within the system. In 1963, the ISO issued its first set of "international standards for containers." These were standards for "10-, 20-, 30-, and 40-foot containers."[10] Standardization requires the proliferation of rules for all aspects of a technology, ranging from the terminology to the identification markers to dimensions to practices for their administration and use. For example, ISO R-668 defined the terminology, dimensions, and ratings of containers;[11] ISO R-79 defined identification markings; and ISO R-1161 made recommendations about corner fittings in order to standardize the practices and processes associated with the containers.[12] Traditionally, the "ISO's practice, wherever possible, was to decide how a product must perform rather than how it should be made."[13] In addition, it was the competition between American, Asian, and European transport companies that undermined the ISO's efforts to harmonize different dimensions for the container. Although the ISO successfully harmonized production standards, quality controls, and security features for the intermodal container, it failed to establish a single standardized size. Differences in infrastructure and demands of the market undermine attempts to create a single container size.

Two major factors determine the demand for different-sized containers and regional preferences. The first is the mode of transportation that comes after the container ship. Global transportation infrastructure varies greatly across different continents. Trains can easily accommodate forty-foot containers, but trucks are less likely to be able to maneuver effectively with such large containers attached. Regional preferences for whether to use trains or trucks determine preferences of regional companies. The exception to this rule is the United States, where trucks with fifty-three-foot containers seem to be the norm rather than the exception; this can be attributed to the interstate highway system, which provides easy navigation for such trucks. The second factor is the market demand. Not every market can sustain forty-foot containers. The demand factor is not simply about the wasted space. As anyone who has used a container to move things around can attest, the container is useful only if you can fill it up completely. Even a small amount of empty space within it can undermine the safety of the whole cargo stored within; loose space allows freight to move around, and this results in (easily preventable) cargo damage. To prevent such mishaps, once again regional companies are forced to take into account different market volumes and demand and to address these factors through different-size containers.

In the long term, these differences in container sizes have proven to be costly for the global transportation industry as a whole. Different standards mean delays at borders and ports and time wasted in moving freight from one kind of container to another. Disagreements over the standardization process delay efforts to further integrate the global transportation grids. As a result of these disagreements, currently we have different regional standards rather than a globally agreed-upon standard for container dimensions. When looked at from the perspective of regional companies that dominate the market, however, even this inefficiency of the system as a whole does not present enough of an incentive to change, for the reasons already explained. Interestingly enough, for the most part, the difficulties associated with this standardization process are due to the land-based modes of transportation and differences in these infrastructures across the regions of the world. The maritime transportation industry has avoided such differences in size mostly because of the safety regulations that are in place. Differences in the standardization process, however, do not spill over into the functioning of the container. In other words, the standards that oversee the intermodality of the container are well developed.

Although the intermodality of the container is what makes the standardization of its dimensions so challenging, it is also what makes it a truly a standardized object. The intermodality of the container refers to its design, which allows for the container to be transferred from one mode of transportation to another. The intermodality of the shipping container transformed transportation practices by bridging the previous divide between the administrative and practical aspects involved in different modes of transportation. This capacity of the container allows a producer in Egypt to pack up a product at its own factory and ship it to the nearest port via truck. At the port, a crane can load the container onto a ship, and within a week that ship can arrive in Rotterdam and be on its way by train to its final destination in France. The intermodal container is central to the way we think about logistics and global production chains.

The intermodality of the container not only led to a dramatic reduction in transportation costs but also changed the way we think about logistics.[14] "Just-in-time" production methods and global production chains have both been made possible as a result of the emergence of the intermodal container as an efficient, reliable, and secure platform. Because of a combination of global standardization and the ease with which a container can be moved from one mode of transportation to another, the conditions of possibility for containerization as a process were met. As I mentioned earlier, the standardization of the container's functions is overseen by the ISO, which ensures that the essential features of the container are universally compatible. Both the standardization and the intermodality of the container contribute to the processes of containerization.

The containerization of global trade has had profound effects. Rather than being delayed at ports or transportation hubs, standardized containers allow for fast, inexpensive, and safe freight forwarding. These three factors are central to the practices and processes of economic globalization. We can assess the impact of containerization from various perspectives. In economic terms, the containerization of global trade has had significant cost-saving consequences. Prior to the intermodal container, "transporting goods was expensive—so expensive that it did not pay to ship many things halfway across the country, much less halfway around the world."[15]

One of the key factors contributing to shipping containers' cost-saving effects is the automation of loading and unloading at ports. Historically, loading and unloading break-bulk cargo was a time-consuming and costly process that required numerous dockworkers to work around the clock. Before cargo began being stored in boxes, dockworkers had to carefully place each item, with not only the safety of the cargo but also the stability of the ship in mind. These days it takes mere hours to unload and reload a sixth-generation container ship with up to 14,500 containers using two state-of-the-art cranes, whereas in the past it took days, even weeks, to load up a cargo ship manually, costing ship owners not only the salaries of dockworkers but also the expense of having a ship waiting to sail. These were all added to the transportation costs of goods being shipped, making global trade a much more costly endeavor.

Reducing the financial costs of international trade comes, in particular, at the expense of those working and living near traditional ports. As containers eliminated the break-bulk practices, the need for dockhands reduced dramatically. Furthermore, the standardization of the container facilitated the development of side technologies, such as container cranes, flatbed trucks, and mobile container scanners, which eliminated the need for a high-volume workforce in ports. The proliferation of containers ravaged the livelihoods of dockworkers, disbanding their once-powerful labor unions and making their profession obsolete. The container as such, then, transformed a whole profession that was once a significant source of employment for thousands of people and the basis of strong unions that were politically active, contributing to the neoliberalization of global economies in multiple ways.

This increased demand for moving thousands of containers every day required the reconfiguration of the existing spatial arrangements of global trade. Such an extensive reconfiguration requires that container ports be built in deep(er) waters, with container-storage areas that can accommodate thousands of containers at once, and that they be connected to transportation grids capable of handling the increasing volume of containers moving in and out of the ports. As a result, where Amsterdam and Liverpool failed, Antwerp, Rotterdam, and Hamburg succeeded. These contemporary transshipment hubs are able to simultaneously accommodate multiple containerships as well as the trains and trucks required to transport their cargo to its final destination in a sustainable way. Similarly, globally we are witnessing a trend of purpose-built container ports located outside of urban centers. The remote location of these ports is, in many ways, enabled by the automation of the port processes that eliminated labor organizations. Similarly, the location of these ports and their distance from urban centers also contribute to the precariousness of working conditions in these ports, and continue the trend of undermining labor. Such ports are being purpose-built in emerging economies such as China, Brazil, and Turkey in order to accommodate their increasing shares in global markets.

Although we cannot attribute economic globalization solely to the intermodal container, the global marketplace as we know it today would not have emerged without the significant reduction of transportation costs associated with the emergence of intermodal containers. Containerization is not the cause of globalization, but it certainly is a condition of its possibility.

The containerization of the international has transformed the global architecture of trade. Experts now think of next-generation container ships when they are planning future ports. The increase in global trade volume, which can partly be attributed to the containerization of trade discussed in this chapter, and the ease with which we can securely move goods around have required this transformation of architecture and infrastructure. Containers not only have reduced the cost of international trade but also have transformed the way we think of logistics. This reconfiguration of the system has not been without resistance. Pressure from corporations, unions, and different regional trade blocs continue to undermine the harmonization of global standards for intermodal containers. This provides certain challenges to the containerization process, but it certainly has not been enough to stop its proliferation. One thing is clear: intermodal containers reduce the cost of international trade, and thus they are here to stay. The limits of the containerization process are not going to be tested by resistance to its implementation. Instead, structural elements such as space and geography will test the limits of containerization.

Containerization Challenges

It seems that when it comes to containers, it is not the sky that constitutes the limit. Instead, it is the bottom of the ocean that is causing a major headache for logistics experts. One of the most profound challenges facing the containerization process is the depth of certain choke points on the global transportation grid. Key passages such as the Suez and Panama Canals and the Malaga and Bosporus Straits are providing yet another structural challenge to the borderless global world myth; these passages are too shallow—and at times too narrow—to accommodate the new generation of mega container ships. Although the current technology allows us to build ships that are much larger than the *Emma Maersk,* the largest container ship at the moment, the geographical limitations of these key choke points limit the potential of the industry. Even the *Emma Maersk* is built with the limitations of these significant passages in mind.[16] Similarly, the depth of international ports is another cause for concern. With dredging operations being costly and paralyzing for ports during the duration of the construction, port managers and ship operators are realizing the material limits of the ideational forces behind containerization processes. Territory, in this sense, is presenting the strongest challenge to the seemingly smooth progress of containerization.

Another challenge comes as a direct result of a certain structural imbalance within the global economy: trade deficits and trade surpluses. For most economists, trade imbalances are numerical problems that refer to the imbalance between imports and exports. If a country imports more than it exports, it ends up with a trade deficit. If it exports more than it imports, then it has a trade surplus. This may seem like a fiscal issue at first, yet, as most port operators on the West Coast of the United States would attest, there is definitely a material side to the trade deficit between the United States and China. China's status as a net exporter to the United States has resulted in the one-way movement of full containers from China to the United States. The problem that stems from this situation is that the United States is left with empty containers. As the United States does not produce enough goods to be exported back to China, storing empty containers in major U.S. ports becomes a major problem; shipping empty containers is both inefficient in cost and unsafe because of the weight imbalance generated by empty containers. Nevertheless, a significant number of empty containers travel back to China to be refilled and reshipped. The ones that are left behind have become a source of inspiration as much as a headache for port managers.

This so-called empty-container problem has led to innovation in the field of architecture. At least some of the empty containers left in ports, or at final destination points, are being refitted to serve as homes in their afterlife. In this process, the standardization that made containers appealing for the transportation field—their durability, integrity, and interlocking systems—has made them appealing for architects who transform them into modern residences. This last point speaks to the significance of the standardization process for the containerization of the international. More so than the durability and the versatility of the container, the standardization of the container contributes to its internationalization.

Notes

1. For more on the Shanghai Containerized Freight Index, see the Shanghai Shipping Exchange website, accessed April 28, 2014, http://en.sse.net.cn/.

2. "Container Port Traffic (TEU: 20 Foot Equivalent Units)," World Bank, accessed April 28, 2014, http://data.worldbank.org/indicator/IS.SHP.GOOD.TU.

3. Ibid.

4. Charles W. Ebeling, "Evolution of a Box," *Invention and Technology* 23, no. 4 (2009): 8–9.

5. Greg Allen, "The Race to Dig Deeper Ports for Bigger Cargo Ships," *National Public Radio,* January 5, 2012, http://www.npr.org/2012/01/05/144737372/the-race-to-dig-deeper-ports-for-bigger-cargo-ships.

6. *Break-bulk* refers to an earlier system of moving cargo around separately rather than in containers. This was the prominent method of transportation prior to the 1950s. See Marc Levinson, *The Box: How the Shipping Container Made the World Smaller and the World Economy Bigger* (Princeton, N.J.: Princeton University Press, 2006).

7. Ibid., 29.

8. Ibid., 36–53.

9. Ibid., 52.

10. Ibid., 138.

11. ISO, "Series 1 Freight Containers—Classification, Dimensions, and Ratings," ISO 668:1995, sec. 1.1, 1995, http://www.iso.org/iso/catalogue_detail?csnumber=24007.

12. ISO, "Series 1 Freight Containers—Corner Fittings—Specification," ISO 1161:1984, sec. 1.1, 1984, http://www.iso.org/iso/catalogue_detail.htm?csnumber=5732.

13. Levinson, *Box,* 137.

14. Brian Slack, "Containerization, Inter-port Competition, and Port Selection," *Maritime Policy and Management* 12, no. 4 (1985): 293–303.

15. Levinson, *Box,* 1.

16. For more information on the *Emma Maersk,* see "*Emma Maersk:* The Largest Container Ship," accessed April 28, 2014, http://www.emma-maersk.com/.

Bicycle

Oded Löwenheim

February 5, 2013

"How much did this bicycle cost you?" the Palestinian worker asks me, in Hebrew. We are standing on a trail in the Luz (Hebrew: "almond") Wadi (Arabic: "valley," "dry riverbed"), on the outskirts of Jewish Jerusalem. I am on my way to Hebrew University's Mount Scopus campus on my mountain bike, riding from Mevasseret Zion (Hebrew: "the herald of Zion"), my town—a suburb of Jewish Jerusalem—to the university in the capital city. I ride not on the main road but on dirt trails and in wadis, in order to avoid the violence of motorized traffic. I know that the trails, the open spaces in these hills, are no less violent, but at least the sight of trees, flowers, and hillsides reduces the emotional pain. But sometimes the sight of the open landscape is even more painful—like today.

The trail is partially blocked with rocks and boulders, and a big excavator digs along the hillside above the two of us. The excavator is too close to the trail, and it keeps turning along its axis, pouring the dirt and rocks it excavates from the slope, its digging arm arching over the trail. At first, I had planned just to sneak through the gap between the excavator and the pile of rocks and dirt, but as I approached the worksite (it wasn't a worksite the last time I rode here! But that was more than eighteen months ago), I decided that it would be more cautious to stop and pass after the workers noticed me.

As I watch the excavator working and try to see if perhaps I can find a bypass behind the pile it made and the thicket, I hear the man question me again: "How much did this cost you?" I switch my attention to the worker, wondering what and how to respond. A few days ago, I returned from an eighteen-month sabbatical leave on Vancouver Island, British Columbia, and there is something in the man's question that annoys me. It reminds me in a very blunt manner that I am no longer on Vancouver Island. There, no one on the trails and the woods asked me such questions, which I interpret now as personal ones. Well, I decide, as the excavator continues to rumble, this man does not truly invade my privacy. And even if he does, this whole worksite, this *Israeli* worksite—

where the workers are Palestinians—invades Palestinian land beyond the Green Line (namely, the 1949 armistice-agreement line that is supposed to be, in some unknown future, the border of an independent Palestinian state—that is, if the "two-state solution" will remain feasible till then). So, knowing that this is not Canada here, and weighing the moral balance of invasions—a supposed invasion on my privacy versus a concrete invasion of land (by me on the bike and by the bulldozer alike)—I finally answer: "About nine thousand shekels." In fact, the bike cost CA$2,300 when I bought it on Vancouver Island, but somehow I easily translate the Canadian price into Israeli shekels. "You mean nine hundred shekels?" the man asks hopefully. His Hebrew, I notice, is basic. "No, nine thousand," I emphasize. Then I repeat in my basic Arabic: "Tis'at alaf." "Tis'a ma'a [nine hundred]," the man insists. "La [no]," I insist too, "Tis'at alaf." He does not seem to understand me. He calls his friend, another worker who stands a few meters away from us. This man speaks better Hebrew. The first man asks him to ask me how much my bike cost. "Nine thousand shekels," I say again, unsure where this exchange is taking us. "Are you certain, nine thousand, not nine hundred?" The second man, too, seems reluctant to accept my reply. I am uncertain whether to become annoyed, amused, or saddened. He translates my reply to Arabic for the benefit of the first worker: "Tis'at alaf." The first worker reaches to grab the front tire of my bike. The excavator by now has stopped working, and its operator, another Palestinian man, watches us above from the cabin. A somewhat uncomfortable feeling spreads through my body as I recall the wave of tractor terror attacks in Jerusalem a few years ago, when several Palestinian construction-tractor operators used their machines to smash into crowded buses or to hit cars and pedestrians. My full-face helmet, shin and elbow guards, sunglasses, and hydration system on my back will not be much help should the excavator operator decide to strike me with the machine's arm. I know that it is mainly fear that overtakes me now and that neither the operator nor the other two workers are likely to harm me. This is just an incidental encounter in the hills. Nonetheless, I don't like the man's holding of my bike's tire—that *is* an invasion of my privacy, of my sovereignty—and I still don't like the fact that the excavator operator stopped his work. No, I must "desecuritize" this situation, I think. I grab hard to withdraw with the bike, to return this machine, this extension of my body, into my full control again. At any rate, the worker solves my dilemma: he takes his hand back.

 I hate what they are doing to this landscape. Wadi Luz, a dry riverbed that runs between two steep slopes just beyond the Green Line below my town, Mevasseret Zion, was, when I left here in August 2011, a pristine valley. There were old olive groves in the wadi, some of them "Roman"—ancient olive trees supposedly from the Roman period in Judea, as the Palestinian *felahim* ("farmers") used to call these trees. Old watchmen huts built of fieldstone guarded the groves, and a bit further up the trail from here, a small and clean spring trickled its water into a round pool. If you rode a few more hundred meters into the wadi, the ruins of an ancient Crusader farmhouse would be revealed standing on the slope, watching over the main road to Jerusalem, a few kilometers away from here, and over the changing conquerors of the land who pass on this highway. But now, the wadi is a worksite of the Israel Railway Company, which is building a new fast-railway track from Tel Aviv to Jerusalem.[1] Part of the track goes in the Occupied Palestinian Territories, or rather, in a tunnel *under* the Occupied Palestinian Territories. Here, in this wadi, the track emerges from the hillside behind us and is supposed to cross the dry riverbed on a high concrete bridge and enter another tunnel. The digging here is part of the excavation of the bridge's legs. I look ahead and see in the distance, coming down the trail in the wadi from the vicinity of the Crusader ruin, a private security company's off-road truck patrolling the wadi, a thick cloud of dust rising behind it. What kind of permits regime will apply here when the bridge is built? Will the Palestinian owners of the olive groves be allowed to work their trees at the foot of the bridge, which will undoubtedly be considered strategic infrastructure by then? Already there are clear indications of abandonment here: high weeds grow among the old trees, sucking the precious water from the ground. The farmers from the neighboring village Beit Iksa have not been here for many months. I become saddened in the face of all this destruction and dispossession, recalling Israeli public debates that preceded this project and the claim that the whole track could have been built underground, avoiding the destruction of this and other beautiful wadis. But colonial construction must exhibit this destructive element, I suppose, to assert ownership and superiority; otherwise, there might be a feeling of wasted potential.

My thoughts return to the conversation about the bike's cost. "Yes, nine thousand shekels," I reply. The two men do not hide their amazement. "But why would anyone pay so much for a bike?" the one who speaks better Hebrew wonders. "You could have bought a car for this price," he adds. Along with the wondering, there is some sense of sadness in his tone. It's not envy or a grudge, only sadness. "Is this some special bike?" the man who previously felt my front tire inquires. "It's a good mountain bike, but there are much better than it," I answer. "You can buy one for even twenty thousand shekels," I add. The second worker translates my words into Arabic. "Well, what can you do with it?" the first worker wonders, and now he's smiling. I'm sure he thinks I'm some sort of a fool. I look around me, and see a narrow single track coming down from the hillside where the bulldozer's teeth have not yet bitten. "Look," I tell them, and I start pedaling up the trail. The trail is rugged and winding, probably a goat trail. In a few hours, the trail will be gone forever. Hundreds of years it took herders, goats, and deer to tread down this path, and soon the bulldozer will bite the slope.[2] The bike's twenty-nine-inch wheels glide over the rocks and quickly bring me up to a small plateau.

I look down to the wadi bed, the excavator, and the workers. The security truck stands in the distance. I then press the pedals and start rolling down the trail. I jump over the rocks and skid along the switchbacks of the trail. I feel the shock absorbers of the bike work, hear the *tst tst* sound of pressurized air and oil working in the hydraulic systems, hear the quiet whoosh of the chain on the gear cogs and the thumping of the tires on the packed, wet soil when the bike lands each time I jump over a drop. Sweet scents of almond and mustard flowers and lush green weeds fill the air all around me. The trees and the weeds, the flowers and the bushes, they will be gone when I come back from campus this evening. There will be just a big scar in the hillside. I try to comfort myself that my riding here now is demonstrating some resistance to this dispossession and destruction, that I store this old trail in my memory, but in less than thirty seconds I have returned to the wadi bed. Should I have stood in the bulldozer's way, actually resisting the machinery of occupation, rather than just postponing the destruction for a few more minutes? There is no doubt that had I really interfered here, the security team would only have been glad to come and take care of me. They would also damage my bike—or the bulldozer would. Or the workers would drive me away for interfering in their work, fearing that the Jewish contractor, their boss, who is somewhere near for sure, would fire them or cut their salary. And this is how the occupation really works, isn't it? By colonizing the consciousness of both the occupied people and their occupiers.

"That was nice," the good-Hebrew-speaker worker tells me, smiling. The first worker and the excavator operator are also smiling. "Well, is it worth the nine thousand shekels?" I ask them. "Tis'at alaf shekel?" I hear the first worker asking his coworker again in dismay. Nine thousand shekels is probably a three- or four-month income of this man, I know. No downhilling tricks that I will perform for them will convince him that this not a terrible waste of money. (The full-face helmet and my other protective gear cost, by the way, another twenty-five hundred shekels. The hydration system—a fancy name for my backpack that has a water bag in it and a rubber straw that enables me to sip water on the ride—five hundred shekels.)

Standing here with these Palestinian workers who are employed by the Israel Railway Company, at the foot of the bridge that is built just a few hundred meters beyond the Green Line, so close to this now-deserted olive grove,[3] seeing the proven method of creeping annexation in action, and sensing the deep inability of the Palestinian men to understand how one could spend so much money on a bicycle, sophisticated and maneuverable machine as it is, it dawns on me that I am back in this country of occupation, segregation, and separation. For me, a privileged Ashkenazi Jewish university professor, this bike represents the spirit of freedom, an extension of my body that enables me to wander, to decommute, in the trails and wadis between my home in Mevasseret Zion and the Mount Scopus campus of Hebrew University, to cross the Green Line at will, to resist for a few minutes the occupation, whereas for men such as these persons, my bike is simply an incomprehensible waste of money and perhaps also a colonial instrument of teasing and provocation, like this bridge they're building. This bike, which crossed more than fifteen thousand kilometers to get here from the far end of Canada, where I bought it, this bike that I love to look at, to carefully maintain, to lubricate, and to wash the mud and dirt from, this bike, the machine that I am happy to spend money to fix and to buy new gadgets for, is alien to these people inasmuch as I am alien to them and they are to me.

February 26, 2013

A narrow and very steep path diverges from the broad dirt road that passes between my town, Mevasseret Zion, and the town of Har Adar, another northern suburb of Jewish Jerusalem, along the separation fence between Israel and the Palestinians. I patrol the fence today with my bike to remind myself to where I have returned, to re-embed myself in the political landscape here. The landscape architects of the fence have done a good job in concealing this megastructure from the Israeli side. (On the Palestinian side, the fence is very visible.) The fence seems to blend into the planted pine forest that borders it. The pines were planted in the 1960s as part of a "security afforestation" effort, a project designed to prevent Palestinian infiltrators from coming to work the lands of depopulated villages and, at the same time, to provide camouflage and shelter to Israeli military units.[4] Now, the pines hide the fence, with its sophisticated movement detectors, its surveillance systems and cameras. I know I am being seen now on some screen in an operation room, winding my way along this trail.

But now, although the broad trail I'm riding on continues for a few more kilometers until it reaches Har Adar, I decide to examine this single track and to leave the path of the fence. I have had enough of the fence and its detection systems for today. The fence lies heavy on me. It keeps reminding me of the unresolved conflicts within myself; I want the fence down. I abhor both the long and winding route it carves through the hills and ridges, hidden as it is by the security forest, and the ghettoization of the Palestinians who are penned behind it. I hate the land grabbing that the path of the fence created, preventing Palestinians from accessing huge agricultural areas, denying them the delight of strolling and walking in these hills without being detected. But I am also aware that the fence has a strong stopping power, that it has reduced to zero the number of suicide-bombing attacks since it was erected. The fence sprawls in the hills for hundreds of kilometers, encircling Palestinian and Israeli villages, settlements, and towns, carving out enclaves and separating people not only physically but also emotionally. The fence, which began as a security project, quickly developed, for the Israelis, into a sacred object in itself, a structure that helps many—too many—Israelis forget the very existence of the Palestinians and their plight and the challenge they pose. It is a technology and device of repression, in the sense of forcing and of disregarding at the same time.

I take the steep single track and sharply descend toward the kibbutz of Kiryat Anavim. The trail is much steeper than I thought. I feel the earth below me flying. Clods of earth are scattered before me, the wind blows in my ears, and a pleasant feeling of fear floods my body. Suddenly, I see in front of me on the trail the trunk of a medium-size pine tree blocking the way. It probably fell during the last snowstorm here, a few weeks ago. These pines never really became naturalized trees here. There is too little rain to help them grow strong, and the earth is too rocky for their roots to go deep. They burn quickly in summer—often after Palestinians set fire to them, or so we're told in the media. And in winter, strong rains or rare snows make them collapse. Seeing the pine on the trail, modest in size as it is, I feel the pleasant fear become an urgent one. I don't have enough time to stop—I'm too fast and the tree is too sudden. I decide to hop over the trunk. Should I brake, I will surely bump into the tree and fall or skid into the gorge below. The trunk approaches quickly; I lean back and pull the handlebars. The front wheel rises and passes the fallen trunk. I lean forward and the rear wheel passes, too. As the rear wheel lands, I give a long "woo hoo," happy for not crashing into the fallen tree. I continue to fly down. I have probably descended two hundred meters in elevation by now from the level of the fence. Suddenly, again so suddenly—I recall in a flash of a second a seminar discussion about the dimension of suddenness in the Israeli landscape, a landscape in which, for example, security forces can appear out of nowhere, as more than half of the country's lands are under some form of control by the military[5]—I see the trail ending. It enters into a graveyard.

I suppose I have made a dramatic entrance to this place, judging by the amazed faces of three Palestinian garden workers who are busy pruning the shrubs here. I squeeze the brake levers and the bike stops in a screech of tires against the stone floor of the place. The noise cuts through the quietness of the graveyard. I look around and realize that I am in a military cemetery, the military cemetery of Kibutz Kiryat Anavim. I have come to this place once before, but not from this direction. This is the cemetery of the fallen soldiers of the Har'el Brigade, a division of the Palmach, which fought in the 1948 war. Before 1948, this was the Jewish militia that occupied and destroyed many Palestinian villages in the Jerusalem corridor, including the villages of al-Qastil and Qalunya, upon the ruins of which my own town, Mevasseret, is built. Without the people who are buried here and their comrades who remained alive, I would not have lived where I live today, would probably not be able to ride the paths I ride today. But because of what happened in that war, with the demolition and depopulation of hundreds of Palestinian villages,[6] because of the constitutive effect of that war on the state of Israel, the literal stamping of violence into the landscape and into so many other aspects of life here, I find it ever so hard to stay in this country.

I slowly ride along the lane, looking at the names of the soldiers embedded in the tombstones. Some of them were so young, the youngest sixteen years old when he was killed. I respect their sacrifice and mourn it, but it also makes me shudder. Benny Morris and Meron Benvenisti, two prominent historians of the Israeli-Palestinian conflict, think the clash of 1948 was inevitable.[7] Perhaps they're right that in 1947–48, when Israel fought its war of independence and the Palestinian Arabs suffered their *nakba* ("disaster"), the historical context was such that it inexorably led to this conflict. It certainly was a disaster and tragedy for the Jews, too, to establish a state in which violence constitutes its basic structure.[8] Today, more than sixty-five years after these soldiers died, the discourse of "no partner for peace" has taken over Israeli society and politics so strongly that even talking about peace, even *hoping* for peace, has become a sign that one is a radical, delusional "lefty" or, worse still, a traitor of the fifth column. The more powerful Israel becomes, and the more it refuses to relinquish its fears and insecurities as well as its heritage of occupation and war, to seriously engage with history and think about whether indeed the events of 1948 were inevitable—at least in part, and whether Israel's policy has truly ever since been defensive and whether its hand has always been outstretched for peace,[9] the higher the chances that military cemeteries will continue to grow in this country.[10]

"Sir!" I suddenly hear a person calling me. I wake up from my lofty reflections and see that one of the gardeners stands next to me on the lane along the tombs. "Sir, this is not respectable, to ride a bicycle here," he says. "This is a sacred place, a cemetery. You can't burst into here like you did, and ride here," he adds. I don't know if this really is disrespectable—indeed, I did not mean to enter like that, materializing from nowhere, so to speak, and skidding with screeching brakes. But now I'm just riding slowly and carefully along the tombs, surveying the tombstones, lamenting the tragic history of this land. Is there something in the bicycle that makes it disrespectable here? Is it my appearance—with the full-face helmet and other body armor—that is foreign and out of place here? And if so, why? Never mind, I think. I apologize, saying that I had no idea that the trail I came from would end here, and turn back to exit the place the same way I entered it. On the climb back on the single track—how different always is the climb back from the descent on these trails—I think that perhaps there is some hope in this country nonetheless: if an Arab gardener can work in a Jewish war cemetery and truly care for the dignity of the people who are buried there—people who participated in the depopulation of the villages that once dotted these hills—guarding the cemetery against "desecraters" such as myself, then perhaps there do exist partners for peace, partners for forgiveness and reconciliation.

A Note on the Materiality of Decommuting in the Hills

My ridings from home to campus began in the mid-2000s, when I stopped using public transportation due to the horrors of buses being exploded by Palestinian suicide bombers (in the early years of this millennium, with the eruption of the second Palestinian uprising, dozens of suicide bombers exploded themselves in buses, malls, and other public locations in Israel, and Jerusalem suffered heavily). As our family car served my wife and children (for practical and security reasons) and we were unable to bear the costs of owning a second car, I had to settle for the bicycle. (I knew very little then about the condition known as upgraditis—the unquenchable craving to upgrade the bike, its components, and the rider's equipment.)[11] Thus, my riding began as a commuting practice out of practical and political constraints. At least, I thought then, I could save on gas, help the planet, and do some exercise on the road.[12]

In addition to being afraid of riding a bus, I was also indisposed to cycle on the highway to Jerusalem (Highway 1); Israeli drivers are not very tolerant toward cyclists, to say the least. I decided to look for alternative routes. Riding a mountain bike (instead of a faster and lighter road bike) on dirt roads and trails thus became the preferable choice. Increasingly, I enjoyed the physical exercise, the open spaces, and the freedom that the bicycle offered. I also rejoiced in the bike's ability to evade "much of the disciplining that other forms of mobility are subjected to."[13]

As the years passed, I noticed that my biking in the hills turned from a practice of somewhat adventurous and sportive commuting to work into a work(out) of experiential and explorative decommuting: I was still riding in order to get to campus, but I also started straying off the trail or doing "longcuts" instead of riding the shortest way. Increasingly, I started exposing, and being exposed to, the landscape of violence and conflict in Israel-Palestine. Among the thickets, behind cactus hedges, and at the end of narrow single tracks that only few people know now, I discovered ruins of 1948 deserted or depopulated Palestinian villages, followed the route of the separation fence, and saw the scars left, along what was once the Green Line, by Jordanian and Israeli military bunkers, outposts, and battlegrounds. Also, I realized that a great deal of my thoughts and observations took shape on the trail and its surroundings.[14] Furthermore, I also thought about the trail. I thought about the emotional burden that riding in this landscape of violence creates in me, and pondered what to do with this burden. Gradually, I understood that I want to share this burden with others, to write about my experiences in the hills in an effort to resist a culture of violence.

Justin Spinney argues, "While technologies are often considered simply as means to meet practical demands, the character of a place depends on 'how things are made' or experienced, and is consequently determined by the technical realization of a place. Ultimately then, our perceptions of our environment are informed by the goals, skills and technologies available to us."[15] In this light, it is worth noting that my experience of the hills of Jerusalem inevitably became highly influenced by the materiality of the mountain bike. The mountain bike, as a machine, enabled me to explore these spaces, established a strong emotional contact between me and the landscape, and thus helped me open up to the political and human meanings of the sights that I saw.

Compared to other types of bicycle, like road or racing bikes, mountain bikes are made of stronger and more durable materials and components, and hence allow riders to negotiate a wide array of obstacles along the trail, to experiment with various kinds of track, and to maintain a high level of control over the machine in changing path conditions. Moreover, the mountain bicycle offers a much more intimate, embodied, and unmediated contact with the terrain and the outside environment than other vehicles do. This engenders a unique travel experience that is usually absent when using more "normative" forms of mobility. The intimacy and connectedness generated by mountain biking also create a more nuanced perception of the places and spaces I pass along and through. Thus, especially compared to the driver or passenger of a car or other form of motorized commuting transportation, and even compared to the road cyclist or to riders of motorized off-road vehicles (such as motorcycles, ATVs, and pickup trucks), the mountain biker is much more "in" the environment.

On the mountain bike, the rider feels the imposing weight of a summer day's heat or the stab of a winter night's chill, tastes the dust of the path, struggles with headwinds, and smells the various odors and scents of the world. The rider's bones and muscles absorb vibrations and blows from rocks and roots on the trail, and one's skin is often bruised or cut if one loses traction and falls on some "element" along the trail. The lesser ability to insulate oneself physically from the environment also arouses a completely different spectrum of emotional reactions and sensory experiences than those felt by motorists, who are usually much more materially protected than riders of mountain bikes.

Yet despite such feelings of exposure, which sometimes might indeed become unpleasant, the physical effort and bodily experience during the ride connects me (and, I believe, many other riders as well) much more and in a direct and close manner to the space I ride in. The close connection between me, the bicycle, and the terrain not only generates a perception of the bike, the machine, as an extension of my body, but also creates an emotional connection with the space by producing a sense of belonging, achievement, overcoming, and satisfaction during and after each ride. Geography becomes a much more felt constraint (whether it is uphill or rugged, technical terrain, for instance) and a source of excitement and joy (say, a fast and challenging downhill ride). In the medium and long run, geographical space even shapes and transforms the bodily "geography" of the rider. The daily trial of the ascents and technical trails builds certain muscle groups in my body and sometimes leaves bruises, wounds, and scars on my skin. Thus, geographical space blends with my body materially and emotionally.

In this sense, an important material characteristic of a mountain bike is its being an extremely maneuverable machine that can cross almost any terrain. Where it cannot roll, one can lift it on one's shoulder or walk and pull it along. There is no necessary element of "path dependence" in riding a mountain bike. For me, this maneuverability translates into many detours and excursions from my regular path into places that I would not have discovered otherwise. Thus, my ride line often looks like a zigzag or a tree with many branches. In such excursions, I often not only see and experience new places and spaces but also get to talk and interact with "marginal" people who seek the privacy and loneliness that the dirt paths around Jerusalem can offer or who find their livelihood in this frontier region. Such conversations and discoveries of new locations, small as they might be, largely constitute my sense of place.

Another material aspect of the blended mountain biking and decommuting I practice is that it usually takes much more time than driving. This longer duration enables a close familiarity and knowledge of the trails and the surroundings. The relative slowness of the ride also highlights the fact that our world, which we increasingly see as a small and connected place, is in fact very big, and that there is plenty of space for everybody in it. It takes me, a skilled, well-equipped, and fit rider, almost an hour to complete the eleven-kilometer ride from home to campus if I ride without straying off the trail (but I often do). In this manner, I get a more concrete, direct, and in fact corporeal sense of distances in larger scale.[16] I also understand the vulnerability of modern society in the sense of its high dependence on fast and comfortable mobility. The relatively long duration of my rides, the direct contact with the terrain, the daily exposure to the landscape, and the ability to investigate it with the bike become for me an opportunity to actually do my job: to open my eyes to what and who is around me and to reflect on what I see in the world.

Notes

1. See Coalition of Women for Peace, "Crossing the Line: The Tel Aviv–Jerusalem Fast Train," October 2010, p. 8, http://www.whoprofits.org/sites/default/files/Train%20A1.pdf.

2. On these trails, see Raja Shehadeh, *Palestinian Walks: Notes on a Vanishing Landscape* (London: Profile Books, 2010).

3. Irus Braverman, "'The Tree Is the Enemy Soldier': A Sociolegal Making of War Landscapes in the Occupied West Bank," *Law and Society Review* 42, no. 3 (2008): 449–82.

4. See Nurit Kliot, "Afforestation for Security Purposes: Spatial Geographical Aspects," in *Studies in Eretz Yisrael: Aviel Ron Book,* ed. Yoram Bar-Gal, Nurit Kliot, and Ammatzia Peled [in Hebrew] (Haifa, Israel: University of Haifa, Department of Geography, 2004), 205–19. Recently, along the separation fence of the Gaza Strip, the Ministry of Defense and the Honey Council (the organization of honey producers in Israel) inaugurated a joint project to plant eucalyptus, carob, jujube, and avocado trees. The trees will hide the Israeli villages and towns behind them and will provide the bees with opportunities to collect high-quality honeydew. See Adi Hashmonai, "The Honey Is Enlisted: Trees against Rockets," *NRG Maariv,* March 20, 2011 [in Hebrew], http://www.nrg.co.il/online/1/ART2/223/708.html?hp=1&cat=459.

5. See Amiram Oren and Reaffi Regev, *Land in Uniform: Territory and Defense in Israel* [in Hebrew] (Jerusalem: Carmel, 2008).

6. On the history of the 1948 Palestinian depopulated villages, see Meron Ben-

venisti, *Sacred Landscape: The Buried History of the Holy Land since 1948* (Berkeley: University of California Press, 2000). See also Noga Kadman, *The Depopulated Palestinian Villages of 1948 in the Israeli Discourse* [in Hebrew] (Jerusalem: November Books, 2008); and Rochelle A. Davis, *Palestinian Village Histories: Geographies of the Displaced* (Stanford, Calif.: Stanford University Press, 2010).

7. See Benny Morris, *1948: A History of the First Arab-Israeli War* (New Haven, Conn.: Yale University Press, 2009); and Benvenisti, *Sacred Landscape*.

8. Compare this to the utopian vision of the Jewish state written by Theodore Herzl in his *Altneuland* (1902). Herzl envisioned a Jewish society without an army, peacefully accepted—welcomed, in fact—by the Arabs of Palestine.

9. For a critique of the Israeli defensive-policy argument, see Oded Löwenheim and Gadi Heimann, "Revenge in International Politics," *Security Studies* 17, no. 4 (Winter 2008–9): 685–724; Amiram Oren, Oren Barak, and Assaf Shapira, "'How the Mouse Got His Roar': The Shift to an 'Offensive-Defensive' Military Strategy in Israel in 1953 and Its Implications," *International History Review* 35, no. 2 (June 2013): 356–76.

10. In the recent wars Israel has conducted, the hierarchy of death, so to speak, puts soldiers' lives before Israeli civilians'. At the bottom of this hierarchy are the Palestinians of Gaza. See Yagil Levy, *Israel's Death Hierarchy: Casualty Aversion in a Militarized Democracy* (New York: New York University Press, 2012).

11. Rachel Aldred argues that cycling involves "a more or less conscious nonconsumption." Indeed, cycling consumes no petrol and hardly any parking space and road space. See her "'On the Outside': Constructing Cycling Citizenship," *Social and Cultural Geography* 11, no. 1 (February 2010): 36. Yet cycling, and especially mountain biking, often involves consumption of many gadgets and accessories. It is also often characterized by frequent replacements of the bicycle itself, which keeps developing and improving. In this context, bike technological development becomes a manifestation of postmodern consumerism; see Paul Rosen, "The Social Construction of Mountain Bikes: Technology and Postmodernity in the Cycle Industry," *Social Studies of Science* 23, no. 3 (1993): 479–513.

12. On cycling and exercise in the biopower context, see ibid. She writes: "Regimes of bio-power construct the responsible citizen-subject who maintains his or her body, with stigmatized signs of failure including obesity. Cycling as a body practice could thus be seen as a means of displaying one's identity as a healthy, low-carbon subject" (36).

13. Justin Spinney, "Cycling the City: Non-place and the Sensory Construction of Meaning in a Mobile Practice," in *Cycling and Society*, ed. Paul Rosen, Peter Cox, and David Horton (Aldershot, Eng.: Ashgate, 2007), 29.

14. See, in this context, these words of Max Weber in his essay "Science as a Vocation": "Ideas occur to us when they please, not when it pleases us. The best ideas do indeed occur to one's mind in the way in which Ihering describes it: when smoking a cigar on the sofa; or as Helmholtz states of himself with scientific exactitude: when taking a walk on a slowly ascending street; or in a similar way. In any case, ideas come when we do not expect them, and not when we are brooding and searching at our desks. Yet ideas would certainly not come to mind had we not brooded at our desks and searched

for answers with passionate devotion." In *Gesammlte Aufsaetze zur Wissenschaftslehre* (Tubingen, 1922), 524–55. Available at "Theorizing Early Modern Studies," University of Minnesota, http://tems.umn.edu/pdf/WeberScienceVocation.pdf, p. 6. I am thankful to Brent J. Steele for this quote.

15. Spinney, "Cycling the City," 29.

16. Sovereignty in premodern times was much constrained and limited by topography and the difficulty in projecting power, which was essentially human power, into hilly or rugged terrains. See James C. Scott, *The Art of Not Being Governed: An Anarchist History of Upland Southeast Asia* (New Haven, Conn.: Yale University Press, 2009). Scott suggests that in order to better understand the various power dynamics between states and nonstate populations and actors in premodern Southeast Asia, which were much determined by the rugged geography and constraining climate, "we would have to devise an entirely different metric for mapmaking: a metric that corrected for the friction of terrain. . . . The result for those accustomed to standard, as-the-crow-flies maps would look like the reflection of a fairground funhouse mirror. Navigable rivers, coastlines, and flat plains would be massively shrunken to reflect the ease of travel. Difficult-to-traverse mountains, swamps, marshes, and forests would, by contrast, be massively enlarged to reflect travel times, even though the distances, as the crow flies, might be quite small" (47).

Boats

Geneviève Piché

Boats have been integral objects in making possible the international, crucial to the very possibility of intercontinental travel, the development of modern cartography, and international trade. Following the materialist turn influenced by the work of Bruno Latour and others, this chapter explores how vessels in maritime-migration incidents are international political objects informing the representations and practices that shape these incidents. Boats make possible the international by permitting a level of autonomy of movement to groups and individuals who would otherwise not be available to them, which highlights the limits of the international refugee regime and state immigration controls. I consider how some aspects of the history of a vessel construct the vehicle's lifetime and shape the responsibilities and roles surrounding such incidents. Transformations that occur across this lifetime, along with the vessel's movement across various spaces and jurisdictions, complicate questions of authority and responsibility aboard the boat. The intersection of various authorities in relation to vessels, including shipmasters and captains, owners and shipping companies, charterers, flag states, destination states, transit states, and international bodies, makes these objects spaces of contestation, where agents challenge each other and struggle for power. The physical structure of a vessel is constructed to respond to the activities foreseen for the boat, but over the course of the boat's lifetime, the boat's architecture is adapted. The architecture becomes central to the boat's use as a tool of power, where passengers use it as a tool for resistance or where operators and states use it as a means to control or subdue passengers.

Using evidence from empirical research on maritime-migration incidents, this chapter argues that the boat is an important international political object. The boat is a unique vehicle for autonomy in intercontinental travel, unencumbered by the limits of travel by land and the regulation of air travel, including security practices and technologies at the airport. The conditions encountered by boat migrants as a result of the vehicle used (such as the duration of the voyage, the jurisdictional spaces crossed, the structure of the vessel, the limited provisions, the sea, and customary practices of seafarers) are unique to this mode of travel, challenging effective border controls and highlighting the tensions between humanitarian and security concerns in ways that distinguish it from other modes of travel.

Political Material

The literature found within the materialist turn explores the place of material objects in the world. These objects are shaped by and shape social relations but do not exist in isolation from discourses, actors, and practices. Objects inform and are informed by discourses and actors, engendering and accompanying the development of other technologies, infrastructure, and rules that govern the use of objects. They exist within assemblages and networks of actors, spaces, and practices, enabling certain relationships between them. Objects play a significant role in making possible the international, and their materiality does not stand apart from how we understand them and their uses, how they are implemented and developed, and how they relate to the discourses and practices that shape and are shaped by them.

Latour has shown that the distinction between the human and the thing, a distinction that has shaped and been promulgated by science and the scientific method, has been integral to the very definition of modernity, but that it becomes untenable and even undesirable in the face of hybrids that challenge the very distinction between nature and society.[1] In their discussion of the social's shaping the technological, Wiebe E. Bjiker and John Law point to the contingency of objects and technologies. The technologies and artifacts we know are not necessarily how they must have been; they did not emerge out of an innate technological logic, necessarily taking the form they have taken. They are instead shaped by decisions, through the social, and in turn they shape the social.[2] In a volume that discusses the material turn among the important approaches in critical security studies, Nisha Shah states that to adopt this perspective is to consider technologies as things that are "composites of the relationships between people, devices, and practices."[3] The work of Latour and others has been integral to the materialist turn in social sciences, one that considers that objects and the social are shaped by material and nonmaterial things and by the relationships between them.

Boats belong to an assemblage of actors and tools that accompany and shape them. Law examines how Portuguese expansion in the fifteenth and sixteenth centuries relied on maritime vessels to succeed in controlling at long distances. He demonstrates that Portuguese expansion was made possible by the particular design of those ships, along with the use of "documents" (records of travel and written instructions/observations), "devices" (astrolabes), and "drilled people" (navigators, experts).[4]

The continued development of boats and other technologies, as well as rules, infrastructure, maritime culture, and governing bodies, has shaped a particular international that is made possible by and accessed through maritime travel. Activities, actors, and responsibilities are guided by the institutionalization of certain rules pertaining to maritime travel and trade, such as the law of the sea, the delineation of territorial waters and exclusive economic zones, licensing for marine vessel operation, and the responsibilities of flag states. Infrastructure was developed to facilitate and improve the use of maritime vessels, including the building of docks, ports, drawbridges, and communication towers. International maritime travel has been further facilitated by the development and implementation of certain technologies, such as recording logs, navigational tools, communication technologies, lights and visual-signaling systems, lifeboats and jackets, and waste-disposal systems. And the use of boats for transcontinental travel has been integral to the development of modern cartography and to the improvement of navigational tools. The use of marine vessels has been accompanied by the creation of both state-run maritime agencies, such as the coast guard and the navy, and international governing bodies, such as the International Maritime Organization. Maritime travel has also engendered the development of a particular maritime culture, which contains a particular language as well as conventions such as those regarding the role of shipmaster or captain and the responsibility to rescue. The rules, technologies, conventions, governing bodies, and infrastructure accompanying and accompanied by maritime travel shape the incidents considered in my research.

Boats as International Political Objects

This chapter examines four incidents of encounter between Canada and boat migrants: the arrival of the *Komagata Maru* to the coast of British Columbia in 1914, the voyage of the *Saint Louis* in 1939, the arrival to Canada of the *Ocean Lady* in 2009, and the arrival of the MV *Sun Sea* in 2010. These incidents occurred in varying historical and political contexts. The earlier incidents examined here remain important to the way boat migration continues to be framed and understood in Canada. They have been recalled in later incidents, with references that have often transformed the representations of the actors and practices involved. The most recent incidents explored here are important to understanding the contemporary discourses, practices, and representations in maritime-migration incidents. These later arrivals were also shaped by discourses and practices by Canadian state actors attempting to distinguish boat migration from other forms of migration, who insisted on a different set of rules for processing migrants. In all four of these incidents, the boat was prominent in the representation of the migrants themselves and central to state and community responses, both within the incidents' immediate contexts and in subsequent constructions of historical memory. The purpose of this work is not to conduct a comparative analysis but rather to understand through empirical research the significance of the boat to discourses, practices, and representations in maritime-migration incidents and the transformations that occur therein.

The vehicles in the incidents studied here permitted travel that would otherwise likely not be available to those attempting to migrate. As commercial and passenger air travel began to grow only following the Second World War, the *Komagata Maru* and *Saint Louis* incidents occurred in a context in which maritime travel was the only available option for crossing oceans. Thus, without a boat, migrants from India could not have reached Canada in 1914, nor could German refugees have sought asylum in North America in 1939.

International migration and travel in the present context are largely conducted by air and by land. With airspace heavily monitored and linked to state jurisdictions, and thorough security measures and technologies at airports, international travel by airplane by persons who do not meet certain requirements (including proper identification and documentation authorizing entry into a country) has become increasingly difficult. The movement of 76 people attempting to migrate together from Sri Lanka to Canada aboard the *Ocean Lady* in 2009 without prior authorization from the Canadian state, and of 492 people attempting the same aboard the MV *Sun Sea* in 2010, was made possible only by boats.

The autonomy of migrants and organizers has challenged state control as to who arrives at a state's borders and under what conditions. Even in cases where the coast represented the limit of this challenge, the rejection of migrants aboard vessels was rarely simple or without consequence. In the most recent incidents explored here, the arrival by boat represented a bypassing of the normal avenues for migrant and refugee selection, challenging the state's ability to fully regulate migration and asylum through offshore selection. Contrarily, these incidents were used to further reinforce border controls and the conception of sovereignty over territory and access.

Incidents

The *Komagata Maru*, a Japanese steamship that left Hong Kong, making stops in China and Japan, arrived in Canadian waters carrying 376 Indian migrants on May 23, 1914. The ship arrived in Canada in a political context shaped by racial hostility toward Asian immigrants, an oversaturated labor market and high unemployment in British Columbia, and a Canadian state-building process that sought to demonstrate and exercise its sovereignty over immigration. Laws were in place to make Asian immigration more difficult through restrictions on the immigration of general laborers and stipulations requiring that immigrants arrive by continuous journey. Despite the legal tools available to the Canadian state, it was only on July 23, 1914, after two months of negotiations and legal battles, that the Canadian government was successful in its attempt to remove the *Komagata Maru* and its passengers from Canadian waters. The visibility of the vessel moored off the coast contributed to the attention of media outlets and of the Indian community in Vancouver, whereas control over the ship itself was contested, with the passengers refusing to allow the captain and crew to get up steam upon Canadian orders to leave, eventually resulting in a violent clash with Canadian police and special forces. Control was turned back over to the captain and crew only when the Canadian government agreed to concessions regarding provisioning the vessel. Though the Canadian state was ultimately successful in turning away the vessel and the migrants, the arrival of the *Komagata Maru* on the coast of British Columbia signified the migrants' challenge to what the Canadian government considered its sovereign control over its borders and access to its territory, claiming that they had the right to land as British subjects. The challenge became an opportunity for the Canadian government to solidify its claim to sovereign control over its borders, signifying an important moment in Canadian state building.

The *Saint Louis* arrived in Havana's harbor, carrying 937 Jewish refugees from Germany, on May 27, 1939. The passengers believed they had the authorization to enter Cuba. Upon their arrival, with the exception of 28 passengers who were allowed to land, the passengers learned that the landing permits they had obtained in Germany had retroactively been nullified by a decree signed by Cuba's president, Federico Laredo Brú, on May 5, only eight days before their departure from Hamburg. Despite the hospitality of Cuba and other North American countries that provided most Jewish refugees either temporary asylum or permanent resettlement, the limits of that hospitality were highlighted when this ship's passengers arrived without authorization to land. The hospitality of asylum and resettlement countries was contingent on migrants' obtaining permission elsewhere prior to arrival on their shores. The changes to Cuban law and the fears of the passengers upon the threat of their return to Germany brought to light and engendered a very public debate about the failings of a system meant to assist refugees. This debate occurred while the passengers were still aboard the boat, highlighting the particularity of an incident in which migrants find themselves suspended between territories, waiting for arrangements to be made on their behalf by others on land. Advocates for the passengers presented their situation aboard the *Saint Louis* as an indication of the urgency to secure disembarkation before the vessel's inevitable return to Europe.

The *Ocean Lady* arrived off the coast of British Columbia on October 17, 2009, carrying 76 Sri Lankan migrants, who would make refugee claims. The vessel arrived following the end of a twenty-six-year civil war in Sri Lanka. The country's Tamil minority maintained that persecution continued despite the end of the war. The ship sailed from Indonesia, stopping in Thailand and the Philippines before arriving in Canada. Less than one year later, on August 12, 2010, the MV *Sun Sea* was intercepted off the coast of British Columbia, carrying 492 Sri Lankan asylum seekers. The subversion of a Canadian immigration system predicated on a process of foreign selection on the scale seen in the *Ocean Lady* and MV *Sun Sea* incidents was made possible only by the vehicles used by these migrants.

These two boats challenged Canadian migration controls, leading even to the creation of legislative changes that distinguished migrants and refugees arriving by irregular means, thus creating a particular category of migration on the basis of the mode of travel in an attempt to salvage and reinforce the existing framework for migrant- and refugee-selection processes. The Canadian migration and refugee systems were developed on the basis of the preference of Western governments for processing and selection of people from "over there."[5] The movement of unauthorized migrants demonstrated their refusal to accept that their applications and/or claims be assessed at a distance. It was an active refusal to remain in the place and the alleged circumstances from which they had fled while the system processed their claims.

Conditions

In addition to challenging the very possibility of effective border and migration controls, boats in maritime-migration incidents highlighted tensions between humanitarian concerns and security. Those aboard the boat may have sought refuge, and even without formal, institutionalized obligations, the state might have been compelled to assist the passengers or at the very least not return them to dangerous circumstances. Even where asylum was not the purpose of the voyage, the vehicle and mode of travel created possible dangers of their own requiring action (such as repairing or provisioning the vessel, or even allowing the passengers to disembark). The conditions of boats and the sea were prominent in discourses both promoting the acceptance of the passengers and vilifying the organizers and/or operators.

In the case of the *Saint Louis*, the conditions of the vessel compromised the perception of passengers as being in urgent need of asylum. The passengers were discussed widely as refugees in the media as well as among advocacy groups and government officials. The circumstances in Germany from which they were fleeing were condemned internationally, and Jews leaving Germany, even before the full outbreak of World War II, were considered refugees. But the representations of the passengers were deeply tied to the ship and to class in which the passengers traveled. The *Saint Louis* was a cruise ship, and the passengers aboard traveled in either first class or tourist class, depending on the tickets purchased. The ship contained sleeping quarters that were divided into private rooms for couples and families, as well as a dining room, a projector and screen for showing films, and spaces for a variety of activities, including dancing and parties. The passengers were also discussed in news media as affluent professionals (often lawyers and doctors). Although this discourse promoted the image of the passengers as ideal potential migrants to American countries, it also jeopardized the perception of the passengers as refugees in urgent need of resettlement, contributing to their failure to obtain refuge in North America. The representations of the passengers as affluent professionals conflicted with the idea that they were refugees. The structure of the ship and the amenities aboard the vessel contributed greatly to these representations.

The *Saint Louis* incident was also shaped by the urgency created by a power available to the passengers linked directly to the ship. While the ship was anchored in Havana's harbor, one of the passengers, Max Loewe, cut his wrists and jumped overboard. He was rescued by Cuban police in the patrol boats that were responsible for keeping the ship and its passengers offshore. Loewe was brought to hospital in Havana, ultimately securing his own life, as he would not be returned to continental Europe when, later, asylum for the passengers could not be secured in North America. Of course, suicide attempts and self-mutilation are not unique to boat migration, yet the boat was significant to how this attempted suicide unfolded and to its outcome. Loewe not only cut his wrists but also jumped overboard into Cuban waters, where, instead of remaining under the responsibility of the ship's captain, a certain Captain Schroeder, and crew, he was rescued by Cuban authorities who brought him to hospital onshore. The boat permitted Loewe to act in a way that was both dramatic and visible, resulting in a great deal of attention and concern from refugee advocates, fellow passengers, the captain and crew, and news-media outlets. This suicide attempt became a part of the pleas of the passengers and Captain Schroeder to potential asylum countries to grant refuge to the passengers. It was cited to demonstrate the desperation of the passengers and the urgency of the situation.[6] The fact that the passengers found themselves aboard a ship, fearing return to Germany, meant that in taking to sea, where the vessel was not surrounded by patrol boats, suicide attempts would likely be successful and the ship's operators would be powerless to intervene.

The conditions of the voyage at sea and the conditions aboard the boats were prominent in the representations of the passengers, the organizers, and the operators of the *Ocean Lady* and MV *Sun Sea* voyages. The passengers were depicted as desperate, and the difficulty of the long voyage upon rough seas aboard ships in poor condition was cited as demonstrating the gravity of the situation from which they must be departing and their commitment to reaching Canada.[7] But these same conditions were also used to criminalize and demonize the operators and organizers of these voyages. They were depicted as money-hungry smugglers who did not care for the passengers, as they had subjected them to such a dangerous journey.[8]

Transformations

In the maritime-migration incidents studied here, there were important transformations in the meanings and functions of the boats. In different moments, a boat was a vehicle for escape, a detention center, and a vehicle of deportation or exclusion. The boat also served as a space of various and, at times, competing powers and jurisdictions.

Boats are designed to respond to particular conditions and requirements with regard to physical structure and capacity. But in some of the incidents explored here, there were evident transformations between design and actual function. Whereas the *Saint Louis* was a cruise ship, a liner designed and used primarily for leisure, it was used in those few weeks in 1939 to transport not vacationers but refugees. The *Saint Louis* remained a cruise ship, a fact made evident by its scheduled commitment to pick up passengers in New York City traveling to Europe. When the German refugees could not be disembarked in Havana, the Hamburg-America Line faced losses that resulted from the failure to transport travelers from New York. This was, however, not an unforeseen event, as the ticket price for those traveling to Havana from Germany contained a surcharge as insurance for the company against such an event. Though this was a one-way journey out of Germany, passengers were required to pay a deposit for the return journey, according to a circular distributed by the Hamburg-America Line.[9]

Some of the most dramatic transformations in maritime-migration incidents have been not the adaptation of the boat by those operating and traveling aboard it, but rather a transformation occurring through the encounter of the boat with outside authorities, most notably destination or transit states. The *Komagata Maru* and the *Saint Louis* were transformed into floating detention centers when they encountered state authorities at their destinations. The *Saint Louis* was refused access to Havana's port and held offshore by patrol boats. The passengers were not allowed to disembark, and only a few visitors were allowed aboard. Effectively, this transformed the *Saint Louis* into a floating detention center in much the same way as the encounter with the Canadian state transformed the *Komagata Maru*.

The *Komagata Maru* was prohibited from docking in Vancouver and was held offshore by patrol boats. The passengers were not permitted to leave the ship. When the resources provided by the local Indian community were exhausted, the lack of provisions after a long voyage put the passengers at the mercy of the Canadian state, which could decide whether or not to provide provisions to those held aboard.[10] Permission for access to a lawyer in order to appeal the Canadian government's decision to deny entry to the *Komagata Maru*'s passengers was granted only after lengthy negotiations. Even then, communications with legal counsel and with the Indian community in Vancouver that was advocating on behalf of the passengers were limited and monitored by Canadian state officials and police. A vessel was thus transformed into a floating detention center.

The *Ocean Lady* was also subject to transformations over the course of its lifetime and in its encounter with the Canadian state. The name *Ocean Lady* was a fraudulent name given to the Cambodian-flagged *Princess Easwary*. The vessel was also suspected to have been used by the Liberation Tigers of Tamil Eelam (LTTE), an organization designated by the Canadian state as a terrorist organization, for smuggling weapons from North Korea to Sri Lanka. The suspected illicit activities only further marred the image presented in the media of the passengers aboard and of the organizers of their voyage. The history of the vessel contributed to suspicions that passengers and organizers were linked to terrorism, according to one document released by the Royal Canadian

Mounted Police (RCMP) through an Access to Information release stating that the passengers and operators were arriving aboard "a vessel whose history remains murky."[11] The ship that arrived in Canadian waters painted with the name *Ocean Lady* saw a further transformation through its encounter with Canadian state agencies. In its lifetime, it had been not just a vehicle for smuggling weapons to the LTTE but also a mode of transportation for 76 people attempting to migrate to or seek asylum in Canada from Sri Lanka and, upon its arrival in Canada, a crime scene through its encounter with the RCMP.[12]

Power

Boats in maritime-migration incidents were spaces of and tools in power struggles between competing agents and authorities, including passengers, operators, states, charterers, and shipping companies. Power relations were further complicated as the boats crossed jurisdictional spaces.

The *Komagata Maru* was the site of various competing authorities. The passengers aboard the ship, who believed they were entitled to entry to Canada on the grounds of their status as fellow British subjects, were denied entry by a state that wished to exercise and demonstrate its autonomy from the British, particularly on the issue of immigration. The passengers were led by the ship's charterer, Gurdit Singh, who found himself disputing Canada's decision to send the ship back to India, and who cited the responsibilities of the captain and shipping company to abide by his demands as long as his charter was valid. The captain found himself ordered by Canadian state agents to get up steam and return across the Pacific, pressured by the shipping company to abide by Canada's laws but also to respect the conditions of the charter signed with Singh. The captain faced, on the one hand, interference and threats of physical violence against himself and the crew by the passengers and, on the other hand, Singh's demands that he not obey Canada's order to leave until provisions were sent aboard by the Canadian government.[13] The captain faced competing responsibilities to the charterer and to the passengers of the voyage, as well as to the shipping company employing him, to the crew, and to the state in whose jurisdiction his vessel was located.

The captain of the *Komagata Maru* indicated to Canadian officials that he had lost control of the vessel to the passengers, who refused to allow the boat to depart Canadian waters, and thus he would not be able to obey orders to return the ship and its passengers to India.[14] In an effort to return control to the captain and to force the departure of the vessel, Canadian officers and special agents boarded the *Sea Lion,* a tug, on July 19, 1914, and approached the

Komagata Maru, attempting to board it. A clash ensued between the passengers and the officers, the outcome of which was shaped largely by the physical juxtaposition of the two boats. The *Komagata Maru* was much taller than the *Sea Lion,* and so the officers and agents aboard the tug found themselves positioned below the decks of the *Komagata Maru*. Though the officers and agents were equipped with firearms, they opted to use the hose on the passengers who lined the decks above, in an attempt to subdue them and to board the vessel. The passengers responded by throwing bricks, coal, and other debris onto the officers below. The *Sea Lion* eventually retreated as a result of reported injuries. Canadian agents were thus unsuccessful in returning power to the captain and in forcing the vessel's departure.[15] The more important result, however, was that the passengers were then described by state officials and news media as violent and angry and, above all, undesirable. Only when an agreement was reached regarding the provisioning of the ship did the passengers agree to return control of the vessel to the captain. A much larger Canadian naval vessel, the HMCS *Rainbow,* was used to escort the *Komagata Maru* out of Canadian waters. And in order to assist the captain and crew, who still feared mutiny, two Japanese vessels accompanied the *Komagata Maru* on its journey back across the Pacific.[16]

The *Saint Louis* incident also highlights the ship as a center for power struggles. The captain faced pressures from the passengers, who apparently were desperately seeking refuge and threatening suicide, from his employer, the Hamburg-America Line, and from the Cuban government, which refused to allow the ship to dock or to disembark passengers. Captain Schroeder continued to manage these competing demands while advocacy groups in the United States and in Europe, most notably the American Jewish Joint Distribution Committee, negotiated with different states to secure asylum for the passengers. Upon returning to Europe, Schroeder even physically slowed the ship to allow time for arrangements to be finalized for the disembarkation of his passengers in England, France, the Netherlands, and Belgium.[17]

The use of a ship by dozens of migrants to arrive on Canada's shores without prior authorization in the hopes of making refugee claims, as in the cases of the *Ocean Lady* (*Princess Easwary*) and the MV *Sun Sea,* was a direct challenge to a system composed of highly routinized offshore migrant- and refugee-selection processes. Upon the arrival of Sri Lankan migrants on the coast of British Columbia, briefing notes were prepared for all relevant Canadian agencies, outlining the responses to anticipated media questions. An important question that Canadian agencies were ordered to be prepared to answer was why the Canadian state did not intervene with the ships upon learning that they were destined for Canada and why they did not simply turn the ships away. The response in the media was that the Canadian state had a responsibility to uphold international laws on the high seas, as well as its own commitments to international humanitarian standards, when the boats arrived in Canadian jurisdiction. Canada had no power to interfere with the ships before they arrived in Canada's exclusive economic zone, and once they were within this zone, Canada could not return people to a place where they might face torture or violence without first investigating claims through Canada's refugee-determination process.[18]

Toward a Genealogy of the Boat

Attention to the vehicles in maritime-migration incidents as political objects shows us how material things shape and make possible the international. By considering these objects in relation to relevant actors, technologies, and representations, I have shown that boats act as spaces of contested power, providing a level of autonomy to migrants that would otherwise be unavailable to them, but also challenging the effectiveness of migration and border controls. I have demonstrated further that the boats themselves were transformed by their encounters with various actors and technologies, creating greater nuance in the power relations within these incidents.

Maritime-migration incidents, however, represent only one node in a genealogy of the boat. There are many facets to this object that must be further explored, including the market of maritime transportation, the long history of maritime travel and trade, the development of different types of vessels and technologies as they relate to their use, and the legal frameworks and detailed infrastructures developed for this mode of transportation. A materialist approach offers a promising lens through which such a genealogy can be undertaken.

Notes

1. Bruno Latour, *We Have Never Been Modern*, trans. Catherine Porter (Cambridge, Mass.: Harvard University Press, 1993).

2. Wiebe E. Bjiker and John Law, "General Introduction," in *Shaping Technology/Building Society: Studies in Sociotechnical Change*, ed. Wiebe E. Bjiker and John Law (Cambridge, Mass.: MIT Press, 1994), 1–14.

3. Nisha Shah, "The Internet as Evocative Infrastructure," in *Research Methods in Critical Security Studies: An Introduction*, ed. Mark B. Salter and Can E. Mutlu (New York: Routledge, 2013), 186–90.

4. John Law, "On the Methods of Long-Distance Control: Vessels, Navigation, and the Portuguese Route to India," in *Power, Action, and Belief: A New Sociology of Knowledge?*, ed. John Law (London: Routledge & Kegan Paul, 1986), 234–63.

5. Jennifer Hyndman, *Managing Displacement: Refugees and the Politics of Humanitarianism* (Minneapolis: University of Minnesota Press, 2000), 27.

6. "Fears Jews Will Enter Death Pact," *Globe and Mail* (Toronto), June 1, 1939.

7. Jill Mahoney and Jane Armstrong, "Sri Lankan Migrants Suffered Grueling Journey," *Globe and Mail* (Toronto), October 22, 2009.

8. Citizenship and Immigration Canada, Access to Information release file no. A-2010-12770. See also Stuart Bell, "First Arrests in Ocean Lady Human Smuggling Investigation," *National Post* (Toronto), June 14, 2011.

9. Hamburg Amerikanische Paketfahrt Aktein-Gesellschaft (HAPAG), Hamburg-America Line, "Notice to All Interested Parties for CUBA," April 15, 1939.

10. Malcolm R. J. Reid, dominion immigration agent, to William D. Scott, superintendent of immigration, June 3, 1914, *Komagata Maru, 1914–2014*, Collection, Canada Department of Immigration, p. 1, http://komagatamarujourney.ca/node/11187. See also "Hindus Barred by the Decision of Courts," *Toronto Daily Star*, July 7, 1914; "Hindu Ship Passengers Can't Enter Canada," *Globe* (Toronto), July 7, 1914; "Hindus Quit Fight: Will Return Home," *Globe* (Toronto), July 8, 1914; and "Hindus in Bad Way," *Globe* (Toronto), July 11, 1914.

11. RCMP, Access to Information release file no. GA 3951-3-03544/11, p. 29.

12. Transport Canada, Access to Information release file no. A-2011-00478/RB, p. 1.

13. William C. Hopkinson, immigration inspector, to William W. Cory, deputy minister of the interior, June 10, 1914, *Komagata Maru, 1914–2014*, Collection, Canada Department of Immigration, p. 1, http://komagatamarujourney.ca/node/11203.

14. William C. Hopkinson to William W. Cory, June 17, 1914, *Komagata Maru, 1914–2014*, Collection, Canada Department of Immigration, p. 1, http://komagatamarujourney.ca/node/10910; E. Blake Robertson, assistant superintendent of immigration, to Sir Robert Borden, prime minister, July 20, 1914, *Komagata Maru, 1914–2014*, Collection, Canada Department of Immigration, p. 1, http://komagatamarujourney.ca/node/10724.

15. "Furious Hindus Beat Back 200 Police and Specials," *Globe* (Toronto), July 20, 1914.

16. "Obstinate Hindus Make Fresh Demand," *Globe* (Toronto), July 23, 1914; "Hindus Leave Quietly, Convoyed by Rainbow," *Globe* (Toronto), July 24, 1914.

17. American Jewish Joint Distribution Committee, *The Voyage of the "St. Louis"* (JDC report), June 15, 1939.

18. RCMP, Access to Information release file no. GA-3951-3-03544/11, p. 2.

Ballast

Charlie Hailey

Joseph Conrad often weighed in on ballast. This material, intended to stabilize ships, stirred the sailor-author's maritime narratives. Like the sea itself, ballast could delay journeys, vex captains, hide treasure, ruin profit, and destroy the vessels it helped balance. One of Conrad's early autobiographical stories tells how storm-flung ballast nearly sank his ship and postponed departure for Bangkok by more than a month. Ballast shifted leeward, and the imbalance left the vessel "tossing about like mad on her side."[1] All hands descended below deck to shovel sand windward to right their ship. Ballast binds its characters' fate: what material, other than gold—or perhaps Conrad's claret—might draw together captain and ordinary seamen? Like a sailor's bend knot, it ties ship architecture to narrative space: where else but in a ship's "gloomy" and tallow-lit "cavern" could such inert and burdensome material divine the sea's invisible forces? And ballast collapses distance: how else can we reckon thousands of miles of geography, terrain, city-states, nations, and natures? But a ship—whether narrated or tallied—can hold only so much. Ballast also measures profit, and its stabilizing presence to balance ships can signal trade imbalance.

Balance

When Conrad writes, "[T]here are profitable ships and unprofitable ships," he mixes the empirical nature of a ship's balance with the economic reality that ballast is an unwanted burden to the bottom line.[2] Ships travel either "in cargo" or "in ballast." In the former condition, they are delivering goods, whereas in the latter, they freight only the necessary load for balanced navigation on the return voyage or to the next port of call. Loading and unloading the material was expensive and chronically inefficient, and merchants wished away, sometimes even ignored, the need for ballast. But Conrad cannot fathom such a vessel, and he distrusts advertisements for ships that sail without ballast as maritime paragons that displace a ship's character with "excess of virtue and good-nature" and betray laws of buoyancy and stability.[3] He bemoans the modern seaman's loss of knowledge about such skills as stowing, trimming, and ballasting. Balancing the craft was itself a craft, and ballast was a moral imperative, because the art of sailing could not always account for economies of trade that left a ship's hold empty of valuable cargo. As Conrad narrated his adventures, at the turn of the twentieth century, aqueous ballast was replacing solid ballast, and now, at the opening of the twenty-first, as global marine trade increases exponentially, the necessity for ballast can again be tied to virtue and to those excessively good natures. More ships travel farther distances in ballast, bearing the economic, social, political, and cultural multivalences that Conrad and his fellow seamen recognized long ago. Always an agent of dislocation, ballast is now a more fluid, and less visible, agent of internationalization. Because water carries invasive organisms, it is ballast that measures a ship's value as well as its virtue, and it is ballast that might tell us where the global ship is heading: an unlikely but illustrative material link between economic and environmental concerns.

Design

Ship design entails both buoyancy and balance. Legendarily climbing into his tub, Archimedes recognized the principle that floating bodies displace water equal to their weight. When ships float, they link material properties of water and solids. Water's forces work upward, and a vessel's mass presses downward. On the open sea, a ship must contend with lateral forces of wind and wave, and it is ballast that controls buoyancy, stabilizes the ship, and affects its maneuverability. In the South Pacific, outriggers created stable watercraft, but elsewhere, vessels required sheer weight, the bare load embedded in ballast as an integral part of ship design. Until the latter half of the nineteenth century, when pumps began to harvest the convenient weight of seawater, ballast was solid material: iron, stone, gravel, and sand. The heavier the better. Each nautical manual offered its own rule of thumb, and Conrad's favored source, *Stevens on Stowage,* advised that ballast weighing half a ship's tonnage would stiffen and trim a cranky vessel.

But ultimately, balancing a ship is more art than science, and each vessel has its own physics and chemistry. As pataphysicians who study exceptions to rules, captains not only account for what they see, as their ships list from port to starboard or need trimming between fore and aft, but also must imagine particular arrangements: the invisible matrix of cargo and ballast within the vessel. As alchemists, they order this internal array of matter. Eighteenth-century captains first stowed iron ballast fore and aft, followed with a layer of shingle ballast on which cargo might be secured before another level of the coarse gravel or sand provided purchase for additional strata. But solid ballast proved to be a volatile shipmate. Sand clogged pumps, stones caused spontaneous combustion, shifting pig iron crushed cargo as well as sailors, and shingle lodged between floor timbers and pried them apart, in many cases causing ships to founder. Facing these dangers, shipwrights soon transmuted stone to seawater for ballast, though the need for bare weight remained. Only very recently have twenty-first-century engineers found solutions that might have eased Conrad's concerns about ballast-free vessels, with designs that engage seawater's mass and materiality more fluidly. But ecosystems have already been internationalized, and ballast is the vector.

Labor and Landscape

Ballast is a well-traveled term. As fifteenth-century innovations refined the technics of balancing ships, ballast was born along the Hanseatic League's trade routes. A family of words for this bare load and its deadweight provided common ground—and stable vessels—for merchant guilds across the Baltic's and the North Sea's roiling waters. Swedes loaded ships with *barlastadh*, Danish stevedores said *barlast*, Norwegians called for ballast in Hamburg's docks, and further afield, Spaniards loaded *balastro*, just as Venetians once heaped postclassical Latin *ballastra* onto cogs bound for the Black Sea. Ballast later came to identify other materials that stabilized infrastructure as well as systems: gravel for road bases, railway lines, and flat roofs, and electrical currents for lighting fixtures. But its core meaning began with the labor required to move this material that offered weight for its own sake. Among the lowest in an already poor laboring class, ballast heavers—like *Oliver Twist*'s coal whippers—loaded this material, ballast getters—like George Eliot's cigar-end finders—retrieved ballast from river and harbor bottoms, and lightermen barged stone, lime, mud, and iron from ship to ship and from ship to dock. Ballast-men oversaw this process from agencies such as Dublin's Ballast Office, with its landmark clock, which inspires epiphany in *Ulysses*: "Mr. Bloom smiled O rocks at two windows of the ballast office."[4] James Joyce no doubt connected the ordinary routines of office and clock with the mundane ballast rocks. These other wandering rocks are banal things that reference distant horizons, rouse sublime beauty, and bring into focus "inscrutable" systems and strange "transmigrations."[5]

Ballast was a workaday thing, but its changes to the landscape did not go unnoticed. Because it was throwaway material in the wake of more valued cargo, its accumulations became shipping hazards. Regulation soon followed, acknowledging early on that ballast transforms places where it is discharged. On June 1, 1746, Great Britain's King George II mandated "in what manner ships may unlade ballast" with laws that were unprecedented and anything but ordinary.[6] You must unload at high tide and onto a barge if the main ship draws more than eleven feet of water. You are not allowed to extract shingle ballast from shorelines, and you must discharge materials from the side of the ship nearest the shore. And you must use a tarpaulin to prevent ballast from falling into the sea. Noncompliance meant fines of as much as ten pounds and prison for two months. Despite such rulings, discarded ballast spawned landscapes born of displaced materials from far-flung lands.

New landforms emerged. Ballast islands, ballast hills, ballast mounds, and ballast banks marked coastal landscapes as shipping hazards, material reserves, and windbreaks. Such waste often became infrastructure, with these unwitting spoils of trade repurposed for buildings, roads, and railways. Historians still speculate about the transnational provenance of Charleston's cobbled streets and Baltimore's masonry buildings. Builders of the latter adopted "bastard" bricks, discarded as too thin or irregular on one side of the Atlantic but perfectly good for transatlantic weight and equally acceptable for thrifty construction in the New World. Ballast also internationalized the Pacific region. When Seattle's coal exports outpaced imports, ships returned to Puget Sound with chunks of Australia and the Sandwich Islands offloaded to form Ballast Island, where outcast American Indians camped between the wharves.[7] Topography became topological, mixing material and folding it together in other places, as in the example of Bristol, England, and New York City. During World War II, empty American supply ships returned home laden with "blitz ballast" from the English city's rubble, which became part of East River Drive's substrate. In this way, ballast tells time. Both maritime and terrestrial archaeologists have learned to read its deposition. Rome's Santa Croce rests on ballast from Jerusalem, and divers recognize centuries-old, otherwise vanished shipwrecks through traces of obdurate ballast. Art also follows displacements of ballast. Goods sent from Thailand to China weighed more than products that returned, and the Chinese sculptures at Bangkok's Temple of Wat Pho made up the difference as necessary burdens that doubled as gifts between rulers.

In some ports, whether remote or heavily regulated, ballast was hard to come by, and necessities for stabilizing material subtracted from the landscape. In the middle of the nineteenth century, John Ruskin linked the need for ballast with the "necessity for destruction" of architecture that should not be restored: "Accept it as such, pull the building down, throw its stones into neglected corners, make ballast of them, or mortar, if you will."[8] Such tactics quarried ballast from the exigencies of time and the vulnerabilities of nature. Two centuries earlier, ruins became spoils of colonial trade as Dutch sailors plundered sites for ballast along the Arabian Sea's coastline. In one case, after the East India Company ship *Discovery* offloaded cargo from Bombay in Gombroon, sailors hunted for stones across the strait on the island of Ormus. On January 30, 1627, with a day off during this search for ballast, master's mate David Davies looked southeast, trade winds fluttering his journal's velum pages and morning sun warming his face, to witness some of Persia's finest marble disappear belowdecks.[9] The need for ballast marked landscapes with resource extraction: captains harvested sand, gravel, and stone from Boston's harbor islands, just as industrialists mined New Jersey's Palisades for landward ballast under railway tracks. But when Davies sketched his eroded city, not only did he document a historic site irreparably mined for raw material, but he also traced the contours of an emergent mode of trade that sacrificed environment for economy, exchanged destruction for profit, and consequently made ballast an international villain.

Water

Ballast is not a neutral thing. Simply weight, it is bare load. But complexities of global exchange also make it "bale load." Ballast can be bad, and *bale* carries this wicked etymological weight of an indeterminate but nonetheless potent evil.[10] Moving ballast was back-breaking work, its shifting mass foundered ships, and extraction ruined its source. And yet, as earth and stone, ballast was a relatively tangible component of trade, and its solidity tallied with commodity. But with water for ballast, the legible displacement of solid matter gave way to less visible, more fluid dislocations. Pumped in here, pumped out there, moved seamlessly by mechanical hydraulics rather than manual labor, hidden in ships' tanks and veiled by the sea's surface, water for ballast would have a pernicious effect on its places of deposit. It was no longer an identifiable, finite landscape that was made international by ballast. It was instead the entirely fluid, limitless matrix of transport—the oceans, seas, gulfs, bays, lakes, and rivers—that served as both global resource and compromised environment. Water became vehicle and vessel. More than an indicator of transnational trade dynamics, water ballast has emerged as an agent of internationalization.

Ships first harnessed water for ballast around 1850. Centuries earlier, innovators had proposed water to weigh down submarine prototypes. And Giovanni Alfonso Borelli's submersible design in 1680 deployed water-filled goatskins for an early version of ballast tanks. But, unlike subs' closed volumes, ships were open trays that could be routinely shoveled full of material. And great effort had gone into making ship hulls watertight. Bringing water into such a vessel was counterintuitive. Shipwrights tried canvas rather than animal skin, but these bags still broke, and leaking ballast tanks could ruin dry cargo. Meanwhile, steel hulls and mechanical pumps transformed ship design, and iron steamers began to cross the Atlantic in the 1830s. With rising costs and formidable delays related to solid ballast, ship owners could not compete with rail transport. In 1852, the first iron-screw collier carried coal for cargo and water for ballast, and mercantile ships soon followed. Water ballast cut costs by more than 5 percent and averted delays that, with solid ballast, could add more than a week in domestic ports and as much as two months internationally. A ship's water ballast was "so valuable . . . in the course of a year [as] to make the difference between its being a profitable or an unprofitable ship."[11] Water was all around, it was free, it was much easier and faster to pump than to shovel, and its discharge was mostly unregulated. Though Lloyd's of London did not routinely insure water-ballasted ships until late in the century, by the 1870s, stabilizing vessels with seawater was well established.

Shipwrights continued to refine water ballast's proportion and placement, and ship design shifted to twin hulls for professed unsinkability and for ready-made water storage. The *Great Eastern* was not the first such vessel, but at the time it was the largest, and it tapped the transatlantic zeitgeist generated by London's 1851 exposition. In January 1858, the ship famously exhibited this system of water storage. After an ignominious christening as *Leviathan* and three months of inching its way to the Thames, crowds watched river water pumped between its hulls to ground the vessel while the crew prepared for the long-awaited launch. Two weeks later, on January 31, 1858, another crowd witnessed fountains of water pumped out so the seven-hundred-foot-long, twenty-thousand-ton ship could demonstrate its thirty-thousand-ton displacement and commence its four decades of unrivaled scale. Only water—six hundred thousand gallons—could ballast such a monster.

And water ballast's efficiency increased with a vessel's bulk. The bigger the better. Captain Francis L. Norton made this argument in support of his design for a lifeboat that could be neither capsized nor sunk. Ballast was at the core of his project, which Norton professed to have been "thoroughly examined and tested by the greatest nautical experts of modern times . . . and pronounced by many of them to be the greatest invention of the age."[12] His system streamlined the ballasting process by automatically filling air chambers with water, and the double-hull design increased stability and speed with what Norton called "live ballast." The inner hull floated on imprisoned ballast water that bore three-quarters of the ship's weight so that the outer hull could ply its watery medium with less friction. According to this logic, ballast water increased buoyancy, more water yielded greater efficiency, and larger ships thus meant increased performance. What better material for safety, buoyancy, and economy than water itself? Ballast came alive in Norton's plan, and his patent—registered after his ironic and tragic death at sea aboard the *Neversink*, in 1892—marked the next century's ship designs, which would be measured not so much by technical advancement as by scalar progression.

Vector

Each year, ships move one trillion gallons of ballast water. Traveling in ballast, it would seem their watery cargo might make them one with the sea, melding vessel with medium. Indeed, these ships, logging more than four thousand trips per year, are stabilized by their origin's seawater and travel on a lens of familiar sea until they reach their destination. But they must trade water for valuable goods. Borne across vast stretches of ocean, in correspondingly massive holds, ballast water makes its own environment: one that is conducive to stowaway organisms. When it was solid, ballast carried primarily weight, but now, as liquid, ballast carries other things as well. Ballast freights its own passengers. The pathogens and organisms that thrive in ships' water tanks threaten global biodiversity. Imagine emptying three Monterey Bay aquariums, complete with their exotic contents from around the globe, into the Pacific each day.

Ships themselves have long been understood as vectors of change. When rats and fleas disembarked from Black Sea ports and carried Black Death to Venice, late in 1347, trading vessels transformed public health; and a network's burgeoning economic connectivity became its vulnerability. Solid ballast has also transported its share of fugitive cargo, but in smaller quantities and with more measurable results than today's ballast water. Seeds in gravel and dirt sent native plants overseas: poppies spread from England to New Jersey's Meadowlands, and seeds from everywhere else traveled back to London and Bristol. In 2012, artist Maria Thereza Alvarez unearthed the latter city's history of ballast seeds to build *Seeds of Change*, a floating garden planted with imported species, including figs from Cyprus, tassel flowers from the Caribbean, and love-in-a-mist's flowering buttercups from northern Africa. Also in soil, night crawlers came to North America, and European shore crabs clung to ballast rocks on their way to New England. In addition, though not ballast, ship-borne fouling has always vexed captains and redistributed sea life. Barnacles, mussels, clams, oysters, crabs, tube worms, leeches, anemones, brown algae, and bacteria have all made various global journeys attached to ships' hulls. Their designation as fouling organisms connotes their offensive nature. Acorn barnacles slow ships with added friction and restrict flow with reduced pipe diameters, and Eastern Hemisphere anemones have displaced Western Hemisphere species. Dockyard workers in Copenhagen could curse these visible signs of disruption, but Danish botanists were only beginning to diagnose the effects of less discernible, though equally unwanted, imports in ballast water. In 1903, from Denmark's coastline, a young Carl Hansen Ostenfeld watched an exceptional bloom of Asian phytoplankton algae in the North Sea. Five years later, he would make the first connection between ballast water and the introduction of nonnative organisms.[13]

It is estimated that 90 percent of global trade occurs with ships. Ballast water and its living contents are now carried with unprecedented efficiency and scale, with an increasing rate of invasion that has only recently been acknowledged. Following Ostenfeld's observation, seven decades elapsed before direct samples of ballast water were taken. On May 15, 1973, J. C. Medcof pulled out his magnifying glass to spy swimming crabs, darting marine worms, and seed shrimp floating in ballast water carried across seventy degrees of latitude from Japan to Australia. The marine biologist made these discoveries in Eden, New South Wales, but the waters of this *locus amoenus* would prove far from idyllic, and Medcof tapped ballast water as a vector of change, including the latest mysterious appearances of Pacific oysters in New Zealand and concurrent dangers posed to native rock oysters.[14]

Meanwhile, the rate of marine invasion increased exponentially with more than two hundred exotic species introduced to each bioregion during the last three decades of the twentieth century.[15] Mitten crabs went from northern Asia to the Baltic Sea and to North America's West Coast. Round gobies swam in ballast water from the Black Sea to the U.S. coastline. Asian kelp went global, fanning out to southern Australia, New Zealand, the U.S. West Coast, Europe, and Argentina. The green crab left Scotland for South Africa, North America, and Japan. In 1951, blue swimmer crabs moved from the western Atlantic's Chesapeake Bay to the Venetian lagoon's Porto Marghera, where intensive oil-shipping activities further broadcast this species to the rest of the Mediterranean region. Zebra mussels wreaked five billion dollars' worth of damage after their introduction to the Great Lakes from the Black Sea in 1988, when a freighter from eastern Europe transferred ballast water with mussel larvae to Lake Saint Clair. Tankers had already carried the North American comb jelly the other way, pushing anchovies and sprats to near extinction and collapsing major fisheries in the Black and Caspian Seas. Pathogens in ballast have also spawned late twentieth-century epidemics. A strain of cholera previously reported only in Bangladesh reached Peru, where it killed more than ten thousand people between 1991 and 1994.

Maps and Risk

Such a complex network of interactions is difficult to track. By some estimates, ten thousand species move daily across the globe in ballast water.[16] Mapping ballast poses challenges to geographers, environmentalists, biologists, and cartographers. Some maps rely on vector diagrams with arrows drawn across continents to show the lines of invasion and circles of infestation highlighting overlaps between trade pathways and exotic intrusions. Others ink areas of influence, as in charts of the Great Lakes that show zebra-mussel colonies radiating out from shoreline points. Maps also trace shipping lanes and global cargo-ship movements, but lines of travel end up meshing in a solid mass. Such meshes of data easily confuse modes of transport, because specific types of ships move in particular ways. Container ships traverse repetitive paths, whereas dry-bulk carriers and oil tankers move less predictably between ports. To illustrate trade, maps must also register the betweenness of ports and their position as both nexus and connector: a topological condition that mirrors the folding of alien ballast water into local water. Other maps demonstrate connectivity by isolating ports and marking areas by ballast load and discharge. This material-flow analysis seeks to quantify flows of materials and energy through the economy. Still other documents attempt to map risk, correlating ballast-water volume and the number of ships in particular ecoregions to potential environmental degradation.[17]

Risk has always been a part of marine trade, and ballast water adds environmental peril to an already fragile system. As insurers evaluate delays, weather, war, inspections, politics, and mechanics, regulators seek to manage ballast water's dangers with equipment, best practices, and redesigned vessels. But intake strainers, easily damaged and often left unrepaired, can corrode and fall off. And completely purging a ship of ballast water is not possible. As a result, the designation NOBOB (no ballast on board) is a misnomer. Residual ballast becomes unpumpable. As much as one hundred thousand gallons of "dead water," along with nutrient-rich sediment, provides favorable, if tenuous, habitats for fugitive marine organisms. Designs for ballast-free ships have been realized, but changing an industry with more than forty thousand vessels in current use is like halting the tanker *Knock Nevis*. Now decommissioned, the world's largest ship needed almost six miles to make a full stop. Enforceable regulations have moved just as slowly. After a decade of discussion, the International Maritime Organization (IMO) adopted the Ballast Water Management Convention in 2004, but eight years later it still lacked ratifications to reach the 35 percent of world merchant-shipping tonnage. By July 2012, thirty-five states had ratified the convention, more than the thirty required by the 2004 conference, but they represent only 28 percent of world shipping tonnage, less than the required 35 percent. In the meantime, the IMO has set up GloBallast, the Global Ballast Water Management Program, to help countries understand the problem. Since 1993, the United States has required arriving ships to choose from three options: retain ballast water for the entire voyage in sealed tanks; execute an "environmentally acceptable alternative" such as the use of onboard treatment systems and delivery to onshore treatment plants; or transfer ballast water out at sea. Known colloquially as "swish and spit," the latter choice requires ships to exchange ballast water at the midpoint of the trip outside of the exclusive economic zone (EEZ).[18] In international law, the EEZ identifies a state's rights to marine resources in an area that extends two hundred nautical miles from its seaward boundary. But neither native nor ballast-transported organisms abide by such political zones, making an EEZ's boundaries capricious for ballast regulation.

Waste and Agency

"Swish and spit" suggests that ballast water is waste. U.S. regulations also reflect this idea: ballast water is included with oil, noxious liquid substances, garbage, and municipal or commercial waste under Subchapter O "Pollution." By these accounts, it seems that oil spills are preferable to bio-invasion. Though still unresolved and outpaced by rates of invasion, regulation has reframed what ballast means within networks of trade. Ballast water's destabilizing effect has consequently informed a new infrastructure that engages the politics of water and waste. With ballast as intermediary, oil can now be traded for wastewater. Tanker ships leave from Japan in ballast, not with seawater but with wastewater, a cargo that can be exchanged in Qatar for crude oil or liquefied natural gas. Ballast becomes a valuable thing in newly minted trade networks where water and waste are both commodities: Japan needs energy, Middle Eastern states lack water, and ships need steadying in between. Water is increasingly central to a complex web of environmental concerns; and, with changing dynamics of trade, water-treatment systems such as ClearBallast now target ballast water along with sewage treatment and the separation of oil and water. Other waste exports have similarly transformed trade and ballast. After export economies deliver products to the Western Hemisphere, scrap metal, newspapers, and recyclable plastics provide weight for return voyages to China, where a labor force recycles this waste ballast. According to the carbon-balance sheet, shipping beats landfilling, and a ten-thousand-mile journey still represents savings compared to inadequate recycle centers and landfill hazards. Ballast fills the trade gap, and liquids other than seawater now fill ballast tanks.

Trading with ships still requires ballast. Ships move faster, but shipping changes slowly. Speeds of forty knots will continue to improve trade connectivity but will also increase success rates of bio-invasion, with more species surviving shorter travel times. In the past, the need to balance ships indicated imbalanced trade and influenced natural balance, but not on today's global scale. As a mobile and now fluid disturbance, ballast inextricably links trade practices to the environment. As a material whose properties have changed from solid to liquid, ballast remains an object that balances ships, stabilizes economic interaction, and consequently maps patterns of global trade. At the same time, it can destabilize ecological balance. By engaging the sea's common ground, ballast in essence has internationalized environmental debates and dilemmas. Though latent and paradoxical, this arrangement also suggests new means for balancing trade and understanding global ecosystems. More than a witness to these changes, ballast water, as its own vector and vessel, acts to produce economic and environmental effects.

As an agent of internationalization, ballast defines its own underlying trade network. Down below in ships' holds and dispersed in its own fluid matrix, ballast was long an unidentified actor, previously intangible in its fluid materiality and invisible in its effect. With these properties, it cuts across boundaries and provides a necessary substrate for the physics of ships and the logistics of shipping. With bigger and faster ships, export trade will continue to increase, and ballast makes this infrastructure international. As an agent of dislocation, ballast is a major force in what the United Nations has termed the "globalization of nature." If, as the UN also notes, the sea is "nature's free highway," then ballast has been its base.[19] Freed from the labor costs of solid ballast, this infrastructure requires very little investment. Given this relatively open road, transnational corporations have elided political boundaries with fragmented production that has in turn made for more voyages and more ballast. When world shipping passed the one-billion-ton mark for deadweight tonnage, in 2007, this number also bore the weight of its ballast, which continued to disrupt ecosystems. But ballast is also an indicator of global connectivity, and it gauges new interactions.

Though it also threatens the very context of these relations, ballast now inscribes a critical network of instability that allows for the visualization of global connectivity. From Lycurgus to Edmund Burke to Michel Serres, ballast has often been used as a metaphor for political stability. Plutarch narrates how Lycurgus, the ancient Greek lawgiver, balanced Sparta's government with an elected senate: "For before this, the civil polity was veering and unsteady; but now, by making the power of the senate a sort of ballast for the ship of state and putting her on a steady keel, it achieved the safest and the most orderly arrangement."[20] Senates balance absolute power just as objects stabilize systems, particularly in terms of their relations. Ballast is the "silent thing" that might steady a twenty-first-century economy and its epistemology.[21] It floated quietly but heavily in the House of Commons when Burke delivered his famous address, arguing for conciliation with America's colonies while striving for a principled response that would "ballast my conduct."[22] Metaphorically and materially, ballast negotiates these social—many times international—contracts, mediates knowledge, and, as Serres contends, affects natural contracts.[23]

Keystone

Like metaphor, ballast bears alternative meanings and uses. This tendency forms the paradoxical core of its ability and agency to disrupt as well as to stabilize. It holds different cargoes. Ballast makes room for waste to become useful, for unconventional payloads to move within a global economy's network. When ballast becomes a commodity, the environmental imbroglio starts to make sense. When Bruno Latour cited an apostle's dictum "it was the stone rejected by the builders that became the keystone," he opened up the possibility that waste objects might build worlds.[24] With ballast, what was once considered a useless thing—a true burden—has now become valuable, necessarily useful despite its onerous environmental weight. Although it has not always been overtly traded, ballast has been a consistent token of exchange. Intermediary to global trade's economic imperatives and ecological concerns, ballast is Lycurgus's senate in Latour's "parliament of things." In this context, Latour again deploys ballast: "Natures are present, but with their representatives, scientists who speak in their name. Societies are present, but with the objects that have long been serving as their ballast from time immemorial. . . . The imbroglios and networks that had no place now have the whole place to themselves. They are the ones that have to be represented; it is around them that the Parliament of Things gathers henceforth. 'It was the stone rejected by the builders that became the keystone' (Mark 12:10)."[25]

Ballast carries moral weight; it is a moral thing that binds us together. Conrad knew this. And this shifting and fluid paragon might actually help balance the global ship.

Notes

1. Joseph Conrad, "Youth," *Blackwood's Magazine*, September 1898; and *Youth* (London: Blackwood & Sons, 1902), 3.
2. Joseph Conrad, "Her Captivity," *Blackwood's Magazine*, September 1905; and, later, in *The Mirror of the Sea* (London: Methuen, 1906).
3. Joseph Conrad, *The Mirror of the Sea* (Garden City, N.Y.: Doubleday, 1926), 40.
4. James Joyce, *Ulysses* (New York: Vintage, 1990), 151.
5. James Joyce, *Stephen Hero* (New York: New Directions, 1963), 211.
6. Richard Burn, "Rivers and Navigation," in *The Justice of the Peace and Parish Officer*, ed. Joseph Chitty, 26th ed. (London: Stevens & Sons, 1831), 3:307–10.
7. See Arthur Warner, "Ballast Island," *Pacific*, April 24, 1983.
8. John Ruskin, *Seven Lamps of Architecture* (New York: Grove, 2008), 180.
9. See John Knox, "Mr. John Nieuhoff's Voyages to the East Indies," in *A New Collection of Voyages, Discoveries, and Travels* (London, 1767). Gombroon is modern-day Bandar-e-Abbas, and Ormus is the island of Hormuz.
10. See the entry for *ballast* in the *Oxford English Dictionary*, 2nd ed. (New York: Oxford University Press, 1989).
11. Benjamin Martell, "On Water Ballast," oration delivered August 30, 1877, in *Transactions of the Institution of Naval Architects* 18 (1877): 336–48.
12. Francis L. Norton, cited in "To Protect All Vessels," *New York Times*, July 22, 1888.
13. Carl Hansen Ostenfeld, *On the Immigration of* Biddulphia sinensis *and Its Occurrence in the North Sea during 1903–1907*, Plankton Series, vol. 1, no. 6 (Copenhagen: Medd Komm Havundersogelser, 1908), 1–25.
14. John Carl Medcof, "Living Marine Animals in a Ship's Ballast Water," *Proceedings of the National Shellfisheries Association* 65 (1973; repr., Easton, Md.: Waverly, 1975), 54–55.
15. J. T. Carlton, *Introduced Species in U.S. Coastal Waters*, Pew Oceans Commission, Arlington, Virginia, 2001.
16. NOAA Coastal Services Center, National Oceanic and Atmospheric Administration, "Ballast Water: Michigan Takes on the Law," *Coastal Services*, July–August 2007, http://www.csc.noaa.gov/magazine/2007/04/article2.html.
17. Pablo Kaluza et al., "The Complex Network of Global Cargo Ship Movements," *Journal of the Royal Society Interface* 48, no. 7 (January 2010): 1093–1103; Susanne Kytzia, "Material Flow Analysis as a Tool for Sustainable Management of the Built Environment," in *The Real and the Virtual Worlds of Spatial Planning*, ed. Martina Koll-Schretzenmayr, Marco Keiner, and Gustav Nussbaumer (Berlin: Springer-Verlag, 2003).

18. See United States Code of Federal Regulations, Title 33, Navigation and Navigable Waters, Part 151, Subchapter O, Subpart C, "Ballast Water Management for Control of Nonindigenous Species in the Great Lakes and Hudson River."

19. See Board of the Millennium Ecosystem Assessment, *Living Beyond Our Means: Natural Assets and Human Well-Being*, (Washington, D.C.: World Resources Institute, March 2005).

20. Plutarch, *Lives*, trans. Bernadotte Perrin, vol. 1 (New York: Macmillan, 1914), "Lycurgus," 221.

21. When Bruno Latour discusses Serres's "archeology of stones" in the latter's *Statues*, he brings together ballast and Serres's stone witnesses: "Serres ballasts epistemology with an unknown new actor, silent things" (Latour, *We Have Never Been Modern*, trans. Catherine Porter [Cambridge, Mass.: Harvard University Press, 1993], 83). I like to think Serres—a sailor who studied at France's Naval Academy and the son of a fisherman, gravel dealer, and bargeman on the Garonne River—considered ballast stones to be a part of this archaeology, perhaps even quasi-objects in their own right.

22. Edmund Burke, "On Moving His Resolutions for Conciliation with the Colonies," March 22, 1775, from *Select Works of Edmund Burke: A New Imprint of the Payne Edition* (Indianapolis: Liberty Fund, 1999).

23. See Michel Serres, *The Natural Contract*, trans. Elizabeth MacArthur and William Paulson (Ann Arbor: University of Michigan Press, 1995).

24. Latour, *We Have Never Been Modern*, 144.

25. Ibid.

PART II
Bodies in Motion

Symptoms

John Law and Wen-yuan Lin

What does it mean to be international? What if the way things are international could have been different? Can we imagine other modes of international? These are our questions.

Let's start by thinking counterfactually. Here's one of the great "what-ifs" of history: What if the Chinese had dominated the world from the fifteenth century onward? This might have happened; indeed, it almost did. In 1400, Ming dynastic China was in expansive mode. A huge expedition sailed from China in 1405. Its 317 vessels and 28,000 crew members were commanded by Zheng He (zhèng hé, 鄭和). (Compare and contrast the "voyages of discovery" by Christopher Columbus and Vasco da Gama—3 vessels and 4, respectively). Over the next thirty years, there were six more expeditions to Southeast Asia, to India, and to present-day Sri Lanka. A trilingual stone tablet—in Chinese, Tamil, and Persian—was left at Galle in Sri Lanka. And the fleets went further, too, to the Arabian Gulf, the Red Sea, and then to the Horn of Africa, to Mogadishu and south to Mombasa.[1]

This was exploration on a grand scale. But then again, it wasn't. The fleets were simply following trade routes well known to the Song Dynasty (960–1279 CE) and the Three Kingdom periods (third century CE). So Zheng He and his people knew where they were going. There are excellent charts that show this. And they knew what they were up to, too. These expeditions were about trading, but they were also about diplomacy and extending Chinese influence. As a part of this, they were about saber rattling: there were plenty of armed men on that first expedition, and Zheng He wasn't afraid to use them. Pirates received short shrift, and he wasn't above fighting the odd land war (Sri Lanka in 1410). And then it all came to a stop. After 1433 CE, there were no more expeditions.

So why wasn't a Chinese version of international created in the fifteenth century? How come the marginal Spanish and the even more impoverished Portuguese instead set the gold standard for what it was to be international? There is no shortage of stories: imperial succession; the size of China; court intrigue; Confucianism; a general distaste for foreigners; a Mongolian invasion;

a sense that the periphery of the world was unimportant. Those debates don't really matter here, because we're interested instead in what we might think of as different "modes of international." We're interested in how international has been *done* in the world since the Spanish and Portuguese expansion. Our argument is that this has happened in a very specific way. It implies particular ways of acting and fighting, and specific versions of the person, the organization, the body, and healing—and not least, of understanding itself. In particular, it clings to the idea that realities and truths are universal (that, so to speak, they're already international). As a part of this, it takes it for granted that there's just one world, and its order and its structure are out there waiting to be revealed. So here's the counterfactual in which we are most interested: *What might a different mode of international look like?* No doubt there are all sorts of possibilities, but let us anticipate our conclusion. Even though we live in an international world whose parameters were established by the European expansion, it isn't that there aren't alternatives, *so long as you know where to look for them and how to see them*. And (let's add) if you're also willing to be fuzzy around the edges about what it means to be "international." One of these alternatives is alive and well in Chinese medical practices.

To look for this alternative, we're going to come home. Home is in the north of Taiwan. It's Hsinchu, the "Windy City." We're going to visit a GP's surgery and consult with a doctor (we'll call her Dr. Lee) who's doubly trained: as a biomedical doctor and as a practitioner of Chinese medicine. But first, an initial word of caution: all the books tell us that big categories are dangerous. We've done it already, but it's better not to talk about "the West" or "the Chinese," because neither is a great monolith. Indeed, such categories are endlessly heterogeneous. However, we're going to have to lose some subtlety in order to tell the story quickly, so we will need to say that there are two modes of international at work in this clinic. There's the one we first learned from the Portuguese and the Spanish, the one that is "Western." We'll also say that this is "analytical." And then there's a second that is "Chinese" and is "correlative." In what follows, we're going to try to show that they're quite different. But now for the story.

Dr. Lee is well known, and she's popular. You may have to wait hours to see her, but it's worth it. So you go in through her storefront door. What do you see? It's not "traditional," this practice. It's not in a dark room with herbs and dried fungi hanging from the ceiling. It's clean, it's light, and there's a Formica counter where you register once you've arrived. There's a small waiting room, and there's a pharmacy—you can just see it down the corridor. And you know, because you've been there before, that Dr. Lee's got a small consulting room behind the pharmacy. You're going to have to wait, but take the time to look around the waiting room more carefully. Ignore the fish tank. Look at the display cabinet instead. It's filled with bottles. Look carefully and you'll see an array of traditional Chinese medication: ginseng, deer horn, shark cartilage, and tortoiseshell. These are just for looking at, not for use: endangered species are out of bounds.

Now look at the notice boards. There are Ministry of Health licenses and certificates. Dr. Lee's qualified, and we learn she was trained at the Chinese Medical University. There's a certificate about the electronic patient-record system. Why? That's a little unclear. Then we see the qualifications of the nurses pinned up, too. They've got degrees in biomedicine. Now look at the other notice boards. You can read about diabetes and osteoporosis. You can read about the quality control of "scientific" Chinese medicine. There's been a scandal recently, and the posters are there to reassure you that you will not be ingesting pesticides or heavy metals when you take your medicine. And then you can read about tests. What will the surgery test you for? You've seen the list before in other surgeries: allergies, possible forms of cancer, sexually transmitted diseases. It's the same in many GPs' surgeries around the world: diseases are international, more or less.

Though hang on a minute. There's a whole lot of social science that says this isn't right. It insists that the idea that things like diseases are international doesn't happen automatically. It tells us instead that it takes a lot of effort to *make* things international. Perhaps Vasco da Gama or Zheng He would have agreed. It's not trivial getting from A to B and holding oneself together in one piece on the heaving seas.[2] But let's bring the argument right home and up to date, too. Let's talk about a particular "medical thing." Here's a question: Is hypoglycemia (that's a "medical thing") a symptom from which everyone might suffer, anywhere in the world? It sounds as if the answer should be yes. But it isn't, at least according to science and technology studies (STS), though the argument sounds really strange if you were raised in the "Western" mode of international. STS says that the answer is no, not necessarily, and not everywhere. Hypoglycemia *isn't* a universal condition. Yes, you may feel poorly. You may collapse, and indeed you may die. But do you actually get hypoglycemia? And the answer is no, you don't, not necessarily. Instead, whether you do or don't depends on *where* you are, *who* is looking at you, and probably on a bunch of *technologies*, too.[3]

Here's the argument. Hypoglycemia is possible only in contexts created by biomedicine: you must be biomedically trained and have the right equipment. "Blood sugar level 3.3 mmol/l? Quick, get some carbohydrates down her! Fast!" That's a version of hypoglycemia in practice in biomedicine. But now we need an instant clarification. "Biomedically trained" doesn't mean that you're necessarily a doctor. It means you've learned to think and act in a generally biomedical way; that you're caught up in a biomedical world. Perhaps you're a doctor, a nurse, a patient, a partner, a friend, a medical statistician, a coroner, or a passerby; it doesn't really matter. All it needs is someone who might say, "No, she's not drunk, she's got a hypo"; someone who gives her a glass of orange juice; someone who knows the real cause of the blurred speech; someone who knows that the person who's collapsing has type 1 diabetes.

Strange though it may sound, here's what STS is saying: *Hypoglycemia is hypoglycemia* only *if it is done in biomedical practices*. And then we can add this: *Hypoglycemia is international* only *if it is done in lots of biomedical places and practices in different parts of the world*. This means that it isn't a universal reality. STS is saying that it takes *effort* to turn bodies and their conditions into specific diseases, and then to make them international by making them transportable. And it adds that this is what biomedicine is all about: its fights against quackery, all those laboratories, the years of training, the clinics, the insurance policies, the technologies, and the treatments. It's about creating and practicing disease conditions, pathologies, and remedies, and doing this in different places. It's about creating practices that will hold steady across time and space. This means also that it's about creating *similar kinds of bodies* in different places, too, in the clinics and the hospitals and the laboratories and the streets and the dwelling places of the world. The argument is that it takes work to create the condition (type 1 diabetes) that is caused by autoimmune destruction of the pancreatic cells that produce the insulin that allows the body to metabolize glucose—a lot of work, expertise, and gear. Things *can* be the same everywhere, but only if you put the effort into making them so. Here's the problem: The "Western" mode of international doesn't see this at all. It thinks of itself as a way of uncovering universal truths about a single reality: the diabetic body. This is why the STS argument sounds so weird in the first place.

Sorry about the digression, but back in the Windy City, this is going to become important. Just hold the thought that "it's about creating similar kinds of bodies in different places," and sit down in Dr. Lee's waiting room. Actually, you're lucky. Dr. Lee is calling you already. You walk down the corridor past the pharmacy, and you go into her surgery. There she is. She's sitting at her desk in her white coat. She smiles. You sit across from her, and then your consultation starts. You roll up your sleeve. You lay the back of your right wrist on a firm foam pillow. Now Dr. Lee presses the tips of her middle three fingers on your radial artery. She presses down very gently. She's feeling your pulse. Then she does the same with your left wrist. She does it a couple more times. And then she picks up a little flashlight. She wants to look at your tongue. She doesn't tell you to "say ah" (you don't do that in Chinese), but she's looking at the top of your tongue and then at the bottom. Now she's putting the flashlight down, and she's starting to ask you questions: "How are you sleeping? How's your appetite? What are you eating? How's your digestion?" You answer: You've been having difficulty getting to sleep, have eaten irregularly, and your poo is a bit liquid in the morning. "How's your emotional life?" Hmm, not so good. You've been working too hard and getting irritable. You tell her you've got a pain in your shoulder, and you add that the backs of your legs have been hurting, too.

Dr. Lee is making notes. She's typing into the computer. There it is, the keyboard's handy. Then she calls the nurse. "Take his blood pressure, please." A nurse comes in with the hematinometer. She wraps the cuff around your upper arm and pumps it up. She waits as it deflates, and then she says, "102 over 78." Dr. Lee types the figures in, and then she speaks: "Your pulse points to the root of your problem. It's always deep at the 'chi' position [ch, 尺] in both your wrists. But it's stronger on the left. You're constantly drawing energy to keep going." Now she's moving her fingers between your wrists again, to and fro, comparing the pulses. Then she speaks again: "Your pulse is strong and deep [chén xián, 沈弦] as usual, but it's a bit faster and stronger." She's typing again: "white, thin coating." That's your tongue. Now she's talking about your legs. So what's the problem? Why do they hurt? When you told your biomedical doctor, he sent you off for a scan and a blood test. He was looking for signs of tendon injury, arteriosclerosis, or neuropathy, but the tests came back negative. Dr. Lee takes the results seriously, because they help her to eliminate possibilities, but at the same time, she isn't surprised. She thought that the problem was to do with the kidney meridian (shèn jīng, 腎經) all along. "You've been working too hard," she says. The end of the autumn is the time to restore *yang qi* (yáng qì, 陽氣) to the meridian. But you're not doing that. You're *extracting* it. This is bad news, but it explains why your legs hurt. They're close to the kidney meridian, and the fact that they hurt is telling her that there's not enough *qi* to warm your body.

Step back from this. We've got computers here, we've got biomedical tests, we've got biomedical technologies, and we know already that Dr. Lee and her nurses are trained in biomedicine. So we're in an international world, biomedical style. Neuropathy and blood pressures, these have traveled from Europe and North America, and now they've been carried to a clinic in Hsinchu. This is the biomedical body on the march. If you look at history, you can tell a story that pushes this home, and it's all about domination. Taiwan was first named Formosa and was mapped and made international, Western-style, by the Portuguese in the 1540s. Then, between the 1620s and 1660s, the Dutch and the Spanish occupied parts of northern and southern Taiwan, respectively. Much later, Taiwan became a Japanese colony—from 1895 to 1945. Indeed, it was a "model colony," but this applied to health care, too. Chinese medicine? The modernizers among the Japanese said that it wasn't scientific, it wasn't properly regulated, it wasn't effective, sometimes it was dangerous, and it made no kind of biomedical sense. Meridians? Could you point to them in the anatomical body? The answer was no, you couldn't. Or *qi*? What on earth was that? So the history tells us that for fifty years Chinese medicine was pushed to the periphery. It was suppressed, persecuted, and outlawed.[4] And the result? By the time Japanese left, in 1945, there wasn't much of Chinese medicine left, yet all the time the biomedical body was on the march. It had been installed in the clinics and the hospitals and in the daily health-care practices of Taiwan and was practiced in those places—indeed, just like type I diabetes and its penumbra of blood-sugar levels and hypoglycemia. Chinese medicine pretty much stayed marginal after the Japanese left, but it wasn't quite squeezed out. It had hung in there in "folk practices" and "backstreet clinics," and after 1945, very slowly and very partially it started to inch back in from the cold, such that in due course it became possible to train as a Chinese medical practitioner. Indeed, it became possible to put Chinese medicine and biomedicine together. The Chinese Medical College (which later became the Chinese Medical University) was founded in 1958, and the college, which was intended to revive Chinese medicine, gradually started to include biomedical diplomas and departments, too; as we have said, Dr. Lee was a product of that double training. In short, there was a pushback, although even now more than 96 percent of Taiwan's health-care budget goes toward biomedicine.

So that's the story, told historically. It's a story about making biomedical things like bodies international on a grand scale. It's about the movement of those things. It's a story about a small pushback from Chinese medicine, with emphasis on the "small." It's also another version of how international has been done in a "Western" and analytical way since the Spanish and Portuguese expansion set its parameters. This is a specific mode of international with its particular ways of knowing, healing, and self-understanding, in which realities and truths are taken to be universal: they were always *already* international. That's how it works.

So what is Dr. Lee up to? The answer is given in her training. She's putting biomedicine (the hematinometer and all that) *together with* Chinese medicine. She's not uninterested in the causal pathways of biomedicine. She's happy enough to take test results from the biomedical laboratory. "Ah, good," she's thinking. "We're not looking at neuropathy." And she's perfectly happy with the idea that there is a biomechanical body with its own specific anatomy. But there's something else going on, too, or something different, because what she's doing doesn't fit with biomedicine's understanding of itself. This is because when Dr. Lee gets to work, biomedical realities and truths are no longer the only realities and truths. Suddenly they are lying down with kidney meridians and seasonal imbalances in *yang qi*. So how should we think about this? In the end, the answer is going to lead us to another and quite different mode of international, one that is "correlative"—a term we are borrowing from sinology, where it has a specific meaning.[5] It's going to take us on a difficult journey to a place where things look and are quite different, because we're going to have to give up on biomedicine's understanding of itself and the habits embedded in the "analytical" mode of international we've inherited from the Spanish and the Portuguese. To see what this might mean, let's make a short historical detour.

There's a surviving classic text in Chinese medicine called *The Yellow Emperor's Inner Canon* (huáng dì nèi jīng, 黃帝內經).[6] Did the Yellow Emperor really exist? At least as a single historical figure, the answer is probably not. Most likely we're back in Chinese origin myths about proper rulers, proper states, proper statehood, and proper forms of culture; but if he did exist after all, then we're talking around 2500 BCE. However, the *Inner Canon* was compiled much later, in the Han period (206 BCE–24 CE). It tells us about yin and yang (yīn yang, 陰陽) and how they relate, and it collects together a bunch of theoretical principles about pathology and physiology in terms of seven emotions and six evils (qī qíng liù xié, 七情六邪), twelve meridians (jīng luò, 經絡), and five *zang* (zàng, 臟) and six *fu* (fǔ, 腑) visceral systems. In practice, it's an attempt to assemble the knowledge and practices of a whole series of partially connected medical traditions, including the five schools of ancient medical practice from the Warring States period (475–221 BCE) and the five-phases (wǔ, xíng, 五行) theory that was added and adapted during the Qin Dynasty (221–206 BCE) and the Han period. Indeed, archaeological and textual evidence suggests that there have been different theories about meridians and that the *Inner Canon* is just one of them. But for now, the historical details don't matter. What's important is that the *Inner Canon* is *syncretic*. It's a collection of different thoughts, approaches, practices, and forms of intervention. Even more important, its syncretism sets a pattern that has characterized Chinese medicine throughout its entire extended history, because subsequent classics have also been hybrid classifications. They've taken the form of collections of medical records (yī àn, 醫案) and reinterpretations of earlier classic texts. Indeed, most recently they have more or less happily taken up and absorbed facts emerging from biomedicine. So here's the bottom line: The *qi* body and the theory of meridians have been heterogeneous from the beginning. From a Western point of view, they don't look coherent, but heterogeneity or hybridity has never been a problem, and biomedicine is just the most recent addition.

Here's something else: these ideas have been diverse, too. This is because different practitioners have always worked in different ways. More important, they have always taken it for granted that this was the right thing to do and indeed probably *necessary*. Why? Because those who have practiced Chinese medicine have always taken it for granted that the *qi* body is one that is relational and contextual. They have assumed that diagnosis and intervention need to include what in Western terms we would think of as (Chinese versions of) bodies, diet, symptoms, emotions, social relations (including practitioner–patient relations), the passage of the seasons, location, the daily round, and stage of life. All of these are both irretrievably contextual and irretrievably variable from one place to the next, or indeed from one consultation to the next. At the risk of oversimplification, we might say that in the Chinese scheme of things, there is no fixed knowledge and there are no fixed causal relations. Everything varies.

Now we have moved back to the realm of self-understanding. Remember we said that the self-understanding of biomedicine takes it for granted that, in theory, realities and truths and (let's add) causal mechanisms are universal. In theory (the practice is different), there's just one world and one set of (okay, very complex and only partly known) mechanisms out there.[7] Well, none of this applies to Chinese medicine. It knows that it is hybrid; it knows that there aren't fixed mechanisms; it knows that everything is contextual. That's how it understands itself. It includes whatever it finds syncretically, which means that it's also diverse, because the same size doesn't fit all.

This detour into history makes it easier to see why Dr. Lee works in the way she does.[8] If *The Inner Canon* tells us that *qi* flows in the meridians with daily life and that it ebbs and flows with the seasons, then this is why she asks you questions about your diet, your emotions, your sleep, and your bodily functions. It is why she makes sense of your problem by thinking about the interrelations between the liver, the spleen, and the kidney meridians. But it is also why she doesn't find it difficult to absorb parts of biomedicine, too. We see this again when she starts to think about treatment. What does she say? The answer is all sorts of things: "You should eat carefully. You should go for an hour's walk every day. You should make sure you're in bed before eleven." And then she prescribes a modified version of Wendan decoction (wēn dǎn tang, 溫膽湯). What she wants to do is to warm your body. She wants to increase your *yang qi*. So she is talking about yin and yang propensities, the dynamics of the five phases, and how *qi* and *xue* (blood) are out of balance between the visceral systems of *zang* and *fu* in the meridians. And then, vacillating between biomedical and Chinese medical concepts, she also reminds you that your liver fire is not real fire, and that the liver in Chinese medicine is akin to the neurological system in biomedicine. Maybe you don't recognize this (perhaps you are too much a child of biomedicine), but she is being utterly thorough in a particular Chinese mode. She is systematically exploring correlations by drawing on some of the endlessly rich resources for thinking about patterns of relations within Chinese medicine. She's working *correlatively*. There is a moment when all this becomes clear, though, which is when she turns to her computer to record her diagnosis. Looking over her shoulder, you discover that she's not writing about *qi* or meridians. Instead she's using the WHO's International Classification of Diseases (ICD). She types "30742, 5649, 7291," and then the ICD diagnostic entries flash up on the screen: "Persistent disorder of initiating or maintaining sleep," "unspecified functional disorder of intestine," "myalgia and myositis, unspecified." Finally the certificate about the electronic patient-record system in the waiting room makes sense. The National Health Insurance scheme will pay Dr. Lee only if she uses the ICD. She has no choice, she has to do this. But as she does so, she's further weaving biomedicine and Chinese medicine together, and she's not doing this reductively. She's not attached to the baggage of a single body with a single set of (complex) causal relations. Instead she's working *functionally* and syncretically. Indeed, she's being *knowingly* syncretic. It's no problem laying bits of biomedicine alongside *qi* or the meridians. This has nothing to do with causes. She's trying to find ways of understanding how they are correlated together in the particular circumstances to hand. And to understand the pattern of those correlations, she's drawing on the corpus of Chinese medicine. The latter is an almost indefinitely rich resource for thinking about patterns. This is what it is to think correlatively rather than analytically.

So now we have what we want to say in place. If we set it up as a binary then we've got the analytical on the one hand and the correlative on the other. In the correlative, there are no big reductions or general explanations. It's site specific, and it shifts and moves as contexts change. It's a grammar or a syntax for sensing and interpreting the endless possibilities in the weave of things, *qi* and all the rest of it. It is a set of metaphors for coming up with a story line that makes sense of the pattern of things as they are in a place. Compare and contrast this with the analytical of the biomedical. This sees symptoms, and it goes looking for causes. It's into mechanisms. (Think of hypoglycemia again.) That's the theory, even if biomedicine in practice often has more to do with tinkering than with causes.[9] Nevertheless, that's how it likes to imagine itself. It works in a world, a single world, of mechanisms, and it lives in the hope—and quite often the reality—that if we can understand those mechanisms, then we'll be able to intervene and cure or manage the disease. Sit down, chew on these glucagon tablets, drink an orange juice, keep quiet, and hopefully you'll come around in a few minutes. And then, alongside this, there is the commitment to universality: that because things are pretty much the same everywhere, all we need to do is to transport what we know about those things, because our knowledge will work in other places, too. Biomedicine in Taiwan under the Japanese? Of course. There's your analytical mode of the international. If it can overcome prejudice such as the attachment to Chinese medicine, then it will work in the farthest corners of the globe. There's room for only one reality because there *is* only one reality. End of story.

So that's the binary: analytical versus correlative. Let's underscore our earlier warning. The division is too simple, and it's a whole lot more complicated in practice. But if we stick with it, then we learn something interesting about different modes of international. In the world of analysis, Chinese medicine gets squeezed: biomedicine is dominant. But in the world of correlativity, or at least in the version we're seeing in Dr. Lee's practice, things look quite different. The biomedical doesn't dominate here; it's just being added. It's being absorbed into the great, lumpy, heterogeneous weave of possible correlations that makes up Chinese medicine. And now we're back to our opening question: What does it mean to be international? We've seen that it takes hard work to make things transportable and to hold them steady. This is a tough way of doing international, but it's pretty successful even so, and biomedicine's very good at it. Along with imperialism, colonialism, and European gunboat diplomacy in various shape and forms. And massacres. And opium wars. And Japanese imperialism. And economic domination. And technological subordination. And colonization by "Western" ideas, such that a Taiwanese university campus looks very like one in Michigan, with the same kinds of departments and concerns: journal publication, rankings, the Science Citation Index, and all the rest. But the making of similarities like this has been going on since Vasco da Gama in the Western mode of international. Yes, it comes in various guises, as modernization, or Westernization, or globalization, or development, or even civilization. But it's been done on a massive scale since the sixteenth century. And it has always been *analytical* and *dominant*.

So think again about Zheng He's voyages, the world of "what-if." What if the Chinese had expanded to fill the world instead of the Spanish and Portuguese?

What might such a world have looked like? What form might a correlative international have taken? Now that we have been in Dr. Lee's clinic, the ghost of an answer begins to take shape. We have four thoughts on this.

First, *things* would look different. Dr. Lee does so much for *you*, but she is doing almost nothing to your diseases. Instead she's looking for symptoms and circumstances. She's worrying about correlations. And she's trying to patch the clues she sees—bodily, emotional, and contextual—into correlative patterns and figuring out their imbalances. She is telling you that the patterns might be better balanced if you lived in a slightly different way, and then she's prescribing a decoction that might be added to the pattern and help to shift it. So where's the disease in all this? The answer is that it's not clear that she's working on a disease at all—even when she's working with the ICD. Instead she's dealing with the complexity of symptoms. There's no idea of a magic bullet that can go everywhere and work in all contexts. All of this suggests that as things moved from place to place in a correlative mode of international, they wouldn't hold their shape. More strongly, the extent to which it would make sense to talk of "things" at all is quite unclear. Propensities, relations, patterns, flows, and knowledges or forms of expertise for sensing and working upon them—this is how a correlative international would frame the places it joined together. But things themselves? Perhaps they would still be there, but they would turn into shape-shifters. They'd respond to patterns and propensities. Indeed, *they'd look more like symptoms than objects.*[10]

Second, the *relations between places* would change. Chinese medicine is syncretic in each location, but also *between locations*. Yes, you attend to the flows and meridians and the visceral systems and to the balances and movements of yin and yang, but these work differently each time around. It's not even that these principles are *applied* differently. Rather, it is that they don't count as analytical principles at all. Instead they are sensibilities that draw contextually on an endlessly rich set of metaphors for knowing correlativity. Everything shifts from place to place and time to time, and this explains why Chinese medicine was mostly traditionally taught in apprenticeships rather than medical schools. It also explains why practitioners don't (simply) pass exams but gather a following. Finally, it explains why Chinese medical practice is diverse and distributed and why it has (as it were) been "correctly" practiced in endlessly many ways. Places cannot be reduced to one another. They aren't abstract, spatially coordinated locations, sites to which universal truths naturally apply. Instead, specific things-in-place make the particular propensities and patterns of that place. It is almost as if *different places are in different worlds*. And this implies a corollary: centers and peripheries in a correlative mode of international would look quite unlike those of the analytical international bequeathed to us by the Spanish and the Portuguese.

Third, the mode of *doing* international—what it actually takes—would look different, too. It would be more supple or subtle than its analytical counterpart. Think again of the relations between biomedicine and Chinese medicine. As we saw, the former almost entirely displaced the latter in Taiwan. In an analytical mode of international, there's just one right way of doing things, because there's just one version of reality. It's powerful, it's focused, and, to use a religious metaphor, it's monotheistic. This contrasts with the dispersal of Chinese medicine. But now we need to add that there is also a kind of practiced *ease* to the conduct of Chinese medicine. This is because a practitioner like Dr. Lee isn't trying to press everything into an analytical framework. Instead she is drawing on sensibilities educated by the vast metaphorical resources in Chinese medicine for knowing and recognizing relational patterns, and she is gently working on those patterns to improve the imbalances that she detects. She works with propensities to make small maneuvers at strategic points. She shapes things by following and working with things as they flow and return. And then she is inviting you to work with her strategy. The *Dao De-Jing* of Lao-zi (lǎo zi, 老子) tells us that it is better to "do-nothing" (wú wéi, 無為): "The softest thing in the world / Rides roughshod over the strongest. / No-thing enters no-space. / This teaches me the benefit of no-action."[11] This is what it would be to do or to act in a correlative mode of international. It would be flexible, it would be subtle, it would "work with," and it would be minimalist. To be sure, its critics would say that it was manipulative.

Finally, what of *international itself*? The answer is that it would dissolve. The notion of international implies nations, territorial patches, borders, and differences; insides and outsides; lines drawn between places; and the creation of connections between those places. As we have suggested, it also implies *a single space within which all this takes place*. The United States is here on the map; China and Taiwan are over there. They can trade. Let's hope they're not going to fight a war. But here's the problem: All of this is *analytical*. There's one world, and then there are nations within that world. That is how we get to "inter-national." But this doesn't work for correlativity. Think of Chinese medicine. The extent to which it lives in a single world and, importantly, *imagines* that it lives in a single world is very limited. That practice over there? It's in a different place, and the things that make that place are also irreducibly different. Yes, we might learn from looking somewhere else in a correlative world. We might find that we share the *Inner Canon* and the subsequent medical classics. We might apprentice ourselves in a different place and do things differently. But the world is irreducible variably, composed, as the Chinese put it, of the ten thousand things (wàn wù, 萬物). There's no larger context—spatial or otherwise—in which we or our nations could be lined up to join the ranks of the international.

Making things international. In this chapter, we've played with the hypothesis that there are many modes of international. One of these, the analytical, seems to be everywhere, and it occupies most of the space and most of the talk. But we've gone looking for alternatives, and we think we've found one. It is alive and well in a GP's surgery in Hsinchu, and it is correlative. If we are right, this tells us that other internationals are possible. What would the world have been if the Chinese had not withdrawn from the Western, or Indian, Ocean in 1433? The answer is that we will never know.

Notes

We thank Ping-yi Chu, Judith Farquhar, Daiwie Fu, Casper Bruun Jensen, Sean Hsiang-lin Lei, Shang-jen Li, Annemarie Mol, Kuo-li Pi, and Hugh Raffles for dialogue, friendship, and support. We are also grateful to the Taiwanese National Science Council for its generous financial support.

1. Louise Levathes, *When China Ruled the Seas: The Treasure Fleet of the Dragon Throne, 1405–1433* (Oxford: Oxford University Press, 1996).

2. Latour makes this argument by talking of immutable mobiles, and Law explores it for the case of Portuguese imperialism. See Bruno Latour, *Science in Action: How to Follow Scientists and Engineers through Society* (Milton Keynes, Eng.: Open University Press, 1987); and John Law, "On the Methods of Long Distance Control: Vessels, Navigation, and the Portuguese Route to India," in *Power, Action, and Belief: A New Sociology of Knowledge?*, ed. John Law, Sociological Review Monograph 32 (London: Routledge & Kegan Paul, 1986), 234–63.

3. For further discussion of hypoglycemia, see Annemarie Mol and John Law, "Embodied Action, Enacted Bodies. The Example of Hypoglycaemia," *Body and Society* 10, nos. 2–3 (2004): 43-62, also available at Heterogeneities.net, http://www.heterogeneities.net/publications/MolLaw2004EmbodiedAction.pdf.

4. Ming-cheng Miriam Lo, *Doctors within Borders: Profession, Ethnicity, and Modernity in Colonial Taiwan* (Berkeley: University of California Press, 2002).

5. David L. Hall and Roger T. Ames, *Anticipating China: Thinking through the Narratives of Chinese and Western Culture* (Albany: State University of New York Press, 1995).

6. *The Yellow Emperor's Classic of Internal Medicine*, trans. Ilza Veith (Berkeley: University of California Press, 2002).

7. It is indeed important to note that the practice is different. See, for instance, Annemarie Mol, *The Body Multiple: Ontology in Medical Practice* (Durham, N.C.: Duke University Press, 2002).

8. Our account draws on a large body of work about contemporary Chinese medical practice, and in particular on Judith Farquhar and Qicheng Zhang, *Ten Thousand Things: Nurturing Life in Contemporary Beijing* (New York: Zone Books, 2012); and Volker Scheid, *Chinese Medicine in Contemporary China: Plurality and Synthesis* (Durham, N.C.: Duke University Press, 2002).

9. Annemarie Mol, Ingunn Moser, and Jeannette Pols, eds., *Care in Practice: On Tinkering in Clinics, Homes, and Farms* (Bielefeld, Ger.: Transcript-Verlag, 2010).

10. It may be that the Western analytical mode misunderstands itself and is also fluid. See Annemarie Mol and John Law, "Regions, Networks, and Fluids: Anaemia and Social Topology," *Social Studies of Science* 24 (1994): 641–71.

11. Lao Tzu, *Tao Te Ching* (Boston: Shambhala, 2007), 43.

Corpses

Jessica Auchter

The primary objects of study of international relations (IR), namely, conflict and war, produce dead bodies en masse, yet the corpse itself has remained outside IR's purview. International relations then, is literally built on the backs of dead bodies, yet fails to examine dead bodies in their complex potential. This lack of attention to conceptualizing dead bodies in a framework of the international leads to a skewed perception of the fundamental problems of international relations. In short, analysis of world politics shied away from considering the thing as a political entity, but it also suffers from a vitalist bias: a prejudice toward the living and against the dead. New approaches that focus on materiality and the body often promise to resuscitate the agency of the body, but they do so by looking at the way it engages in forms of resistance. But what happens when we look at a material corpus that exists outside of the normally recognized living-and-breathing-based bounds of politics: the dead body?

This intervention considers the corpse in its material and political context through two examples: the question of corpse management after natural disasters, and the abuse of dead bodies by coalition forces in Afghanistan. The politics of the dead body in international relations are illuminated through the context in which the body is rendered visible or hidden from view, as well as through how the dead body is managed and disposed of. Assessing the way in which the body, even when dead, can disrupt the story the state tells about security, I explore how the dead body is integral to conceptualizing the international.

Corpses suggest the lived lives of complex human beings, yet they are ultimately material things. However, as things, they are managed very differently from other things considered to be strictly objects. As Tiffany Jenkins details, "[H]uman remains hold a social category as a 'person' (human, body), but are also a 'thing' (remains, corpse, cadaver, skeleton)."[1] Dead bodies are considered a problem for governance in that they require some kind of management, yet rarely are they considered a problem of governance in that they rarely cause us to reflect on structures of authority and power. By engaging with the material status of dead bodies, we see what work dead bodies do in sustaining a particular conception of the international.

Death itself sustains a particular conception of the state as an entity that, through causing the deaths of others, is able to sustain and nurture the lives of its own citizens. Death is key to the authority of the state. Stories of dead bodies complicate this conception of the state, which comes to be taken for granted as a natural circumstance of affairs. In looking at bodies, I theorize corpse politics through two stories: first, that of international corpse management after natural disasters, and second, that of the abuse of corpses by coalition forces in Afghanistan.

Jane Bennett has theorized a vital materialism that seeks to explore a vibrant assemblage that results in conceptualizing a nonhuman agency. She essentially theorizes the object and argues that objects appear to be objects to us simply because "their becoming proceeds at a speed or a level below the threshold of human discernment."[2] Dead bodies, though not alive, are vital materiality. They are engaged in a becoming that differs from traditional conceptions of subjectivity, in a materiality and corporeality that are mobile; the dead body is not itself a fixed object despite its material (and dead) status. Then what are the dead body's politics? To answer this question, I explore the complex assemblages in which it is enmeshed in two specific contexts, which form a relationship that invokes particular questions of visibility: When is the dead body visible? When is it rendered hypervisible, and when is it rendered invisible or hidden from view?

The Corpse, the Thing

What does it mean to talk about the dead body itself? What is the corpse, and in what ways is it a political entity? The corpse is a dead body. It is a thing, yet at the same time it is human. It is a body, and there are assemblages and interrelationships that come along with the fleshy being that we typically privilege as the subject of politics. It does not possess the self-awareness and vitality that we refer to when we conceive of members of a political community, but it still retains the sacred status of once having been imbued with self-identification, an attribute that sets the corpse apart from other things. This is based on the notion of the body itself as sacred that is often derived from religious myths. As Walt Whitman writes in "I Sing the Body Electric," "If any thing is sacred, the human body is sacred."[3]

Human remains in the context of scientific analysis and museum exhibition are considered to be sacred objects in terms of how they are treated in display, even if they are hundreds or thousands of years old. The Human Remains Working Group Report states that "human remains, irrespective of age, provenance or kind, occupy a unique category distinct from all other museum objects. There is a qualitative distinction between human remains and artifacts."[4] As Jenny Edkins argues, rationally the corpse may be an inanimate object, but our cultures tell us otherwise: "[T]he body may not be alive, but it is grievable."[5] Dead bodies are not objects or subjects, and in many ways they invoke that line between life and death by reminding us that they are our loved ones, yet at the same time they are not fully anymore. Still, they certainly remain imbued with some sense of the identity they held while alive, because we make pilgrimages to their gravesites to visit them. Dead bodies matter to us precisely because they are not simply dead bodies. All living human beings have bodies, and all will eventually become dead bodies. The corpse, then, posits particularly interesting questions for this collection, because it must interrogate not only the relationship between the thing and the international but also the thing-as-thing itself.

The body has been theorized in relation to its materiality, though mainly in reference to the living body. Michel Foucault has discussed the way in which the emergence of biopolitical technologies have placed the body at the center of political life focused on ensuring the spatial distribution of individual bodies through separation, alignment, serialization, and surveillance.[6] As he states, "[T]he body is also directly involved in a political field; power relations have an immediate hold upon it; they invest it, mark it, train it, torture it, force it to carry out tasks, to perform ceremonies, to emit signs."[7] Foucault emphasizes the importance of techniques of visibility in control over bodies. Lauren Wilcox, as in this volume, has theorized bodies as processes, both material in themselves and revolving around other assemblages.[8] Elsewhere, she has focused on other materialities, such as food, as in the case of force-feeding at Guantanamo.[9]

These questions of visibility and control come to the forefront: What might it mean to display certain bodies in certain contexts and not others? Why are some corpses in some spaces rendered invisible and others hypervisible? Monica Casper and Lisa Moore also emphasize the importance of visibility when it comes to bodies, arguing that not all bodies are equally visible. Some bodies are hyperexposed and magnified, others hidden or missing.[10] Dead bodies themselves are significant for politics, especially because, as Henry Giroux lays out, "cadavers have a way of insinuating themselves on consciousness, demanding answers to questions that aren't often asked."[11]

Katherine Verdery focuses on the corpses of elites to demonstrate the politics of dead-body movement and interment.[12] Yet the corpse of the everyday individual is just as political. Indeed, the statement of presence Verdery alludes to is not simply a declaration of being but a declaration of being that relies on a visible presence or trace in order to declare ownership. That is, dead-body politics more generally are worth looking at if we focus on corpses as features in larger assemblages.

Dead bodies are the traces of political hauntings: both the invocation of death by the state in order to legitimize specific policy making, and the focal points we study as the key features of international conflict: war, genocide, economic inequality, famine, disease. How can we study these without looking at the dead bodies they create? Dead bodies are thus the material targets of power, but at the same time, the way in which dead bodies insinuate themselves into the project of radical questioning plays with the question of agency, which we traditionally associate with living-and-breathing activity.

Dead bodies are ambiguous. Bruno Latour's argument that we should take things seriously, Dingpolitik, hinges on the unnecessary distinction between human and nonhuman when it comes to agency, the attribution of agency to things, and a rethinking of the notion of actor. Similarly, Bennett's theorization of vital materiality rests on acknowledging nonhuman materialities as participants and as agents. It enmeshes human beings in complex assemblages along with nonhuman participants in order to re-envision what it means to act and effect. But is the dead body human or nonhuman? What about the character of a being that was once vital in every sense but now falls into a much more complex categorization than nonhuman? What about the character in the interstitial space: the dead body? The dead body poses particularly interesting questions for a materialist retheorization of agency, as Latour and Bennett have both undertaken.

That is, if the dead body does not act in the way action is traditionally defined by our political communities, what are its politics? What does it mean to look at things from the perspective of the dead body? Because dead bodies do not see, they disrupt traditional conceptions of visibility and self-recognition that form the basis of claims to agency and political viability. Dead bodies are enmeshed in material assemblages of human and nonhuman, but there is something unique about dead bodies, because they are neither entirely human nor entirely nonhuman. In this sense, corpses are "ontologically multiple."[13] Yet dead bodies are not these things that Bennett associates with her notion of ontological multiplicity: not lively, not alive, not vegetable, not verbal. They are a different complex ontology or ontologies: human yet dead; hypervisible yet invisible; silent yet questioning; grievable person yet thing; subject yet object; subject of power, subject to power, and object all at the same time. Assessing dead-body politics becomes clearer by looking at two instances of what we might consider to be the complex multiple ontologies of the corpse.

Managing Corpses: What to Do with Dead Bodies after Natural Disasters

The way international organizations deal with questions of what should be done with dead bodies after natural disasters speaks to the difficulty in pinpointing what these bodies are. They are not human beings, yet they are not things, because we must be concerned about questions of dignity in corpse management. They are both and neither at the same time.

Dead bodies are generally considered a problem of management after natural disasters, simply because they appear in places where they have hitherto not been, and at times in places where it is difficult to access them, including being buried in rubble after disasters such as earthquakes. What to do with unidentified corpses poses a particularly salient problem, because it is often infeasible to leave them indefinitely or to preserve them for a later identification that will likely never come in most cases. One suggestion is burial in unmarked or anonymous gravesites, which can provide some sense of closure and alleviate the presumed health effects of lingering dead bodies. This option also provides the possibility of exhumation down the road. But in societies where burial is not the cultural norm, cremation often provides the key option for disposing of corpses. Yet cremation also indicates some sense of permanence in relation to identification status, because once cremated, the bodies will never be able to be identified definitively. Yet neither burial nor cremation forms the basis of policies related to dead-body management. Indeed, first responders are encouraged to not cremate bodies whenever possible, and to avoid immediate burial.

One of the biggest concerns expressed by the general public in relation to body management is the spread of disease. However, the World Health Organization (WHO) and its subsidiaries indicate that the risk of disease is relatively low. Steven Rottman, director of UCLA's Center for Public Health and Disasters, notes that the idea of dead bodies as dangerous is a myth. Melinda Wenner puts his comment in context: "If the disasters have struck populations that have, for the most part, been vaccinated against major communicable diseases like measles, 'the risk of dead bodies following natural disasters being a source for spreading infectious diseases is very, very small,' Rottman says."[14] After all, the deaths have been caused by the natural disaster itself, not by a contagious disease. Wenner goes on to quote the Pan-American Health Organization's former director, Claude de Ville de Goyet: "[O]n rare occasions when victims of a disaster are carriers of communicable diseases, they are, in fact, a far lesser threat to the public than they were while alive."[15] Indeed, viruses and parasites die shortly after their host has died. So how, then, does the corpse come to be discursively constructed as a threat, a risk that must be managed?

The issue is not the infectious nature of dead bodies but rather the fact that the visibility of dead bodies is distressing to survivors. Dead bodies are indeed dangerous, but not because of the health risks they pose. Rather, they are dangerous psychologically. They must be managed precisely so that human mortality is hidden from view, which is not simply a psychological move but also a political one. When we are faced with the notion of our own mortality, certainly it is psychologically disruptive, but it also breeds concern over the way in which our lives and deaths are in fact already managed by the state in a variety of contexts. It is my argument not that corpses should not be managed but rather that the way their visibility is managed in particular contexts is important, and that we need to pay attention to dead bodies and their management as a political notion, not simply a forensic issue.

Corpse-management framework documents illustrate my argument. The WHO puts out a manual for first responders entitled *Management of Dead Bodies after Disasters: A Field Manual for First Responders*.[16] The manual begins by softening the blow of its content with a cover image of red autumn leaves against a smoky background image of some jutting rubble. What meaning does this have for conceptualizing the international health response to corpses? What work does this do to paint corpses in a specific light or, in this case, darkness? Corpses, or dead bodies, the very subject of the manual, are not shown on the cover. A euphemistic image of fallen leaves is shown instead, reinforcing the idea that we can talk about dead bodies without having to view them, that they are not something that is intended to be viewed, even when we are directly dealing with the problem their visibility has thrust upon us. Even before we have begun discussing what is to be done with corpses, we are told that they are not to be made visible, and that, when visible, they should be rendered invisible again as soon as possible. Death, then, becomes a problem that must be dealt with and corpses things that must be managed, and quickly, before anyone realizes that they are complicated political entities.

Further into the manual, several features stand out. First, the authors feel the need to address why they refer to the subject as "dead bodies" rather than, as they put it, "the more respectful and technically correct term 'human remains,'" an issue of translation and language.[17] The manual provides a framework for managing dead bodies through photographic identification, including the use of reference numbers, and it instructs how to take effective photographs that will help with identification down the road. It offers up an example of effective photography of a dead body, with a man laid out with his reference number in each photo. A note accompanying the photo specifies that "for the purposes of demonstration, photos were taken of a volunteer and not of a deceased individual."[18] This notation speaks to the same questions of display and visibility as the front cover. It seems odd that in a manual that provides instructions specifically for those dealing with dead bodies, there is still the sanitation of images of dead bodies. One might argue that there is no need to show a real corpse, yet the issues of photographic identification that arise with a real dead body, including how to capture features when a body has begun to decompose, are addressed in the manual as key issues.

A brief example is illustrative. After the Haiti earthquake in 2010, there were nearly two hundred thousand dead bodies that needed to be dealt with. Immediately the question of dead-body management came to the forefront; it was considered a necessary step before reconstruction could begin. The biggest concern was not necessarily identification, which was considered likely infeasible given the circumstances of the deaths, but burial. However, simple burial was not the objective. International Medical Corps psychiatrist Lynn Jones remarked in the Haitian context that "if deaths are not dignified—that is, lacking proper burials or mourning ceremonies—this denies people the means to accept and come to terms with their loss."[19] Dignity is depicted as something that comes along with the manner of burial, not the manner of death itself. Death is dignified if it is solemnized in cultural rituals or mourning and burial regardless of the nature of the death itself. This revolutionary perspective seems to conceptualize death as defined by the way others respond to it rather than by the circumstances under which it takes place. Dignity, then, is not something for the dead body itself, not simply a thing to be managed, but something for the survivors, precisely because the dead body is not a thing. This paradox is what makes the corpse-as-thing so interesting for conceptualizing the international politics of the dead body. The root of this idea is Latour's notion of representation. The way in which the dead body is represented in disaster-management discourse is precisely through re-presentation. That is, it is presented in a manner that involves a complex layer of assemblies, translations, and negotiations of meaning.[20] And it is this re-presentation that enables a schema of visibility that tells us what corpses are to us.

Managing Corpses: What to Do with Dead Enemy Bodies

The dead bodies of disaster victims are often consigned to invisibility as a political necessity, but the opposite is true of abuse of enemy corpses. Posing with dead bodies or shattered limbs makes a very specific political statement, both about what can be done with dead bodies of those who are labeled terrorists, who occupy a status beyond simply enemy combatant, and what is allowed to be visible.

Two instances from 2012 are illuminating. One is images and video of American soldiers urinating on Afghan corpses. The merging here of two biological processes—urination and death—emphasizes the significance of the material body. I will explore not the act itself, which has been discussed widely, but rather the depiction of the act in the media. The reporting of the act focused on the soldiers themselves, not on the unnamed Afghans in the video. We do not know whether the dead men were militants or civilians. They are simply dead bodies, props supplementary to the main attraction in the film. Most of the websites reporting the story blurred out parts of the images, although the full video can still be found online without any blurring. Most common was the blurring of the crotch area of the soldiers.[21] Some outlets blurred the soldier's faces,[22] whereas others blurred the soldiers from the chest down.[23] Not one Western news source that I encountered blurred the faces of the dead bodies on the ground. This indicates a widespread discursively constructed consensus that the showing of the Afghans' bodies and faces was the norm. Because they were rendered visible, a political statement was made about their lives and deaths; they were both biologically and ontologically dead. Their visibility, at times contrasted with the invisibility of those enmeshed in multiple biological processes with them, rendered them the objects of American military agency even in their death.

The other instance is images of American troops posing with dead Afghan bodies, including that of one soldier holding one corpse's middle finger up in the air. Although these images might be objectionable in themselves, discomfort also arises from the fact that we are perhaps not as uncomfortable as we thought we might be with pictures of dead bodies. That is, we might find the poses offensive, yet the dead body does not render us nearly as uneasy. This reaction itself is then uncomfortable, because we have been taught that viewing dead bodies is taboo except in very specific circumstances. So is there something about the context of Afghan bodies that makes their visibility acceptable to those viewing these photographs? That is, whereas there was shared agreement among American media outlets not to show photos of the falling bodies of 9/11, why did a paper such as the *Los Angeles Times*, which received the middle-finger-pose picture from a secret army source, choose to display these photos without blurring the faces of the dead Afghans? Surely there is a case for showing these photos to the public as a way of raising awareness as to the abuse of corpses by several members of the military. But if images of corpses are usually blurred—sanitized as a matter of dignity—then why are these corpses depicted as matter in these images, as material background, rather than as individuals with complex histories who have key social meanings for their loved ones and communities?

Other photos include poses with only body parts, including those of a suicide bomber whose body was blown apart. The soldiers typically offer smiles and sometimes a thumbs-up. In some of the photos, they can be seen holding the bodies in particular poses. Michael Sledge details a long American history of maltreatment of the corpses of enemy soldiers. Some include boiling the flesh off Japanese skulls to make table ornaments for sweethearts back home, carving bones into letter openers, playing soccer with the head of a dead German soldier, displaying Vietcong bodies out in the sun so that U.S. soldiers could get used to the sight of blood and corpses, and stealing dog tags from dead Iraqi soldiers so that their families would not be able to identify them.[24]

The depiction of decapitation, more specifically, follows on a long history of posing with corpses of the enemy, including one infamous case of a soldier posing with a Japanese decapitated head during WWII, and several cases of American soldiers posing with the decapitated heads of Vietnamese during the Vietnam War. Indeed, this has recurred in the Afghan context as well. War photographer Danfung Dennis was embedded with U.S. troops in Afghanistan in 2008. He tells his story: "During a patrol, the unit came across a decapitated head sitting on a market stall with a note on the forehead that read: 'This is what happens if you work with the Americans.' The young soldiers playfully packed up the head, took it back to base, and then took pictures with it—holding it up and grinning with their war trophy."[25] Dennis was able to photograph this as well, and he brought the photos to *Newsweek*, which elected not to publish them because of their incendiary nature. Dennis notes that these types of events involving desecration of corpses are relatively common from his observation, and speculates that in the Internet age, increasing numbers of these photographs will be circulating among the general public.

This phenomenon can be usefully contrasted with debates over photography of coffins of dead American soldiers, or the controversial display of five executed American soldiers on Iraqi television in March 2003. The official U.S. perspective at the time was that showing their remains on television violated Geneva Convention rules for prisoners of war.[26] This, juxtaposed with the display of enemy dead bodies, leads one to believe, as Sledge concludes, that one kind of display is criticized whereas the other is justified using an Orwellian logic dictating that "all bodies are equal, but some bodies are more equal than others."[27]

One of the more interesting features of the abuse of corpses is the U.S. Army's response to the photographs of American soldiers posing with dead bodies. Army spokesman George Wright said in response to the publication of the photographs, "It is a violation of Army standards to pose with corpses for photographs outside of officially sanctioned purposes. Such actions fall short of what we expect of our uniformed service members in deployed areas."[28] One might wonder what the "officially sanctioned purposes" might be whereby soldiers may legitimately pose with corpses.

Invisibility and Hypervisibility: Conclusions

When one normally thinks of dead bodies, one does not desire to look at them. Dead bodies are consigned to the realm of invisibility in the service of an illusion that paints those still living as resistant to the vulnerability that rendered the dead into corpses. Yet there are other circumstances in which dead bodies are hypervisible, where there is a desire to look upon the corpse, whether it be the dead body of Mu'ammar Gadhafi, which numerous spectators paraded past in images distributed globally through the media, or the bodies of Afghans subsumed under a discourse of the U.S. presence in Afghanistan. There is also a desire to look upon the corpse in the context of dead-body management in natural disasters. Indeed, the primary focus of dead-body management as laid out in the WHO manual described earlier is identification, which is frequently done visually because of the lack of infrastructure for forensic identification in disaster circumstances. Thus, we are both encouraged to look and discouraged from looking at dead bodies. We have used these two examples to explore the complex ontologies of the dead body; there is no hard-and-fast rule when it comes to dead-body management, and this is precisely what makes what is done to and with dead bodies by the living such a key political question. The way we, the living, navigate our relationships with the dead, whether it be those we identify as part of our political community, as loved ones, or as others, speaks to the way in which identities, lives, and deaths are all managed in a larger schema of visibility that dictates the degree to which one is or should be visible, and hence that dictates the terms of membership in a political community. Statecraft, then, is not simply the overt acts of the state but also the stories we circulate ourselves about what death constitutes, what is appropriately done with a dead body, whether we can or should look on it and how, and ultimately what are the thing-politics of the corpse itself. Dead bodies are objects, but dead-body politics cannot ignore the history of the dead body, that it was once living. There is a need, then, to map the geographies of the corpse, to trace the politics of its visibility and agency, and to explore the complex assemblages in which it is enmeshed, which I have begun to gesture to here.

There is more work to be done in exploring the politics of the corpse. Dead bodies are complex ontologies that posit uncomfortable and disruptive questions about the functioning of international politics and the theoretical frameworks utilized to conceptualize international relations. They are both human and thing, persons imbued with the social meanings of their community and social ties, but at the same time material evidence of explicitly political events; they are both humanity and raw materiality. By exploring the assemblages in which dead bodies operate, as I have demonstrated in the two cases I have explored here, it is possible to trace not only how dead bodies are managed both materially and visually but also how this management constructs a specific understanding of international politics as something that is inscribed upon dead bodies precisely because of the way it manages their visibility. That is, dead bodies should be considered not for their forensic status as evidence of an event, but for the way they blur the boundaries of what counts as vital or material, and the way they provide evidence, in some sense, of how vitality and materiality are managed in international politics. The dead body, then, is not the constitutive outside to IR but rather a complex ontology that forms one piece of the intermeshed international.

Notes

1. Tiffany Jenkins, *Contesting Human Remains in Museum Collections* (London: Routledge, 2011), 107.
2. Jane Bennett, *Vibrant Matter: A Political Ecology of Things* (Durham, N.C.: Duke University Press, 2010), 58.
3. Walt Whitman, "I Sing the Body Electric," in *Leaves of Grass* (Philadelphia: McKay, 1900), retrieved from Bartleby.com, http://www.bartleby.com/142/19.html.
4. Department of Culture Media and Sport, *Report of the Working Group on Human Remains* (London: Department of Culture Media and Sport, 2003), 166.
5. Jenny Edkins, *Missing: Persons and Politics* (Ithaca, N.Y.: Cornell University Press, 2011), 126.
6. Michel Foucault, *Society Must Be Defended* (London: Picador, 2003), 242.
7. Michel Foucault, *Discipline and Punish: The Birth of the Prison* (1975; repr., New York: Vintage, 1995), 25–26.
8. Lauren Wilcox, "Practices of Violence: Theorizing Embodied Subjects in International Relations" (unpublished manuscript).
9. Lauren Wilcox, "Dying Is Not Permitted: Sovereignty, Biopower, and Force-Feeding at Guantanamo Bay," in *Torture: Power, Democracy, and the Human Body*, ed. Shampa Biswas and Zahi Zalloua (Seattle: University of Washington Press, 2011).
10. Monica Casper and Lisa Moore, *Missing Bodies: The Politics of Visibility* (New York: New York University Press, 2009).
11. Henry Giroux, "Reading Hurricane Katrina: Race, Class, and the Biopolitics of Disposability," *College Literature* 33, no. 3 (2006): 174.
12. Katherine Verdery, *The Political Lives of Dead Bodies* (New York: Columbia University Press, 2000).

13. Bennett, *Vibrant Matter*, 8.

14. Steven Rottman, quoted in Melinda Wenner, "How to Handle the Dead in Haiti," *Popular Mechanics,* January 21, 2010, http://www.popularmechanics.com/science/health/4343222.

15. Claude de Ville de Goyet, quoted in Wenner, "How to Handle the Dead in Haiti."

16. Oliver Morgan, ed., *Management of Dead Bodies after Disasters: A Field Manual for First Responders* (Washington, D.C.: Pan-American Health Organization, 2006), available at the International Federation of Red Cross and Red Crescent Societies website, http://www.ifrc.org/docs/idrl/I967EN.pdf.

17. Ibid., 1.

18. Ibid., 15.

19. Lynn Jones, quoted in "Haiti Shows Importance of Dealing with Dead Bodies When Disaster Strikes," *Guardian*, November 1, 2012, http://www.guardian.co.uk/global-development/2012/nov/01/haiti-dead-bodies-disaster-strikes.

20. Bruno Latour, "From Realpolitik to Dingpolitik; or, How to Make Things Public," in *Making Things Public: Atmospheres of Democracy*, ed. Bruno Latour and Peter Weibel (Cambridge, Mass.: MIT Press, 2005).

21. See Daniel Bates and Lee Moran, "US Troops Urinating on Dead Afghan Bodies Used as Taliban Recruitment Tool," *Daily Mail* (London), January 12, 2012, http://www.dailymail.co.uk/news/article-2085378/US-troops-urinating-dead-Afghan-bodies-video-used-Taliban-recruitment-tool.html.

22. See Jim Miklaszewski and Courtney Kube, "Marines: Video Shows Troops Urinating on Corpses," *NBC News*, January 11, 2012, http://usnews.nbcnews.com/_news/2012/01/11/10120072-marines-video-shows-troops-urinating-on-corpses?lite.

23. "US Marines Urinate on Dead Bodies in Afghanistan," *Huffington Post*, January 11, 2012, http://www.huffingtonpost.com/2012/01/11/afghanistan-marines-urinating-video_n_1200324.html.

24. Michael Sledge, *Soldier Dead: How We Recover, Identify, Bury, and Honor Our Military Fallen* (New York: Columbia University Press, 2005), 243.

25. Danfung Dennis, cited in Harry Siegel, "How War Can Make Dehumanizing and Horrific Events Seem Normal," *Daily Beast*, April 18, 2012, http://www.thedailybeast.com/articles/2012/04/18/how-war-can-make-dehumanizing-and-horrific-events-seem-normal.html.

26. Sledge, *Soldier Dead*, 12.

27. Ibid., 243.

28. George Wright, quoted in David Zucchino, "US Troops Posed with Body Parts of Afghan Bombers," *Los Angeles Times*, April 18, 2012, http://www.latimes.com/news/nationworld/nation/la-na-afghan-photos-20120418,0,6471010,full.story.

Virus

Melissa Autumn White

> *In the societies of control . . . what is important is no longer either a signature or a number, but a code. . . . Individuals have become "dividuals," and masses, samples, data, markets, or "banks."*
> —Gilles Deleuze

Virality has become a dominant metaphor in the contemporary moment, one that is marked, in Jan van Dijk's terms, by "too much connectivity."[1] It is perhaps, then, no surprise that the virus, an infectious agent that remained unidentified until the invention of the electronic microscope, in the 1930s, has conceptually leaped across genres, from the material realm of the biological to the virtual realm of the digital, drawing attention to the precarious binary drawn between "natural" and "artificial" life. No matter the medium, the virus overwhelmingly signifies a bit of bad news, whether wrapped in the passing drag of a protein coat or disguised by a loop of programming software. Signaling a "ubiquity of epidemiological encounters in the so-called age of networks,"[2] viruses are in fact masters of undetectable mobility across highly invested borders on multiple scales, from those drawn between bodies to those that demarcate species and nations. These multiplying and unruly mobilities of the virus are what make the virus such a source of fascination and fear, not only because of their initially unapparent and promiscuous movement across body-species-nation boundaries but also because of the virus's attendant capacity to set into motion continual flux, rapid mutation, and transformation. In other words, it is not merely viral "contamination" or infection per se that is frightening, it is also the capacity that the virus holds for "uncontrolled and unstoppable diffusion throughout all the productive nerve centers of our lives."[3] Emerging "at the edge of life,"[4] viruses challenge integrities of all kinds, unmaking or dissolving the boundaries between bodies, species, and nations.

This chapter follows the virus around to explore the processes, systems, and assemblages the virus enacts and participates in as lively matter. The emergent effect of the chapter offers an intuitive feel for how the cross-species virus contributes to making things international at the cellular level.

Uncontrollable Diversity

In the societies of control, the virus has become associated with what Luca Lampo describes as "uncontrollable diversity." Describing the computer virus specifically as an "uninvited stranger" or "other," Lampo asks us to think about the virtual bug as "digital sans-papiers" in the machine of the social.[5] But biological viruses are also metaphorically figured as unauthorized migrants. Before identification documents can be issued, the virus has shape-shifted through a genetic recombination or shift, allowing for the movement on to another host, another body, another species, proliferating horizontally and vertically beyond all the border walls. Risking the romanticization of the virus, can we not say that the virus reflects, as a mirror image, some of the most powerful and impossible desires at play in the topographies of control? Given their propensity for continually emerging as other-than-themselves, viruses are among the ultimate escape artists, eluding capture, the violence of representations, and ultimately control. Viruses escape control by becoming other to that which might have been anticipated, decoded, or vigilated against.

As Jussi Parikka has suggested, the virus, whether biological or digital, is overwhelmingly understood as a bit of "bad matter." Biological viruses typically capture our attention only when they are "de-powering in a Spinozan way," generating encounters that "reduce the vitalities of material assemblages."[6] Such a visioning of the virus has had powerful effects on our cultural and political imaginations. For instance, the threat of an impending global viral apocalypse delivered through a bioterrorist attack has become an anticipated certainty at least since the Cold War. The question of when, rather than whether, the pending viral war would emerge has only intensified since the "Amerithrax" transport of anthrax spores through the mail following the attacks of September 11, 2001. As Roberto Esposito suggests, "[T]he type of terrorist attack most feared today, precisely because it is the least controllable, is a biological one . . . released into the air, the water, the food."[7] Indeed, the univocal objectification of the virus as bad matter has mobilized ideologically produced fears around bioterrorism and a collective future of the uncontrollable diffusion of viruses, such that desires for preparedness take precedent over assessment of risk. The virus and its capacity for rapid, unpredictable movement have effected an international scramble for control of circulation and mobility—including the stockpiling by particular nations of antivirals as a central technology of contemporary biopolitics and governmentality.

In this milieu of hypervigilance, medical research has become increasingly militarized. Indeed, the allocation of funding dedicated to biodefense has produced what Lisa Keränen calls a "bioterrorism 'brain drain' wherein health researchers abandon areas of inquiry with high global mortality rates (say, tuberculosis, malaria, and dengue fever) to study rarely occurring biological weapons agents," many of them viral in nature.[8]

The recent controversy over transmission studies of the highly pathogenic H5N1 avian influenza provides a case in point. With funding from the National Institutes of Health, two major teams led, respectively, by Netherlands-based Ron Fouchier and Yoshihiro Kawaoka at the University of Wisconsin–Madison managed to engineer a strain of H5N1 that, within a mere five genetic mutations, was able to move freely between ferrets, the mammalian species of choice for such studies. The idea was that by engineering in advance a strain of the virus that could lead to a deadly human pandemic (whereby the H5N1 virus would become directly transmissible from human to human), scientists would be able to produce in advance, on an international scale, an effective antiviral and stockpile it. Following a major scientific conference in Malta, the scientific teams submitted their findings to the journals *Science* and *Nature* in August 2011, and a controversial global debate quickly emerged. At the heart of the debate was the question of whether governmentally funded research of such a "sensitive" nature should be subject to redaction and censorship, given the potential for the results to fall into the hands of bioterrorists. After all, if Fouchier's and Kawaoka's labs could produce a highly pathogenic version of H5N1, so could the bioterrorists. As the debate heated up, Fouchier's and Kawaoka's preliminary results were held back from publication.

By the summer months of 2012, the U.S. National Science Advisory Board for Biosecurity, a nongovernmental panel of twenty-three experts, had overturned the full redaction of Fouchier's and Kawaoka's studies, and a partial publication of their research appeared in *Science* and *Nature*, with the exception of certain aspects deemed too risky.[9] In a somewhat ironic twist, the publication of these controversial transmission studies occurred during a self-imposed hiatus of research established by the international network of scientists working on H5N1 transmission studies themselves. The scientists had voluntarily suspended their work on the highly pandemic H5N1 for a period of two months, beginning in January 2012, to facilitate a full-fledged global discussion on the risk-benefit ratio and the ethical dimensions of such research in light of biosecurity and biodefense concerns. The initially-agreed-upon two-month pause in transmission-studies research eventually grew into a yearlong lag, stretching into the early days of January 2013, at which time the international network announced the resumption of such studies in a letter simultaneously published in *Science* and *Nature*. Authored by Fouchier, Kawaoka, and their associated teams, the letter acknowledged that although the controversy over H5N1 transmission studies in 2011–12 had highlighted the need for "a global approach" to such research, "H5N1 viruses continue to evolve in nature." The letter continued: "Because H5N1 virus transmission studies are essential for pandemic preparedness and understanding the adaptation of influenza viruses to mammals, researchers who have approval from their governments and institutions to conduct this research safely, under appropriate biosafety and biosecurity conditions, have a public health responsibility to resume this important research. . . . Because the risk exists in nature that an H5N1 virus capable of transmission in mammals may emerge, the benefits of this work outweigh the risks."[10] Highly pathogenic H5N1 avian influenza is a virus that predominantly affects birds, especially chickens; ducks seem capable of living with the virus and have thus been identified as a "Trojan-horse" reservoir housing H5N1,[11] which can pass readily from them to chickens and, under particular mutation conditions, from birds to humans. According to the World Health Organization, over the ten-year period of 2003–2013, highly pathogenic H5N1 avian influenza has resulted in a total of 620 human infections and 367 deaths, predominantly in Indonesia, Egypt, and Vietnam, where relatively large numbers of people live in close proximity with chickens.[12] Despite the high mortality rate of this particularly lethal strain of H5N1, there is no evidence to date showing that the virus, as a contaminating object, spreads easily between humans. Indeed, the majority of people who have been affected by H5N1 have had consistent and close contact with infected birds. These infections, in other words, were strictly zoonotic because of the virus's agentic capacity to move across the heavily invested species line drawn between bird and human. Despite the relatively low rates of infection (consider, for comparison, that in 2010, there were 219 *million* cases of malaria, leading to 660,000 deaths, with 90 percent of these deaths occurring in continental Africa),[13] highly pathogenic H5N1 continues to receive a disproportionate amount of limited medical-research funding because of its capacity for playing a starring role in a much-anticipated viral war waged either by bioterrorists or by a nature indifferent to the unreflexive privileging of human life.

Generally, only influenza strains beginning with the designations H1, H2, and H3 have the capacity to lead to human epidemics, as these subtypes are highly mobile in human populations. Pigs, however, offer a hospitable environment to both human (H1, H2, H3) and bird (H5) virus subtypes. It is well understood that aquatic birds are the ultimate origin of all influenzas, and these originary viruses must normally be reassorted through an intermediary species before becoming highly mobile among human populations: "By the time [influenza viruses] get into humans, they have generally been assembled from H1, H2, or H3 plus the ten other necessary proteins, some of those in forms borrowed from this or that bird flu or pig flu virus."[14] Influenza thus acts in highly promiscuous ways, sticking together a traveling menagerie of genetic data from multiple species. Although H5 strains occasionally leap from birds to humans, they haven't (yet) acquired transmissibility among human populations. Yet the speculative future of a global H5N1 pandemic has been on the horizon since the first documented case, in Hong Kong in the spring of 1997, which led to the death of a three-year-old boy. This event caused a great deal of alarm in the international scientific community, precisely because it was "the first documented case of a purely avian influenza virus—containing no human-flu genes brought in by reassortment—to cause killer respiratory illness in a person."[15] Later that fall, three more cases emerged in Hong Kong. Six years later, highly pathogenic H5N1 turned up again in Hong Kong, leading to two human deaths and the slaughter of one million chickens, the entire chicken population of Hong Kong. By 2005, the World Health Organization had put the world on global pandemic alert, and the race was on to develop anticipatory viral-response capacities such as early-warning systems, vaccines, and antiviral guidelines.[16]

As I've documented elsewhere, the arrival of the much-anticipated global pandemic in 2009, dressed, as it were, in pig's clothing, generated a complex assemblage of responses, from international information sharing between human- and animal-health organizations to regional and transnational surveillance networks to anachronistic local responses, such as President Hosni Mubarak's order that the entire pig population in Egypt—some three hundred thousand animals—be slaughtered as a "precautionary" measure.[17] Despite the massive cull of living animals as potential reservoirs of pandemic disease, many of the preparedness efforts hinged not on the protection of "life itself" but rather on what Stephen Collier and Andrew Lakoff and others have called the preservation of nonhuman "vital systems," including "transport, water, electricity, and so on."[18]

Vital Viruses

Despite their universal treatment as unambiguously "bad matter" to be prepared for, brought under control, and ultimately eradicated or rendered impotent through the fantasy structures of antiviral research and coordinated medical surveillance, a growing body of evidence shows that the majority of viruses are not in fact pathological. Viruses, by far the most abundant biological entities on earth, are much more likely to be benign. For example, a 2009 study by biologist Dana Willner conducted at San Diego State University found an average of 174 viruses in the lungs of ten research subjects, five of whom had cystic fibrosis and five of whom did not. Prior to Willner's study, the lungs of "healthy" persons were presumed to be sterile, or germ-free. Further, a mere 10 percent of the viral species Willner and her team found in the lungs of her test subjects bore any close resemblance or kinship to viruses that had been previously identified. Willner's study offered empirical weight to the theory that viruses live with us everywhere, ubiquitously; in the words of the late feminist biologist Lynn Margulis, "[W]e can no more be cured of our viruses than we can be relieved of our brains' frontal lobes: we are our viruses."[19]

Yet, despite their ubiquity, very little is known about viruses. What we do "know" tends to be framed in the negative: these active agents are "*not* captured by a filter, *not* cultivable in chemical nutrients, *not* quite alive."[20] Comprising RNA and, less often, DNA, viruses contain the essential building blocks of life, but they are unable to put this genetic information to work without taking over the operational machinery of their host cells. Margulis explains that "viruses require the metabolism of the live cell because they lack the requisites to generate their own. Metabolism, the incessant chemistry of self-maintenance, is an essential feature of life. Viruses lack this."[21] Put in the more poetic terms of virologists Marc van Regenmortel and Brian Mahy, viruses lead "a kind of borrowed life."[22]

The treatment of viruses as nonliving substances or particulate matter has had, until recently, significant epistemological consequences on the study of evolution. It is only in the last decade or so that scientists have begun to appreciate the profound role that viruses, as sources of new genetic information, have played in the evolution of life. Nevertheless, their capacity for rapid mutation and creative production of genetic variation is, without doubt, a source of newness in the world: "Anywhere there's life, we expect viruses. They are the major source of biological material on this planet."[23] From the vantage point of mathematical modeling, media theorists Alexander Galloway and Eugene Thacker have argued that viruses, whether biological or digital, can be understood as being in a continual state of "never-being-the-same" or, put differently, in a perpetual process of "becoming-number."[24] The multiplication of *difference* that viruses produce through the mutating inevitabilities of replication ought to be seen as a vital capacity of the virus, allowing viruses, themselves assemblages, to effect, affect, and infect assemblages of various kinds. The significance of this for evolutionary biology is a paradigm shift from Darwinian models of "survival of the fittest" to the phenomenon of emergent life. Viruses, as transport machines for genetic information, are technologies of emergent life as much as inevitable death.

Beyond serving as vectors of genetic transport, viruses also function as nodal points that bring into convergence multiple networks through their exponential proliferations. Indeed, we can think about viruses as in-formation and in-formational, not only communicating already-extant genetic code across multiple borders—cellular, bodily, species, trans/national—but in the process becoming some other combination of genes. Scrambling multiple borders through their divergent circulations, viruses ought to be considered active participants in creating the potentiality of new conditions of life through their capacity to assemble novel coalitions of genes. Comprised of genetic material but ultimately reliant on the metabolic energies of their host cells, viruses act as bioinformatic transport machines, moving significant genetic freight on a microscopic level. Generating continual mutation, replication, and drift while horizontally moving genetic code across and beyond various cellular systems, viruses disrupt sovereignties of all kinds, in the process making and unmaking "the international."

The rhizomatic movements of cross-species genes by viruses make new coalitions and alliances of data possible, although, as already suggested, the evolutionary effects and genealogies of these code mobilities are little understood. The focus on code sequencing in microbiology has effectively partitioned it from the life sciences, such that microbiology has been developed as an effective information science whose epistemological insights have yet to become central to evolutionary biology.[25] Galloway and Thacker suggest that the rise of microbiology as an information science, with its focus on code sequencing and the mapping of nucleic-acid pairs, corresponds with what contemporary sociologist Michael Fortun has identified as a shift under way in classical medicine's care of the body or Michel Foucault's "care of the self," to a post-Fordist "care of the data."[26] This shift maps onto what Gilles Deleuze, in the epigraph to this chapter, presciently described as the algorithmic tendencies of the societies of control. Life becomes split into its smallest components, such that individuals have become "dividuals," systems of data. At the same time, this shift from number to code might equally be considered expansive in the sense that a focus on data-ist (re)combinations, coalitions, and affiliations allows for an (albeit destabilizing) view of life as radically emergent, in-formational, rather than a stable and static form.

The more they've been looked for, the more viruses have been found: they are everywhere, teeming eclectically in oceans, soil, forests, cities, hospitals, suburbs, malls, mines, isolated caves, and acidic pools. They are coeval with bacteria and other microorganisms, fungi, plants, and animals. Popular-science writer Carl Zimmer explains, "[M]any scientists now argue that viruses contain a genetic archive that's been circulating the planet for billions of years. When they try to trace the common ancestry of genes, they often work their way back to a time before the ancestor of all cell-based life. Viruses may have first evolved before the first true cells even existed."[27] In this way, viruses might be thought of as "mobile living archives" of environmental memory, ecological USB keys comprising sequences of genetic code and a protein coat. Communications theorist Parikka points out that viruses act as microscopic maps of "human migration patterns" that simultaneously track "interactions, movements and spatial distributions of wild animals."[28] Akin to bioinformatic geographic information systems, viruses actively map human–nonhuman interactions genetically, transporting a kind of migratory memory across species and scales.[29]

The active code-swapping, code-switching, and code-mutating that viruses make possible open up potent, if unexpected, philosophical consequences. Indeed, virologist Luis Villareal has provocatively argued that viruses illuminate a need to rethink our very conceptions of life and living. Although viruses in and of themselves fail to autonomously reach the critical complexity required to be considered alive, so, too, do many components of complex systems that we generally attribute with the quality of living. For Villareal, viruses insist that life be thought differently, and specifically "as an emergent property of a collection of certain nonliving things."[30] Villareal describes the virus as a "verging-on-life" through a comparison with brain tissues, which must be in assemblage with a complex of neuronal and biochemical activity in order to be considered living.

Both life and consciousness are examples of emergent complex systems. They each require a critical level of complexity or interaction to achieve their respective states. A neuron by itself, or even in a network of neuronal nerves, is not conscious; whole-brain complexity is needed. Yet even an intact human brain can be biologically alive but incapable of consciousness, or "brain-dead." Similarly, neither cellular genes nor viral proteins are, by themselves, alive. The anucleated individual cell is akin to the state of being brain-dead, in that it lacks a full critical complexity. A virus, too, fails to reach a critical complexity. Approached from this perspective, viruses complicate the distinctions scientists tend to draw between capacities for life. More than inert matter, viruses verge on life. It is worth noting that Villareal's virological insights here resonate strikingly with Deleuze's argument that life designates not a *form* but an *affectivity*—a capacity or power to act, in the Spinozan sense. Life, in other words, provides a shorthand for a process (becoming) rather than a stable entity (being).[31] In still other words, life assembles a "complex reaction between different velocities, between deceleration and acceleration of particles."[32]

What is startling here is that viruses may not simply be bits of "bad matter" that slow down, disassemble, and debilitate complex systems. Rather, viruses might well be approached from another vantage point altogether, as vitalizing forces in complex ecological systems in which humans are not the center. What insights like Villareal's suggest is that viruses are central to the possibility of life on the planet, as movers and shakers of horizontal gene transfers that result in new coalitions beyond vertical filiation. We don't even know what a virus can do.

Viral "Sex"

Viruses, and particularly those that hold the emergent capacity to "go viral" (that is, pandemic on a global scale), open space for a rethinking of the relationships between the global and the intimate, or the ways in which "the global and the intimate constitute one another."[33] The virus carries an intimate touch not only in its ultramicroscopic cellular movements but also in the ways in which it moves genetic material around, across all mediated and mediating boundaries, on a transnational and transspecies scale. Cellular embodiment is intimately tied up with the transnational migration patterns of the virus, the code itself being a complex assemblage of intimate environmental transactions and recombinations. Interpreting "sex" as, in its broadest sense, simply the "recombination of genes from more than one source," biologists Lynn Margulis and Dorion Sagan have argued that "the passing of nucleic acid into a cell from a virus, bacterium or any other source is sex."[34] From this bioinformatic (and heteronormative) vantage point, an influenza infection could provocatively be described as a "sexual act" that creates, through the insertion of viral material into cellular processes, a reconfiguration of the international at the microscopic level, a rewriting of genetic code through global circulations of genetic transport.

Subverting the borders of species embodiment on an international scale, viruses act as polymorphously perverse agents, inserting genetic cartographies of human/nonhuman encounters into the cells they infect. The virus thus carries the capacity to transform relations of scale, bringing species-strangers into microproximity while archiving international circulations of cross-species entanglements. The movements and migration patterns that viruses simultaneously enact and record can thus be understood as a mobile cartography of what Luciana Parisi calls "abstract sex," an inventive force driving "a physical association between different bodies," or the condition of endosymbiosis, "the condition of one body residing within another."[35]

If we wish to consider the (reproductive) performativity of viral labor, Parisi's concept of abstract sex provides a means to explore the entanglement and codependency of the so-called micro levels of cellular activity and horizontal gene transference and the macro levels of vertical sociocultural and politico-economic systems, as well as the indivisibility of biological and technological spheres.[36] Challenging Darwinian notions of evolution as "the survival of the fittest," endosymbiosis, the nesting of different "bodies" one inside the other, is, according to Margulis's theory, "a nonnegotiable requisite for many differing kinds of life."[37] Endosymbiotic proximity, as imperceptible as it is affective, is necessary not only for survival but also for (evolutionary) flourishing.

Viral Politics, Emergent Life

We turn now from the biological to the computer virus. Deleuze has suggested that the latter materializes a site of active resistance in the societies of control, acting as a wrench in the gears of digital capillary power.[38] In these analytics of the societies of control, capitalism itself is understood to be like a virus, continually mutating in response to and effecting new environmental conditions. Uncontainable in or by enclosures of power, viral capitalism, like biological and technological bits of contagious and recombinatory code, can be characterized as "essentially dispersive," deformable, and transformable.[39] Globalization as "universal contagion" has thus become a highly communicable metaphor in the contemporary moment, instantiating epidemiological readings of political economy, bioeconomic circulatory networks, and drastically uneven global distributions of (in)vulnerabilities.[40]

The term *virus*, as it is used today, describes a relatively new taxonomy for subvisual, ultramicroscopic bioinformatic species that verge or edge on life. First defined in its modern sense by Martinus Beijernick in 1897 to define an agent that could pass through a Chamberland filter (designed to capture bacteria) and go on to infect healthy plants with tobacco mosaic disease, the "infectious filterable agent" did not become visible until the 1930s, when its incredibly wide range of diversity began to be systematically examined with the help of the electron microscope. The science of viruses emerged as a distinct subfield of the biological sciences in the 1950s with the publication of the first edition of *Virology*, the textbook that would become central to initiating new ontoepistemologists of the field.[41]

Apart from spectacular materialisms of the depowering work that viruses perform as (dis)assembling agents, there remain great clouds of unknowing around what a virus can do. Scientists have only recently established the extent to which virus genes harbor the greatest genetic diversity of life on the planet, archiving genetic patterns that predate the splitting of life from a common ancestor, close to four billion years ago. In collaboration with evolutionary biologists and climate scientists, virologists are in the early stages of exploring how viruses help to produce much of the oxygen we breathe while contributing to the regulation of the planet's thermostat.[42]

And yet, as the controversy over the transmission studies of highly pathogenic H5N1 reveal, the race to preemptively apprehend viruses that carry the capacity to significantly disrupt both vital systems and human populations as we currently know them is on. As emergent forms of and in assemblage, viruses are, by definition, difficult to manage and control. In this way, viruses have played an important role in the rise of practices and strategies that ensue from what Collier and Lakoff describe as a "rationality of preparedness":

> In place of unknowable risks, the rationality of preparedness deals with unpredictable, future events, *imagined* vulnerabilities. Public health is but one set of actors in a decentralized and distributed set of emergency preparedness which coalesces around a shared investment in being prepared. In addition to the intensification of disease surveillance, preparedness efforts privilege the coordination of "vital systems security" with the involvement of state and non-state agencies.... Protecting them involves the development of warning systems (not only for disease but also for natural disasters and terrorist attacks) and the continuous rehearsal of readiness to respond to disasters through the networking of government and private agencies responsible for their maintenance.[43]

Engineering in advance potentially transmissible viruses such as H5N1, scientists such as Fouchier and Kawaora participate in creating life-likenesses that do not yet exist. The purpose of this is to preemptively configure highly communicative code in order to determine in advance mechanisms to debilitate it and bring it under the control of human knowledge and technologies of power.

Luciana Parisi and Steve Goodman suggest that such regimes of preemptive power go far beyond the biopolitics of "control."[44] They explain that preemptive power, with its focus on potentiality, aims to harness the "not-yet-live."

Preemptive power reaches beyond the biopolitics of life itself to focus on futurity, affectivity, and vital systems. However, preemptive power is not simply an imperialist expansion of the logics of control. The preemptive logics of the viralities associated with contemporary forms of capitalism point us instead toward new objects of "unlife"—like viruses—that cannot be adequately accounted for within the theory of biopolitics. In the desire to preempt that which may yet come to be, preemptive power reasserts the primacy of human life in the face of complex ecological systems that are continually changing.

Preemptive power, rather than the viruses themselves, reproduces the primacy of notions of sovereignty and partition that the international relies upon for its continued salience in what has otherwise been described as a transnational or global world. Even as the unruly mobilities of the virus illuminate the precarity of borders on multiple scales—from the global to the microcellular, mutually entangling both—responses to cross-species viruses by scientists, governments, pharmaceutical companies, and extragovernmental organizations (such as the World Health Organization) tend to reinforce or reassert the primacy of the international. To put it another way, whereas the cross-species virus has been very effective at *unmaking* the international through its proliferation of difference at the microcellular level on a global scale, the responses to these capacities of the virus have tended to (re)make the international.

The international, as re-created through anticipatory and reactionary responses to the appearance of harmful viruses at the level of population, conceptually reasserts, after all, the materiality of bounded nations and the salience of borders, suggesting the continued relevance of colored zones depicted on a map, with crisp lines drawn between "here" and "there." Scientific research on pandemic viruses certainly reproduces the international as a single space of cooperation, competition, and information sharing, despite the vast disparities in the geopolitical distribution of the "goods" of viral research (such as the availability and affordability of antivirals, which are patented and stored predominantly in wealthy nations). Indeed, the surveillance networks that have emerged to track and observe the movements of viruses are concentrated on the northern side of the global North-South divide. During the H1N1 outbreak in 2009–2010, for example, there was very little information available on how the virus was moving through and affecting highly populated regions in much of the global South, simply because this region comprised countries—zones on a map, colonial artifacts—without an infrastructure available that would lend itself to tracking viral movements.

To conclude, I suggest that following viruses around—thinking with and not necessarily against them—opens new space for questions of traversal, transport, and transgression of space. Even though the cross-species virus and its mobilities highlight the radical interconnectivities of "remote" locales, forms of human and nonhuman contact, and differentially embodied interdependencies, I suggest that it takes us beyond the international and into less certain but no less global territories, both conceptual and material. What I hope this thinking with the virus might do is to suggest new somatechnical approaches to ecologies of difference that ultimately challenge sovereignties of all kinds—embodied, taxonomic, international. Themselves mobile living archives of encounter, cross-species viruses draw our attention to the microcellular, but no less global, intimacies of being in relation with many others, human and otherwise, whom we have never met and will never meet but with whom we are nevertheless involved in a deeply ecological sense. Our bodies are neither singular nor a part; rather, they are affective ecologies of capacities and debilities, transfer points for the circulation of viral life.

Notes

1. Jan van Dijk, *The Network Society* (London: Sage, 2006), 187, quoted in Tony D. Sampson, *Virality: Contagion Theory in the Age of Networks* (Minneapolis: University of Minnesota Press, 2012), 1.

2. Sampson, *Virality*, 1.

3. Roberto Esposito, *Immunitas: The Protection and Negation of Life*, trans. Zakiya Hanafi (Cambridge: Polity Press, 2011), 3.

4. Ed Rybicki, "The Classification of Organisms at the Edge of Life; or, Problems with Virus Systematics," *South African Journal of Science* 86 (1990): 182–86.

5. Luca Lampo, "'When the Virus Becomes Epidemic': An Interview with Luca Lampo," by Snafu and Vanni Brusadin, April 18, 2002, quoted in Jussi Parikka, "FCJ-019 Digital Monsters, Binary Aliens—Computer Viruses, Capitalism, and the Flow of Information," *Fibreculture Journal* 19, no. 4 (2005).

6. Jussi Parikka, "New Materialism as Media Theory: Medianatures and Dirty Matter," *Communication and Critical/Cultural Studies* 9, no. 1 (2012): 98.

7. Esposito, *Immunitas*, 4.

8. Lisa Keränen, "Concocting Viral Apocalypse: Catastrophic Risk and the Production of Bio(in)security," *Western Journal of Communication* 75, no. 5 (2011): 464.

9. Sander Hersft et al., "Airborne Transmission of Influenza A/H5N1 Virus between Ferrets," *Science*, May 2012, 1534–41; Masaki Imai et al., "Experimental Adaptation of an Influenza H5 HA Confers Respiratory Droplet Transmission to a Reassortant H5 HA/H1N1 Virus in Ferrets," *Nature*, June 2012, 420–28.

10. Ron Fouchier et al., "Transmission Studies Resume for Avian Influenza," *Science*, January 2013, 520–21.

11. David Quammen, *Spillover: Animal Infections and the Next Human Pandemic* (New York: Norton, 2012).

12. World Health Organization, "Cumulative Number of H5N1 Cases," February 15, 2013, http://www.who.int/influenza/human_animal_interface/EN_GIP_201302013 CumulativeNumberH5N1cases.pdf.

13. World Health Organization, "World Malaria Report 2012 Fact Sheet," December 17, 2012, http://www.who.int/malaria/publications/world_malaria_report_2012/wmr2012_factsheet.pdf.

14. Quammen, *Spillover*, 507.

15. Ibid., 508.

16. Niamh Stephenson and Michelle Jamieson, "Securitizing Health: Australian Newspaper Coverage of Pandemic Influenza," *Sociology of Health and Illness* 31, no. 4 (2009): 525; Melissa Autumn White, "Viral/Species/Crossing: Border Panics and Zoonotic Vulnerabilities," *WSQ: Women's Studies Quarterly* 40, nos. 1–2 (2012): 121.

17. White, "Viral/Species/Crossing," 121.

18. Stephen Collier and Andrew Lakoff, "Vital Systems Security" (ARC Working Paper 2, 2006), quoted in Niamh Stephenson, "Emerging Infectious Disease / Emerging Forms of Biological Sovereignty," *Science, Technology, and Human Values* 36, no. 5 (2011): 622. Also see Dmitris Papadopoulos, Niamh Stephenson, and Vassilis Tsianos, *Escape Routes: Control and Subversion in the Twenty-First Century* (London: Pluto, 2008), 128–29.

19. Lynn Margulis, *Symbiotic Planet: A New Look at Evolution* (New York: Basic Books, 1998), 64.

20. Quammen, *Spillover*, 269 (original emphases).

21. Margulis, *Symbiotic Planet*, 63.

22. Marc van Regenmortel and Brian Mahy, quoted in Luis Villareal, "Are Viruses Alive?," *Scientific American*, August 8, 2008, http://www.scientificamerican.com/article.cfm?id=are-viruses-alive-2004.

23. Mark Young, "Yellowstone Virus Startles Scientist with Ancient Lineage," "Current News at MSU," Montana State University, March 12, 2004, www.montana.edu/news/1711/yellowstone-virus-startles-scientists-with-ancient-lineage.

24. Alexander Galloway and Eugene Thacker, *The Exploit: A Theory of Networks* (Minneapolis: University of Minnesota Press, 2007), 88.

25. See Lynn Margulis, "Microbial Actors in the Evolutionary Drama," *BioScience* 53, no. 2 (2003): 179–80.

26. Michael Fortun, "Care of the Data" (talk given at the symposium "Lively Capital 2: Techno-corporate Critique and Ethnographic Method," University of California, Irvine, October 23–24, 2005), quoted in Galloway and Thacker, *Exploit*, 107.

27. Carl Zimmer, *Planet of Viruses* (Chicago: University of Chicago Press, 2012), 94.

28. Jussi Parikka, "Contagion and Repetition: On the Viral Logic of Network Culture," *Ephemera: Theory and Politics in Organization* 7, no. 2 (2007): 289.

29. Luciana Parisi and Steve Goodman, "Mnemonic Control," in *Beyond Biopolitics: Essays on the Governance of Life and Death*, ed. Patricia Ticineto Clough and Craig Willse (Durham, N.C.: Duke University Press, 2011), 173.

30. Villareal, "Are Viruses Alive?"

31. Jussi Parikka, "The Universal Viral Machine: Bits, Parasites, and the Media

Ecology of Network Culture," *CTheory.net*, December 15, 2005, 7, http://www.ctheory.net/articles.aspx?id=500.

32. Ibid., 3.

33. Alison Mountz and Jennifer Hyndman, "Feminist Approaches to the Global Intimate," *WSQ: Women's Studies Quarterly* 34, nos. 1–2 (2006): 446.

34. Lynn Margulis and Dorion Sagan, *Microcosmos: Four Billion Years of Microbial Evolution* (New York: Simon & Schuster, 1986), 85, 156.

35. Luciana Parisi, *Abstract Sex: Philosophy, Bio-technology, and the Mutations of Desire* (London: Continuum, 2004), 204–5.

36. Ibid., 11.

37. Margulis, *Symbiotic Planet*, 5.

38. Gilles Deleuze, "Postscript on the Societies of Control," *October* 59 (1992): 6.

39. Ibid.

40. Michael Hardt and Antonio Negri, *Empire* (Cambridge, Mass.: Harvard University Press, 2000), 136; White, "Viral/Species/Crossing."

41. Zimmer, *Planet of Viruses*, 5.

42. Collier and Lakoff, "Vital Systems Security," cited in Papadopoulos et al., *Escape Routes*, 129.

43. Parisi and Goodman, "Mnemonic Control," 175.

44. Ibid., 163–76.

Microbes

Stefanie Fishel

International relations (IR) is decidedly anthropocentric, with a focus on human institutions and structures. The state, levels of analysis, and the individual as a rational, unitary actor figure prominently in IR discourse and practice. Be it ideational, a priori, or natural, the state is taken as the primary agent in international relations. Sovereignty—and political agency—is expressed both through the state as a territorially bounded unit and through men and (sometimes) women as self-limiting, self-understanding, self-conscious, and self-representing subjects.[1]

What if this anthropocentric focus is a misleading account of human subjectivity? What if humans are not as limited, rational, singular, and evolutionarily superior as once was thought? What if humans are not at the top of an evolutionary ladder, with other bodies and processes falling below in importance as legitimate objects of study, or what if those bodies and processes are subjects in their own right? How can we understand sovereignty and the state differently? To bring these questions to the fore, this chapter draws inspiration from a school of thought often referred to as critical or new materialism. It is labeled "new" not because of its place on a temporal or teleological line but because of the "unprecedented things that are currently being done with and to matter, nature, life production, and reproduction." This newness compels theorists "to rediscover older materialist traditions while pushing them in novel directions or toward fresh applications."[2]

Materialism gives matter agentic capabilities by bringing the object into a place of its own as a lively actant, rather than downplaying the nonhuman as something static and brought into being only by its encounter with human consciousness. This lively matter does not render human agency an illusion; it merely places it within a complex, interactive environment. Materialism aims to be positive and constructive by "describing active processes of materialization of which embodied humans are an integral part,"[3] rather than setting humans apart from dead matter and other life. This focus on materiality and corporeality does not privilege human bodies and is posthuman in the sense that all bodies have capacities for agency, even if this agency is more diffuse; this is a difference in degree rather than kind. Differences that have previously set humans apart from other species—self-reflection, reasoning, moral reflection—can be seen not as an outcome of evolution but rather as contingent and shared with other species along a continuum of consciousness only recently understood.

If humans take this decentralized and diffuse view of the world seriously, this puts the very categories of politics into question, even the concept of the political itself.[4] This does not reduce humanity to the rest of nature but supports an appreciation, as William Connolly asserts, of the flow of agency from "simple natural processes, through higher processes, to human beings and collective social assemblages."[5]

To investigate this material and object-oriented focus, I take microbes as an object of study to demonstrate that in the relationship between humans and microbial communities, multiple perspectives and objects—beyond human-created institutions and subjects—can be seen as vital and necessary to politics and human survival. This is both a metaphorical and an actual relationship. Humans survive with the help of microbial communities, and these ties can be used analogically to better understand human institutions, politics, and community creation. The addition of the microbes to these relations offers a view of a complex world of interactions between human bodies and their microbiomes, and it unsettles the human's place in the great chain of being and allows the agentic capability of nonhuman actors to be seen. This aids in taking seriously Darwin's assertion that even the most remote organic beings in the scale of nature "are bound together by a web of complex relations."[6] To interrogate how subjects may be formed while thinking about objects, I offer another introductory query: What might human bodies and their microbial relations tell us about global politics and international relations?

This is not to treat IR, cultural theory, or the biological sciences as hegemonic disciplines but to regard them as complex and varied approaches. All three include sets of practices that differ over time and space, the desires and directions of individuals, and varying and diverse epistemological commitments. A dialogue between them brings out the complex nature of who decides, defines, and controls who is a subject; what is an object; and how each is treated politically. These discussions should be open, inclusive, and careful to reflect the values and ethics that are necessary in creating mutual public space.

Specifically, the human microbial community will serve as storytellers for a tale about the somatic bonds that create a human body and how these bonds can connect human relations to a wider—and more fragile—set of relations with the natural world. To break from the macro focus of humans and our political institutions, I use research from the science of metagenomics, or genetic analysis applied to entire communities of microbes and studied in a way analogous to the study of a single genome.[7] Metagenomic techniques reveal complex communities of bacteria that inhabit the intestinal tract of humans. This microbial DNA in and on human bodies outnumbers human DNA ten to one. At the genetic level, we are much less human than previously thought. I solicit these microbial communities, as understood by metagenomic processes, to lead IR's discussion toward ideas of community that allow for a different take on supporting and nourishing pluralism within states and across borders. This, in turn, supports an idea of connectedness, or a "nestedness" of diverse bodies that allows for a reassessment of the ontological and ethical claims of IR, especially those surrounding the state.

Along with our human individuality and uniqueness, human nature includes our inter- and intradependence on multiple species as well as competition and alienation. The numerous constituents at play in these complex relations display agentic capacity at micro and macro levels. This approach hopes to sensitize humans to the presence of things, or objects, at play in our bodies and our politics. This "becoming sensitive," as Bruno Latour calls it,[8] can teach us lessons about both IR and human relations, as well as give us a wider understanding of global existence in which humans and nonhumans are co-constituted. As Stewart Brand asks, "When confronting a difficult problem we might fruitfully ask, 'what would a microbe do?'"[9]

Microbial Agents

Microbes fall into one of four major groups: bacteria, fungi, viruses, and protozoa. There are ten million species of bacteria and more than five thousand species of viruses. Microbes are too small to see with the naked eye but make up half the biomass on the planet and have exerted influence on the course of human evolution and history. Humans also have positive and negative long-term relationships with microbes. Well before we knew of the microprocesses involved, humans used bacteria to ferment beverages and cheeses and to preserve olives. Plagues and diseases have affected bodies, populations, and events across continents, changing the course of human history,[10] but viruses and bacteria are also essential to sea and freshwater regulation and human reproduction.

Commonly referred to as "germs," microbes influence daily life on earth in myriad ways: microbe modulation and maintenance of the environment, microbe suppression of plant disease and support of plant growth, microbe management of fuel leaks, and microbes in human hosts. Yet, except for pathogenic viruses and bacteria, microbes rarely register as objects important to politics. This is partly a matter of scale; humans are the largest organisms and microbes the smallest. In *Living at Microscale,* David B. Dusenbery asserts that it is difficult for humans to imagine the physical situation of bacteria. "Imagine," he writes, "stepping off a curb and having to wait days for your foot to reach the ground."[11] Practically, this means humans have much more experience with and relevant knowledge about larger things rather than smaller things. The next section explains a study of this smaller world.

Metagenomics

Meet *Bacteroides thetaiotaomicron.* This rod-shaped and gram-positive bacterium is found in the human intestine, where it aids in breaking down polysaccharides that the human body would not otherwise be able to process. Approximately ten trillion bacteroides are packed into the gut, and about 15 to 20 percent of our daily caloric intake is accomplished through bacteria like this one fermenting and processing complex carbohydrates that we could not fully digest on our own. Bacteroides also produce heat—in the same way fermentation creates heat—and this is used to regulate human body temperature.[12] These bacteria, and other microbes in the intestine, also defend humans from disease and environmental toxins, and function as a key interface between our bodies and the environment.

Metagenomics is the study of the microbial world—including bacteroides—from an aggregate level, "transcending the individual organism to focus on the genes in the community and how genes might influence each other's activities in serving collective functions."[13] In a report entitled *The New Science of Metagenomics: Revealing the Secrets of Our Microbial Planet,* the National Research Council (NRC) Committee on Metagenomics writes that metagenomics is a new science that combines genomics, bioinformatics, and systems biology to study the genomes of many organisms simultaneously. Most microbes have been essentially invisible until now, as the majority of them cannot be grown in a laboratory or studied with the methods of microbiology.[14] The authors of the report stress that metagenomics is operationally novel and is more than just listing and cataloging sequences and microbial diversity. John Dupré and Maureen O'Malley agree: "The inventories also give unique insights into microbial community structure and biogeography. They enable subtle understandings of ecophysiological characteristics of communities, in which adaptations to different environmental gradients result in different metabolic and

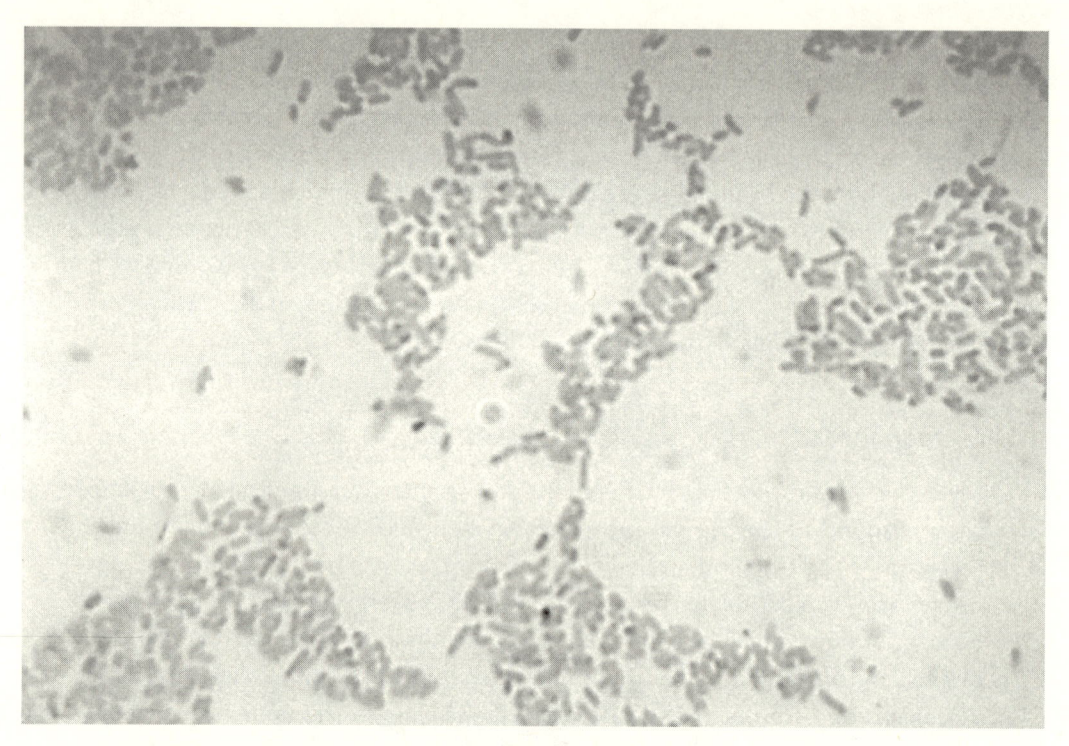

This photomicrograph reveals the bacterium *Bacteroides thetaiotaomicron* grown on blood agar for forty-eight hours. Source: Centers for Disease Control, Public Health Image Library. Courtesy Centers for Disease Control / Dr. V. R. Dowell Jr.

morphological strategies (e.g., capacities for movement) that spread vertically and horizontally through community members."[15] Previously, almost all knowledge about microbes was found in the laboratory, "attained in unusual and unnatural circumstances of growing them optimally in artificial media in pure culture without ecological context."[16] This encouraged the study of "lab weeds," or microbes that did well only in aseptic environments. The ability to study microbes in their natural environment has revealed that microbes are understood better by knowing where they fit within their own community and adjacent communities than by being isolated in the lab. The previous mechanistic framework of the "abstraction of an ontology of fixed entities" may need to give way to the idea that "life is in fact a hierarchy of processes (e.g., metabolic, developmental, ecological, evolutionary) and that any abstraction of an ontology of fixed entities must do some violence to this dynamic reality."[17]

This ability to study entire communities of bacteria opens up new applications and areas of study within biology, medicine, ecology, and biotechnology: "Traditional microbiological approaches have already shown how useful microbes can be; the new approach of metagenomics will greatly extend scientists' ability to discover and benefit from microbial capabilities."[18] Basic ideas in biology will need to be rethought and studied anew: organism-constituted hierarchy, teleological constructions of evolution running from the unicellular (simple) to the advanced (multicellular), and the mechanistic models of fixed entities.[19]

The Human Microbiome Project (HMP) is one of several international projects undertaking studies using metagenomic analysis to study human health by analyzing the human microbiome. It was initiated in 2008 as a five-year project to build and create community resources in the emerging field. The microbiome consists of all the microorganisms that live in or on the human body, and, importantly, the microbiome is an ecosystem in which all the various members maintain balance and equilibrium. These communities inhabit the mouth, skin, guts, and respiratory and reproductive tracts of human bodies. Some of these microorganisms cause illnesses, but many are necessary for good health. Importantly, "these functions are conducted within complex communities—intricate, balanced, and integrated entities that adapt swiftly and flexibly to environmental change."[20]

In May 2010, the HMP published an analysis of 178 genomes in the human body, reporting that, genetically, microorganisms outnumber human cells in the body by a ratio of ten to one. A survey of news, magazine, and journal articles about the human microbiome reveals a wealth of colorful statements about this discovery:

> We think that there are 10 times more microbial cells on and in our bodies than there are human cells. *That means that we're 90 percent microbial and 10 percent human.* There's also an estimated 100 times more microbial genes than the genes in our human genome. So we're really a compendium [and] an amalgamation of human and microbial parts.[21]

> *Outnumbering our human cells by about 10 to one*, the many minuscule microbes that live in and on our bodies are a big part of crucial everyday functions. . . . But as scientists use genetics to uncover what microbes are actually present and what they're doing in there, they are discovering that the bugs play an even larger role in human health than previously suspected—and *perhaps at times exerting more influence than human genes themselves.*[22]

> We're each like a *superorganism*—a unified alliance between the genes of several different species, only one of which is human.[23]

> We are not just the expression of an individual human genome. We are . . . *a genetic landscape*.[24]

> We human beings may think of ourselves as a highly evolved species of conscious individuals, *but we are all far less human than most of us appreciate*. . . . We are only beginning to understand the sort of impact our bacterial passengers have on our daily lives.[25]

> *Maybe we aren't quite as human as we thought.* . . . But however you feel about the "trillions of friends" that make up your microbiome, one thing is sure. . . . You're never alone.[26]

These quotes demonstrate how discoveries like those exemplified in the HMP have affected popular imaginings of what it means to be human, and the intra- and interconnections with nonhuman objects. Taken together, ideas about the relationships between humans and their microbial communities have shifted from the definitional "book of life" to the more expansive "communities of life." Microbes are imagined as friends of and boons to humans rather than enemies. Even more important for this chapter, Brigitte Nerlich and Ina Hellsten have found in their discourse analysis of news and science publications that "[h]umans are no longer conceptualized as the pinnacle of evolution, standing apart from microbes seen as populating the bottom of the evolutionary tree: we are all human-microbe hybrids" and that we should conceive of a new idea of life itself: "[L]ife has to be seen in a much less deterministic and much more fluid and flexible way."[27]

The architects of the HMP and the scientists involved in this research are also aware of the implications of changing metaphors and perceptions. A program funded by the National Institutes of Health (NIH) entitled "Human Microbiome Research and the Social Fabric" gathered thirty interdisciplinary participants to evaluate current ethical, legal, and social issues surrounding microbiome research.[28] The HMP is an opportunity to reexamine a host of issues: biomedical ethics, public health, property law and the human body, and privacy.

The participants in the NIH program discussed the HMP's effect on self-identity specifically. They reported that the HMP is likely both "to transform how we think of the microbes on and in our body, from enemies that must be eradicated to entities that are important in maintaining health," and to "change our concept of the human organism and affect the distinction between us and our environment." They recommended that care should be taken with the way HMP education is approached with the public and that "researchers and clinicians need to be mindful when developing language to describe microbial inhabitants."[29]

New ideas about genomes, organisms, and species, and evolution-prompted discoveries from microbial-community function will lead to changes in how we understand biological functions. The NRC Committee on Metagenomics echoes this interest in new ideas. It heralds a "grandmother" of a paradigm shift in biology: "It will refocus us one level higher in the biological hierarchy (molecules, cells, organisms, species, populations, communities, the biosphere). It will shift the emphasis from individuals to interactions, from parts to processes."[30] As Eric T. Juengst points out, "[T]he proponents of the Human Microbiome Project have much grander ambitions: they to seek to spark nothing short of Kuhnian scientific revolution in our understanding of human biology."[31] This idea supplements and works along with other dynamic-systems approaches in political science and IR, such as complexity theory, object-oriented-ontology approaches, new materialism, and network theory.

Conceptually, metagenomics implies that the human body as a communal gene pool and its "self-extending symbioses" are "highly adaptive and robust against environmental perturbation" and "dynamic, self-sustaining and self-repairing processes."[32] Metaphorical framings built from these understandings of microbial communities will aid in bringing system-based understandings of complex processes to the international realm. Many problems that the state in its current form has been unable to address—warming oceans, pandemics, climate change, flows of immigrants and migrants—may be easier to address. Myra J. Hird identifies this way of thinking as a "micro-ontology" that understands humans to be "enmeshed in a complex web of co-domestications . . . such that natural, social and cultural selection may not be so definitively distinguished."[33]

The language and affect involved in explaining the complexity of metagenomic discoveries to the public and the scientific community's further interest in "the social fabric" and in microbes' relation to humans may demonstrate that there is a space for "creative revision" that these metaphors can open up for shaping politics with objects as actants. By transgressing the boundaries of our autonomous selves and acknowledging the symbiontic colonies in our bodies, we can begin to see the ethical and social implications made possible by relations with diverse objects. Put differently, HMP and metagenomics can provide new concepts that are better equipped to supply IR with different forms of practice for sustainable, ethical, global living with one another and with other life forms as a "bodies politic." These practices can operate on both the micro and macro levels through the way IR understands these bodies and the way these bodies exist together in political communities. This will be the topic of the following section.

Lively Vessels

One of these new forms of practice at the bodily or state level could be to explain human bodies as "lively vessels" rather than passive containers or docile bodies living in isolated individuality from other bodies. These new methods in microbiology and metagenomics find that humans are not intrinsically pure biological individuals but rather, as the earlier popular press quotes show, collective superorganisms, or walking assemblages incorporating multiple species and other organisms. These lively vessels, as seen through the lens of metagenomics, summon a different ontology that aids in demonstrating that an attached and positive view of nature is the best way to proceed. The traditional container metaphors that usually describe the body presuppose that one could stand outside of the world or even be able to distinguish inside from outside as a specific place and time. This supports dualism as natural and unavoidable, both in regard to the world and in regard to the human and nonhuman: "Instead of seeing dualist attachment and domination as a move, a tactic, a ploy, and a very specific way of living in the flow of becoming, we tend to mistake it for the world itself."[34] This dualist detachment and domination of nature, Andrew Pickering and Keith Guzik argue, has veiled the fact that humans live in "the thick of things." This veil needs to be drawn back in order to see the symmetric "process of the becoming of the human and nonhuman."[35]

The idea that life as organized around the pivotal unit of the individual organism, which is traditionally conceived of as an autonomous cell or a group of coordinated cells with the same genome, is being challenged by the idea that "life is not composed in a machine-like way out of unchanging individual constituents."[36] The emerging data (from metagenomics studies) shows clearly that "each organism does not have exactly one genome" and that "this idea of 'one genome, one organism' starts to look like a poorly grounded dogma," when, in fact, "genomes, cells, organisms, lineages are all assemblages of constantly changing entities in constant flux."[37] This cuts directly to the center of the notion of an atomistic individual found in realist and liberal IR theory; the narrowly defined and rational *Homo economicus* is challenged by the refigured and hybrid subject *Homo contaminatus*, whose judgments may be altered by a wealth of unseen bacteria and viruses, commensal processes, and coevolutionary relationships. This hybridity can be used positively to understand how humans interact with, and alter, their lived environments as part of nature rather than as divorced from it.

The tentative and playful rendition of the body as a lively vessel is a potential metaphorical frame for a body wherein it is "not enough to say that we are 'embodied.' We are, rather, *an array of bodies*, many different kinds of them in a nested set of microbiomes."[38] If the "*its* outnumber the *mes*,"[39] a different set of ethical and normative considerations travels to the forefront. Questions such as "What does it mean to be human?," for philosophy; "How do we secure such a human?," for security studies; and "How do we respond?," for policy makers take on new meanings and have new reference points.

Another way to imagine these lively bodies, or superorganisms, at the global level is through the idea of the assemblage. An assemblage is defined as a collective entity made up of a gathering of things or people, but, what is more important, assemblages are created through relationality and never, writes Jussi Parikka, "from a prescribed relation hiding inside it, as if it were a seed."[40] In genomic terms, a gene is not a book or a code that was formerly hidden and that we can now read through its sequencing, but, as shown earlier, a community that includes protein expression, environmental factors, and microbial relationships as well as vertically inherited traits. It is a "folding of the inside and the outside"; this folding is a material connection, so it is "not a question of the body representing drives, forces, or even ideologies but of an intermingling of the world."[41] The organism is understood not as a reaction to a pregiven set of environmental stimuli but as a response to the environment in which it is embedded and with which it must "harmonize" to survive.[42]

In other words, it turns out that organisms (or individuals) are abstractions from larger collective entities that regulate the actions of the organisms, and that genomics-based strategies have fewer preconceptions about what makes a single organism.[43] This brings the biological body more in line with the materialist's and radical empiricist's idea of the body as formed through the environment in which it is embedded. Bodies, in this view, are never a priori but are evaluated as "expressions of certain movements, sensations, and interactions with their environments" rather than expressions of their morphology, or the form that the body takes.[44] As Donna Haraway writes, organisms are made as objects of knowledge, rather than born: "Organisms are biological embodiments; as natural-technical entities, they are not pre-existing plants, animals, protistes, etc., with boundaries already established and awaiting the right kind of instrument to note them correctly."[45]

The international system of sovereign states can also be understood as a site of cultural production, interaction, abstraction, and boundary creation. What is the state but an embodiment of ideas and beliefs about the space and place where politics are enacted and realized? Adding the micro understanding of metagenomics and microbial communities allows for the human's commensal host–guest relations to be used as a dynamic model for human communities at the macro level. This model may demand that humans react differently to issues and scrutinize the effectiveness of current international institutions. Our responses to climate change, for example, have to take into account that our actions affect more than humans and, importantly, that we function in a larger community with actants that support us and offer their own activities toward the preservation of the earth's biosphere. In the words of the NRC Committee on Metagenomics,

> Humans might survive in a world lacking other macroscopic life forms, but without microbes all higher plants and animals, including humans, would die. Not only can many individual systems—for example, the human gut or such processes as the bioremediation of toxic hydrocarbons—be seen to be the tasks of complex and dynamic microbial communities, but these communities are themselves constituents of even larger systems, predominantly microbial, that collectively make up the biggest and most complex functioning system we know: the biosphere. Whatever the causes, extent, and consequences of the global climate change now upon us, the biosphere's response to the changes—and human survival—will depend on its microbes and their activities.[46]

Global Implications

In conclusion, a beneficial exercise lies in answering the question posed at the beginning of the article, "What would a microbe do?," with a short manifesto drawing from Margulis and Sagan's theory of symbiogenesis. I offer the following possible answers as the beginning of a new conversation about how materialism and object-oriented analysis—alongside the recognition of microbial communities in the human body—might change perspectives. First, the human as a superorganism would interact with former strangers frequently to form stable, unique relationships.[47] Second, we would adapt with fluidity and rapidity to changing conditions.[48] Third, it is vital to be a flexible, cosmopolitan virtuosity and to see other objects as an ongoing part of one's transformation.[49] Fourth, we would "form cooperative associations in a plurality of forms" with others. And, finally, we would be disposed to act for the benefit of others.[50]

These answers can lead to two important implications for IR. First, a rethinking of global ethics and responsibility can lead to a "diffuse sort of ontological gratitude for the post-human era, towards the multitude of nonhuman agents" that support us.[51] This diffusing, or flattening, of social action and ties into a continuum of dynamic object interactions between humans and nonhumans, states, and bacteria and biomes makes the nested and imbricated nature of politics inside and between body politics more visible.

The second implication is explicitly political: IR, and human politics more broadly, should organize collectivities and political organizations to reflect these nested and hybrid subjectivities. As Bruno Latour and Steve Wolgar query, "Once the task of exploring the multiplicity of agencies is completed, another question can be raised: What are the assemblies of those assemblages?"[52] These discussions should be open, inclusive, and careful to reflect the values and ethics we feel are necessary in creating mutual public space, and a "critical proximity, not critical distance, is what we should aim for."[53]

This process begins to sensitize humans to the presence of "things," or nonhumans, at play in bodies and politics. This "becoming sensitive," as Latour calls it, can teach us lessons about both international relations and human relations, as well as give us a wider understanding of global existence in which humans and nonhumans are co-constituted. How are we called to respond to a class of beings or to be sensitive to beings in the world, be they human or nonhuman?

If politics are attached to the material in this novel manner and are viewed as future-oriented, ongoing processes of power relations in an environment of shifting agents rather than as only constitutional or normative, this will necessarily involve a shift in political thinking that encompasses both unified and autonomous sovereign bodies *and* complex networks of power relations characterized by decentralized, multiple, and dynamic connections. The human as a hybrid forum composed of nested sets of complex, permeable bodies will lead to a new conception of bodies politic or a set of evolving and interlocking organic systems within systems. This conception can transform our world and the responsibility we have toward the bodies in our bodies and toward other bodies in the world. Processes and techniques that address these deep imbrications and multiple relationships in the world will lead to changes in ethical and normative frameworks.

Along with the philosophical and theoretical approaches described earlier, metagenomics, as a set of scientific research techniques and as a research field, is also beginning to recognize that a detached domination over nature is not the best way of acting in the world. The NRC Committee on Metagenomics writes of our dependence on and interdependence with microbes in complex communities. Microbes transform the biosphere through fermentation and chemical cycles, and the committee recognizes that although humans are not the authors or viewers of these processes, they are intimately interconnected to these cycles, processes, and microbes for survival: "Although we can't usually see them, microbes are essential for every part of human life—indeed all life on Earth. Every process in the biosphere is touched by the seemingly endless capacity of microbes to transform the world around them."[54] In this vision, the world is a joint project between the human and the nonhuman.[55] This is more than the realization that humans inhabit the same world as other objects; it is an acknowledgment that we are composing, and transforming, the world together.

Notes

The author thanks the American Association of University Women for support through its American Fellow Writing Grant, 2010–2011.

1. Achille Mbembe, "Necropolitics," *Public Culture* 15, no. 1 (2003): 11–40.
2. Diana Coole and Samantha Frost, eds., *New Materialisms: Ontology, Agency, and Politics* (Durham, N.C.: Duke University Press, 2010), 4.
3. Ibid., 8.
4. Bruce Braun and Sarah J. Whatmore, eds., *Political Matter: Technoscience, Democracy, and Political Life* (Minneapolis: University of Minnesota Press, 2010), x.
5. William Connolly, *A World of Becoming* (Durham, N.C.: Duke University Press, 2011), 22.
6. Charles Darwin, *Origin of Species and the Voyage of the Beagle*, ed. Richard Dawkins (London: Everyman's Library, 2003), 59.
7. Metagenomics is also called environmental genomics, community genomics, ecogenomics, and microbial-population genomics.
8. Bruno Latour, "How to Talk about the Body: The Normative Dimension of Science Studies," *Body and Society* 10 (2004): 205–6.
9. Stewart Brand, "Microbes Run the World," *World Question Center*, Edge.org, accessed August 7, 2013, http://www.edge.org/q2011/q11_16.html#brand.
10. William H. McNeill, *Plagues and Peoples* (Garden City, N.Y.: Anchor, 1976).
11. David B. Dusenbery, *Living at Microscale: The Unexpected Physics of Being Small* (Cambridge, Mass.: Harvard University Press, 2009), 3.
12. Anne Maczulak, *Allies and Enemies: How the World Depends on Bacteria* (Upper Saddle River, N.J.: FT Press, 2010), 30.
13. National Research Council Committee on Metagenomics, *The New Science of Metagenomics: Revealing the Secrets of Our Microbial Planet* (Washington, D.C.: National Academies Press, 2007), 12.
14. Ibid., 2.
15. John Dupré and Maureen O'Malley, "Metagenomics and Biological Ontology," *Studies in History and Philosophy of Biological and Biomedical Sciences* 38 (2007): 837.

16. NRC Committee on Metagenomics, *New Science of Metagenomics*, 13.
17. Dupré and O'Malley, "Metagenomics and Biological Ontology," 835.
18. NRC Committee on Metagenomics, *New Science of Metagenomics*, 2.
19. Dupré and O'Malley, "Metagenomics and Biological Ontology," 835.
20. Committee on Metagenomics, *New Science of Metagenomics*, 12.
21. "Bacterial Bonanza: Microbes Keep Us Alive," *Fresh Air,* National Public Radio, September 15, 2010 (emphasis mine).
22. Katherine Harmon, "Genetics in the Gut," *Scientific American,* March 5, 2010 (emphasis mine).
23. Ed Yong, "Introduction to the Microbiome," "Not Rocket Science" (blog), *Discover,* August 8, 2010, http://blogs.discovermagazine.com/notrocketscience/2010/08/08/an-introduction-to-the-microbiome/#.UTDGWxnlVZY (emphasis mine).
24. "A Universe of Us" (editorial), *New York Times,* July 20, 2010 (emphasis mine).
25. Robert Martone, "The Neuroscience of the Gut," *Scientific American,* April 19, 2011 (emphasis mine).
26. Kate Becker, "Less than One Percent Human," *Inside Nova* (blog), PBS.org, February 2011, accessed June 14, 2011, http://www.pbs.org/wgbh/nova/insidenova/2011/02/less-than-one-percent-human.html (emphasis mine).
27. Brigitte Nerlich and Ina Hellsten, "Beyond the Human Genome: Microbes, Metaphors, and What It Means to Be Human in an Interconnected Post-genomic World," *New Genetics and Society* 28 (2009): 27–28.
28. Rosamond Rhodes et al., "Human Microbiome Research and the Social Fabric," NIH Human Microbiome Project, accessed June 14, 2011, http://hmpdacc.org/doc/d3.s7.t3%20-%20Rhodes%20-%20Social%20fabric%20implications.pdf.
29. Ibid.
30. NRC Committee on Metagenomics, *New Science of Metagenomics*, 30.
31. Eric T. Juengst, "Metagenomic Metaphors: New Images of the Human from a Translational Genomic Approach," in *New Visions of Nature*, ed. Martin A. M. Drenthen, F. W. Jozef Keulartz, and James Proctor (New York: Springer, 2009), 133.
32. Dupré and O'Malley, "Metagenomics and Biological Ontology," 842.
33. Myra J. Hird, *The Origins of Sociable Life: Evolution after Science Studies* (London: Palgrave Macmillan, 2009), 56.
34. Andrew Pickering and Keith Guzik, eds., *The Mangle in Practice: Science, Society, and Becoming* (Durham, N.C.: Duke University Press, 2008), 4.
35. Ibid., 8.
36. John Dupré, *Processes of Life: Essays in the Philosophy of Biology* (Oxford: Oxford University Press, 2012), 203.
37. Dupré and O'Malley, "Metagenomics and Biological Ontology," 842.
38. Jane Bennett, *Vibrant Matter: A Political Ecology of Things* (Durham, N.C.: Duke University Press, 2010), 112–13.
39. Ibid., 112.
40. Jussi Parikka, *Insect Media: An Archaeology of Animals and Technology* (Minneapolis: University of Minnesota Press, 2010), xxv.
41. Ibid.

42. Elizabeth Grosz, *Chaos, Territory, Art: Deleuze and the Framing of the World* (New York: Columbia University Press, 2008).

43. Dupree and O'Malley, "Metagenomics and Biological Ontology," 835, 841.

44. Parikka, *Insect Media*, xxv.

45. Donna Haraway, "The Promises of Monsters: A Regenerative Politics for Inappropriated Others," in *Cultural Studies*, ed. Lawrence Grossberg, Carrie Nelson, and Paula A, Treichler (New York: Routledge, 1992), 298.

46. NRC Committee on Metagenomics, *New Science of Metagenomics*, 20.

47. Lynne Margulis and Dorion Sagan, *Acquiring Genomes: A Theory of the Origins of Species* (New York: Basic Books, 2002), 90.

48. Ibid.

49. Ibid.

50. Dupree and O'Malley, "Metagenomics and Biological Ontology," 834–46.

51. Rosi Braidotti, *Transpositions: On Nomadic Ethics* (Cambridge, Eng.: Polity Press, 2006), 270.

52. Bruno Latour and Steve Wolgar, *Making Things Public* (Oxford: Oxford University Press, 2005), 253.

53. Bruno Latour, *Reassembling the Social: An Introduction to Actor-Network Theory* (Cambridge, Mass.: MIT Press, 2005), 260.

54. NRC Committee on Metagenomics, *New Science of Metagenomics*, 12.

55. Pickering and Guzik, *Mangle in Practice*, 2.

Breathless

Peter Adey

Political Atmospheres

It is an old assertion, but might politics be, in fact, atmospheric? Might the political be held in the atmospheres that suspend our lives with oxygenated air? Could we find politics in the air that gives us breath and moves us with the primal animation toward expression and the giving of voice? This chapter aims to address what we might call the political atmospheres of breath. It focuses on when breath is shortened, concentrating on when atmosphere is made difficult to live through or survive.[1] The breath splutters or falters.

Let us note that notions of atmosphere have received something of a surge of interest in recent years as both politically "vital" and "affective."[2] Within work that uses this approach, perhaps what marks notions of affective atmospheres most is their ambiguity. This is meant not so much in terms of the widely varying registers of affective atmospheres, in scope or definition, but in recognizing that atmospheres seem to share an essential quality of vagueness.[3] In Peter Sloterdijk, the air becomes a pseudobiological and extrabodily extension of vulnerability for practices of atmoterrorism performed by states and other political movements. The atmosphere, and by extension our ability to breathe it, is explicated into a strategic technological domain of killing and the economic pursuits of consumer spending by modulating the atmospherics of buying environments.

Atmospheres are so strategic because they hold the quality of being around one. Atmospheres are immersive, held in a room one might enter, or sensed as a particular mood or feeling.[4] At the same time, atmospheres are indistinct and unlocalizable. They are difficult to pin down. Atmospheres seem to exist apart from bodies and subjects, yet they are heavily influenced by the movements and excitations of a persona that fills a space with its substance, as Gilles Deleuze notices of Michel Foucault's charismatic and intensified presence.[5] Atmospheres seem both to be entirely about the perceiving subject calling atmospheric conditions into being, whilst they appear to exist and persist beyond our perception long before and after our ever being there.

Approaches to atmospheres, then, share something of a materialist bent that has sought not to disassociate an atmosphere from its simultaneously meteorological material and substantial thingness—even if the thingness of an atmosphere is qualitatively different from others.[6] An atmosphere is a kind of thing registered differently and palpably across different bodies while it hangs in the air, insubstantially. Yet atmospheres *are* substantial.

This is easier to conceive of on a different scale than one we might suppose when we begin to consider the bodies and volumes of air that press in on us, that are altering the planet as carbon is brought back into the air once more.[7] First articulated by Karl Marx (as discussed by Ben Anderson), an atmospheric pressure was one particular expression of class consciousness willing itself to appear.[8] And insofar as atmospheres might encourage a more global and planetary outlook, even some of the earliest scientific study of atmosphere saw figures such as Alfred Russel Wallace positing a great "aerial ocean" through which all life on earth was connected. The atmosphere of the planet is taken into the body (and expelled out of it again) as breath.

As well as being intimate, then, atmospheres speak to more global, more international spheres, outlook, and scales. They perhaps always have done so, as we can see from some of the earliest Enlightenment thought to modern environmentalism that evokes Wallace's notion of a connected aerial ocean of global belonging. Thus, for Frances Dyson, an atmosphere suggests a manner of integration, a sympathetic lining up of subjects and bodies that are spread very far apart across our planet but are nevertheless placed "within, and subject to, a global system: one that combines the air we breathe, the weather we feel, the pulses and waves of the electromagnetic spectrum."[9] In breath and atmosphere, the international is distributed and made palpable. The politics around such scale and vagueness are manifold. Most often, however, atmospheres articulate conditions of sharing, recognition, and the political flattening between subjects and objects. At the same time, atmospheres may register in sensations of difference and unfamiliarity.

How might atmospheres carry and modulate voice through breath? Atmospheres carry voices and the workings of politics. In the wrong climate, some words have been known to be stuck on the air, frozen, or stifled and muffled.[10] Political speech can be lost within the din of too many voices, weakened in the wrong conditions, and sometimes snuffed out entirely. Within design, some political volumes have, unsurprisingly, paid considerable attention to the atmospheres that pervade the chambers of debate and decision making, and to the ways they might be modified and augmented in order to resolve these problems.[11] In other moments, the voice is effectively carried away by the atmospheres of resonant bodies, amplifying and distributing the soundings of political messages. Anja Kanngieser's extraordinary work on the soundings of voice and specifically on the Occupy movement's "human microphone" illustrates just this point, as crowds amplified and repeated the speech of individuals such as Slavoj Žižek, building his speech into a powerful and apparently unifying chant and resonance during Occupy Wall Street in 2011. Kanngieser, however, urges caution and challenges the likely plurality of such political collectives of voice and breath that might not be found or that may be located in the silences or dissonance of different and competing voices.[12] What's more, it is perhaps when such voices go global that, in this instance, their affective freight is lessened. The repetition of Žižek's voice in audio and on YouTube video, especially without his bodily gestures, loses its breathy material liveliness in the drone of collective mechanical repetition.

So what happens to the voice or breath that begins to falter in the space of global politics? The kind of relationship between atmosphere, breath, and voice that I want to tease out within this space is Luce Irigaray's gender politics of assimilation in or through air. The forgetting of air, argues Irigaray, is a masculine assimilation or absorption of the voice of the air of women—the mother or the Other. As Irigaray writes, "The air of voice leaves no mark in Being. The voice's difference does not take place in Being. Their 'presence,' yet unlit and undemonstrated, remains without any possible recall."[13] In other words, to forget air or breath is to forget the agential voice.

In the following section, I select several key moments, some historical and others contemporary, in the history of breath and its appearance and intersection with the voice. These moments deliberately coincide with the Western emergency politics that have played out in the administration and apparatus of aid and security in nineteenth-century emergency medicine and post-9/11 preparedness.

The Sovereignty of Breath

The association of atmosphere and political voicing is not a new relation; neither is it to be associated with the maturity of a political body or, indeed, a body's maturity of years. Irigaray has argued that our first voice is a call, an aspiration for air for the first time in our lives once we are born to the world. Life begins with a muddling between a noise, a cry, and a call. Thus, our first move as political subjects is a clawing for air and a mourning of an absence made present by our entry into the outside. In the baby's cry that haunts the air, we find both the beginning of life and the "boundless immensity of a mourning."[14] Air appears not simply as a medium or a meditation but also as a loss, a need that is voiced even if no language is spoken to utter it.

Air as "that which restores life" has its own conventions, belief systems, technologies, and practices that revolve around the giving of breath in order to restore life to the body. Let us take briefly the work of the Royal Humane Society, the organization first known as the Society for Affording Immediate Relief to Persons Apparently Dead from Drowning, which formed in London in 1774. Insofar as the apparatus of nineteenth-century political-speaking chambers were intended to insulate their inhabitants with a degree of perfect air, shut off from the outside by the control over circulating air and sound and by the dangerous emanations of breath and stench, the ethos of the Humane Society, and that of the other societies and institutions with which it communicated in northern Europe, sought not to insulate but to conduct the restoring powers of breath. Through resuscitation, breath could be given back to bodies that were found drowned or injured.

According to different ideas about air at the time, the breath was viewed as a vital spirit of sorts. In the emergent forms of mouth-to-mouth resuscitation, it was conceived possible to restart the spark of life by reinflating a drowned subject's lungs, even if it was not all that easy to shake off some of the darker tones of snatching a life from death. Notions of "suspended animation" served to repudiate the more ghoulish associations by suggesting that life had not left the not-quite corpse. Theories over resuscitation were wide and varied. For the Scottish physician John Fothergill, the lungs could be compared to a pendulum in a clock, ready to be restarted by human action.[15] In this line of thought, the body's spark is equated to that of a machine's original motive force. The more mechanistic and practical treatments pursued by society sought to combat resuscitation's more ghoulish associations, as well as the idea that people were replacing the role of God as the arbiter of life. As Benjamin Hawes explained during his speech at the forty-seventh anniversary of the society's founding, in 1821, its activities should be understood as a scientific experiment. Its purpose was not "to raise the dead to life, but to snatch the lifeless from an early grave."[16]

Such atmospheric practices of pneumatic resuscitation took decisions of life and death away from their host, the body, to form a different kind of choreography and geography of care, which overcame even early Victorian sensibilities of touch and intimacy, although mouth-to-mouth resuscitation certainly underwent its own period of crisis. The practice was considered by some to be polluting and potentially dangerous should disease be spread by the practice. (There were considerable fears over the spread of consumption.) The society built four hundred receiving stations in London alone, which were constructed according to resuscitative principles and appointed with an apparatus of restoring technologies and techniques as well as iceboats and ladders. Yet the essential philosophy of the society was that resuscitation and artificial respiration could be conducted by anyone.

Rather than the state and other institutions, restorative potential became the property of the breath of the community, institutions like the Humane Society, and those persons willing to administer it. The privilege of restoring life was distributed from the expert oversight of the fused domain of scientific rationalism, medicine, and the Christian church—with their apparent monopoly over life—to a broader space of community, civility, and philanthropy into which wider international networks of scientific, expert, and practical knowledge of breath and resuscitation techniques would be connected. Mouth-to-mouth resuscitation, the society argued, "may possibly do great Good, but cannot do harm."[17]

Just as the sovereignty over life and breath could be distributed far more widely by these techniques, they were helped by a series of technologies. Resuscitation could be augmented by the right climates, and some of these were manufactured indoors. The receiving stations were designed to be conducive to resuscitative thermal atmospheres. Central heating with hot-water pipes lined the two main treatment rooms of the Serpentine building in order to keep a constant equal temperature without smoke. In case of drowning, a body would be placed in the bath and, if revived, taken to the beds, which had copper bottoms filled with hot water to warm the body. A galvanic battery and an artificial respirator would also be provided and could be administered if needed.

And what about the voice? There are many accounts of successful restorations of life to bodies discovered after a crack was heard as the ice gave way underfoot. Signaled by a cry and the splash of a body entering the freezing Thames or Serpentine, potential restorers sprang into action. But men and women were not necessarily equally treated. Thomas Hood's 1844 poem "The Bridge of Sighs" expressed popular representations of women suicides—"fallen" women who had chosen to jump from Waterloo or Blackfriars Bridge. They were figures commonly represented in newspapers, books, theaters, and literature.[18] The forlorn woman, wandering the city in her newfound but transgressive autonomy, was perceived through her leap to be at the end of a slow downward fall.

Although the Humane Society's techniques and ideas essentially remade the corpse into a half-life promising a return to health, in much of the imagery and stories that are associated with "The Bridge of Sighs," as pictured in George Frederic Watts's famous painting *Found Drowned* of 1850, there were no attempts at resuscitation. This is surprising given the popularity of the Humane

Society's pamphlets and advice. Yet if death is believed redemptive and told through a necrophilic excitement, the holding of the corpse in beatific composure cannot be spoiled even by resuscitative air. The face is lovely, written with a placid smile. The skin is luxuriant and holds a melancholy pallor. Any recovery of the drowned, having been purified by the water, due to the dirty breath of a passerby would simply be immoral.

There are, of course, many notable exceptions. One of the most interesting cases for literary connections and emancipatory politics was Mary Wollstonecraft. Forlorn at the breakdown of her relationship with Gilbert Imlay, the stricken Mary chose death over life by similar means. She did not sink nor drown straightaway. She was forced to wrap her evening dress around her in order to lose buoyancy and find suffocation. (This suicide attempt failed.) William Godfrey, later her husband and author of a memoir of her life, came to identify in Mary a "preternatural action of a desperate spirit."[19] And Robert Southey, so enamored with Wollstonecraft, accused Godwin of "stripping his dead wife naked" in his portrayal, which gives us so much detail of Wollstonecraft's suicide attempt.[20] Because it became just an attempt—despite her wishes. Her voice was left in her suicide note. Men boating on the river soon raced to meet her and pulled her body onboard. She was resuscitated using techniques of the Royal Humane Society, which encouraged a view of death as but a providential suspension of life, apt to be restored by breathing air back into the body. Wollstonecraft's botched leap into the Thames was not necessarily a tragedy, because entering the water became a purification of her "sinful habits." In most of these representations, water cleanses. It signifies rebirth, and thus treatment by air becomes unnecessary lest it spoil the redemptive and sexualized portrayal of the still deathly woman's body. In Wollstonecraft's own words, however, the very act was less than charitable, as she was "inhumanly brought back to life and misery."[21] This is not the woman's voice heard, then, because it doesn't appear to be listened to. Instead it is apparently redeemed. There is deafness to the woman's affective voice even as she does not speak, and we can see this repeated in other moments of violence and insecurity.

Call or Cry

Marcy Borders was the "dust lady" pictured covered from head to toe in dust, the settling of an atmosphere choked in a smog of yellow hue, in the immediate aftermath of the collapse of the World Trade Center's north tower following the collision of the technological forms of the airplane and the skyscraper on September 11, 2001. What was left was not simply the nothingness of a gap in the sky but, at least for a time, a huge column of smoke and dust. The cloud was made up of towers, aircraft, inhabitants, and thousands of rescue workers who were turned into an enormous mountain of rubble and ash by collapsing structures that were visible from space. Many of the victims were never accounted for because their bodies were no longer whole. Bodies were pulverized as individuals were made indistinguishable. The result had been anticipated in Rachel Whiteread's photo series *Demolished* (1996), the smoke looking like an arm pushing out from the earth to claw at the sky.

According to Don DeLillo, what appears to have been so startling about the atmosphere that pervaded the scene was the way in which the destruction of the towers marked the decomposition of preexisting geographical as well as social boundaries of class, gender, race, ethnicity, and education.[22] A collapse of the same kind would be portrayed in Hartmut Bitomsky's strange movie *Dust* (2007), which looks at the terrible composition of air following the World Trade Center's collapse, incorporating "mineral, wool, . . . cellulose, . . . polyaromats, . . . metal particles like copper, . . . mercury, chrome, aluminium, . . . cement, lime, . . . arsenic, cadmium, . . . paraffin, [and] airplane fuel."[23] The civil and material crumpling was also geopolitical, as the scale of the event was rearticulated within an international space. The late Neil Smith identified a series of cascading rips through the scalar referents of state and nation as the attacks began to be seen to extend beyond a violence directed toward a city or a country and onto an international stage, a way of life marked by blurring boundaries yet reinforced binaries that would continue into U.S. foreign policy and its web of allies.[24] This climate of dust and distance was not abstract. People breathed it in, they swallowed it, they tried to remove it from their clothes and hair, carpets and sofas and teddy bears.

Almost as soon as the atmosphere began to fall in the following few days, numerous studies began to investigate the composition of the air inhaled by the public and rescue workers. Several of these projects were delayed because of the problem of the human in the atmosphere. As journalists asked just what the World Trade Center's dust cloud was made of, Paul Lioy, one of the chief researchers involved in the study, had to deal with the question of whether DNA could be identifiable from their samples of the cloud.[25] What the question came down to, as well as the sensitivity of the matter of air, was whether the scientists would require approval for the use of human samples in their tests. As it happened, they didn't. Despite the chance of discovering identifiable DNA from the dust samples residing somewhere near the quadrillionth percentile, law required that medical research seek permission only if the sample of human tissue was actually alive.

As with Bitomsky, this is a subtraction of an emotion. But maybe it is a familiar way of coping with an air whose violence becomes overwhelming as the enormity of the billowing matter pushes at the limits of expression. This is not so uncommon when air becomes what has been called a kind of "human smoke."[26] Along with the dramatic flattening of Bitomsky's film and the literal accomplishment of this fact in the grotesque leveling of towers and bodies to nothing but their constituent parts, only to float away in their suspension on the wind; and despite the fact that the events were indiscriminate, a fact that forms the basis of DeLillo's more hopeful politics through a leveling semblance, what else lurks within the atmosphere of 9/11 but the much more familiar narratives of "us" and "them." DeLillo accounts for the atmospheres of terror working as if they were contaminants to perpetrate the local body and the global witness. The atmospheres of the event serve as a reminder:

> There is fear of other kinds of terrorism, the prospect that biological and chemical weapons will contaminate the air we breathe and the water we drink. There wasn't much concern about this after earlier terrorist acts. This time we are trying to name the future, not in our normally hopeful way, but guided by dread.
> What has already happened is sufficient to affect the air around us, psychologically. We are all breathing the fumes of lower Manhattan, where traces of the dead are everywhere, in the soft breeze off the river, on rooftops and windows, in our hair and on our clothes.[27]

In their dispersal and intake as breath, the atmosphere feeds global politics of unease and vulnerability that DeLillo wants to counter, a feeling that intrudes, that cannot be escaped, that lingers and worries—the sense that difference is not accepted so easily. Atmospheres seem to render circumstances as unfamiliar and strange, certainly threatening.

In the midst of this, the reproduced, overexposed image of Marcy Borders stands out for the cry, call, or noise that is wiped from the representation. The ambiguity of her annunciation is outspoken by the texts and commentary that overwrote it, which speak for her in the global press. Her mouth is open, but no sound is ever heard from it that is hers, because the question, following Irigaray, is an "unthinkable one."[28] Borders's image saturates the space of international journalism to portray the inhabitation of events too big to fathom. Her image provokes sympathy. And yet Borders was not heard until much later, when her trouble with drugs, criminality, and family breakdown were told in a more socially comprehensible narrative of redemption once more—and in her own words. She is perhaps the present body articulated in "the voice that recalls who is there—recalls that she can be there even while saying nothing—[that] fades away in the face of what is spoken."[29] Heidegger, writes Irigaray, "assimilates this still-fluid trace of a 'presence' that penetrates into him, and forgets it in light of the truth of the matters to be treated in thought."[30] In this sense, the cry or voice is subsumed into global representation, writing, and thought; Borders's ownership of her encounter and story is lost. How, then, might we recover or store voice in light of its being caught short of breath?

An Archive of Breaths

In the early morning hours of a workweek day in early July, the downtown core of the city of Seattle, the area surrounding it, and the Joint Base Lewis-McChord installation in Pierce County are the targets of an aerosolized attack organized by a terrorist group. Laboratory tests are the first to confirm that anthrax was released. It is later determined that a few nondescript trucks released the deadly suspension while driving through downtown Seattle. Hundreds of thousands of people were exposed, but no one knew.

Time-shift to one week after the attack. The president and the Washington State governor have declared a state of emergency, which has led to significant federal support appearing on the scene. Provisions of the Stafford Act have been enacted, freeing significant federal resources and assistance for the response-and-recovery efforts. It took a few days for the attack to become evident. The contamination first manifested itself as normal flu-like symptoms, ignored by a large proportion of people who failed to present themselves to a doctor or hospital. Those who first developed the signs and symptoms of disease lived or worked in Seattle, and commuters and tourists from across the region, the state, the country, and the world were exposed as well, many now starting to become ill. Time-shift forward again, to one month, six months, five years, and recovery efforts continue.

Toxic events like this simulation have been produced in a now-global but uneven apparatus of emergency-preparedness regimes and scenarios that build and prepare for particular events, supported by circuits of emergency-planning experts, doctrine, training, expertise, and a vast industry. The events themselves had been understood as atmospherically affective in their production of an especially panicky public, as well as the turbulently endogenous conditions of coalescing and unstable associations through which threat might arise unpredictably.[31]

The evacuation measures, instructions, and exercises intended to prepare for just such events are meant to allow metropolitan areas to evacuate their populace to safety while ventilating the threat. There is a common critique of such stifling practices that simultaneously agitate and enroll city inhabitants, because the corporeal and visceral effects of emergency preparedness are noticeably difficult to hold or capture as security events that we can question or interrogate, often from a distance. There are, for example, ways of writing and rewriting such encounters, as Jeremy Crandall has shown in his performative role as an actor in an emergency scenario.[32] For others, a response through speech or writing fails to attune to the register at which such practices agitate the body. Let us dwell on the works of artist Catherine D'ignazio, who performs under the name "kanarinka," of the Institute of Infinitely Small Things.[33]

It Takes 154,000 Breaths to Evacuate Boston saw kanarinka jogging the entire evacuation route system of the city in 2010, an action that now takes on quite a different sense in other juxtapositions of running and violence seen most recently in the events of the same city. It took kanarinka 154,000 recorded breaths in twenty-six runs, which were podcasted and also performed live, turning the calculative metrics of security back onto the apparatus itself. Intending to somehow bottle the fear surrounding the preparedness campaign, kanarinka collected her breaths in an audio archive and sculpture, *An Archive of Breaths,* which was made up of twenty-six jars corresponding to each run along the evacuation route. A speaker in each jar would play the recordings simultaneously so that the sculpture and each "breath" could not be listened to on its own but overlapped other "breaths" from the twenty-five other jars.

What we have in kanarinka's different voicings of one preparedness apparatus, intended to ready emergency responders and a populace for the kinds of events simulated in the Seattle exercise, is a different sense of subversion than the affective politics of emergency and urgency that dominate those practices. Her approach is to perform the technocratic capture of the population's passions and urgencies herself. She bottles up the atmosphere—her breath—before it can escape, even while she does it. Despite the extractive nature of the process that abstracts the respiration of air from the lungs that performed it, which possibly reproduces the technologically mediated security sensorium performed in other cities and places, the sense for the listener is of being somewhere in the middle of respiration, of exertion, of her fatigue.

Conclusion

So we must ask, what possibilities are there for an atmospheric politics of breath in relation to the international? Attention to breath, or its lack, can

kanarinka, *An Archive of Breaths*, 2010. Custom-made table, twenty-six jars, twenty-six speaker components, wire, thirteen CD players, 45 x 72 x 16 inches.

open up politics of life at play in how societies govern the living and the dead, and how they prepare for events and respond in the face of emergency conditions. An attention to breath might tell us who is restored, brought back, or ignored—whose voice is heard, whose is silenced—and attending to breath helps to outline the circulations at play in the undoing of any naturalized sense of the local, global, or international.

Perhaps, in reconsidering breath, we might have another chance at a more plural kind of voicing or a more subtle attention or manner of listening to the gaps between words and meanings. In kanarinka's sculpture, we hear hundreds of breaths, scores of them happening over the top of one another, as if hundreds of people are running away quite silently. But there are no cries or screams. It is as if the securitized and evacuated body could be made to speak for the first time over the sirens or the shouting and the disciplining instructions of an emergency scenario, and reiterated over and over again.

Notes

1. Peter Sloterdijk, *Terror from the Air* (Los Angeles: Semiotext(e), 2009).
2. Peter Sloterdijk, *Bubbles* (Los Angeles: Semiotext(e), 2011); Mark Jackson and Maria Fannin, "Letting Geography Fall Where It May: Aerographies Address the Elemental," *Environment and Planning D: Society and Space* 29, no. 3 (2011): 435–44.
3. Ben Anderson, "Affective Atmospheres," *Emotion, Space, and Society* 2, no. 2 (2009): 77–81.
4. Teresa Brennan, *The Transmission of Affect* (Ithaca, N.Y.: Cornell University Press, 2004).
5. Gilles Deleuze, *Negotiations, 1972–1990* (New York: Columbia University Press, 1995).
6. Derek P. McCormack, "Engineering Affective Atmospheres: On the Moving Geographies of the 1897 Andrée Expedition," *Cultural Geographies* 15, no. 4 (2008): 413–30.
7. Simon Dalby, *Security and Environmental Change* (Cambridge, Eng.: Polity, 2009).
8. Anderson, "Affective Atmospheres," 77.
9. Frances Dyson, *Sounding New Media: Immersion and Embodiment in the Arts and Culture* (Los Angeles: University of California Press, 2009), 17.
10. See my manuscript "Air."
11. Paulo Tavares, *General Essay on Air: Probes into the Atmospheric Conditions of Liberal Democracy* (London: Goldsmiths, 2008); Marsha E. Ackermann, *Cool Comfort: America's Romance with Air-Conditioning* (Washington, D.C.: Smithsonian Institution Press, 2002).
12. Anja Kanngieser, "A Sonic Geography of Voice: Towards an Affective Politics," *Progress in Human Geography* 36, no. 3 (2012): 336–53. See also Mauritzio Lazarato, "Bakhtin's Theory of the Utterance," *Imaginary Property,* http://imaginaryproperty.com/node/146.
13. Luce Irigaray, *The Forgetting of Air in Martin Heidegger* (Austin: University of Texas Press, 1999), 48.
14. Ibid., 35.
15. John Fothergill, "Observations of a Case Published in the Last Volume of the Medical Essays, of Recovering a Man Dead in Appearance, by Distending the Lungs with Air," in *The Works of John Fothergill*, ed. John Coakley Lettsom, vol. 1 (London: Charles Dilly, 1783).
16. Benjamin Hawes, quoted in Royal Humane Society, *Gentleman's Magazine* 91 (1821): 807.
17. Royal Humane Society, ibid.
18. L. J. Nicoletti, "Downward Mobility: Victorian Women, Suicide, and London's 'Bridge of Sighs,'" *Literary London,* 2008, http://homepages.gold.ac.uk/london-journal/.
19. William Godwin, *Memoirs of the Author of "A Vindication of the Rights of Woman,"* 2nd ed. (London: J. Johnson, 1798), 138.
20. Helen M. Buss, "Godwin's Memoirs," in *Mary Wollstonecraft and Mary Shelley: Writing Lives*, ed. Helen M. Buss, D. L. Macdonald, and Anne McWhir (Waterloo, Ont.: Wilfred Laurier University Press, 2001).

21. Carolyn D. Williams, "'Inhumanly Brought Back to Life and Misery': Mary Wollstonecraft, *Frankenstein,* and the Royal Humane Society," *Women's Writing* 8, no. 2 (2001): 213–34.

22. Don DeLillo, "In the Ruins of the Future: Reflections on Terror and Loss in the Shadow of September," *Harper's Magazine*, December 2001, 39.

23. Hartmut Bitomsky, quoted in Ian Cook et al., "Hartmut Bitomsky's 'Dust': A Reaction More than a Review," *Science as Culture* 20, no. 1 (2011): 117.

24. Neil Smith, "Scales of Terror and the Resort to Geography: September 11" (editorial), *Environment and Planning D: Society and Space* 19, no. 6 (2001): 631–37.

25. Paul Lioy, *Dust: The Inside Story of Its Role in the September 11th Aftermath* (Lanham, Md.: Rowman & Littlefield, 2010).

26. Adey, "Air."

27. DeLillo, "In the Ruins of the Future," 39.

28. Irigaray, *Forgetting of Air,* 50.

29. Ibid.

30. Ibid., 48.

31. See Melinda Cooper, "Turbulent Worlds: Financial Markets and Environmental Crisis," *Theory, Culture, and Society* 27, nos. 2–3 (2010): 191–212; and Brian Massumi, "National Enterprise Emergency Steps toward an Ecology of Powers," *Theory, Culture, and Society* 26, no. 6 (2009): 153–85.

32. Jeremy Crandall, "An Actor Prepares," CTheory.net, February 7, 2008, http://www.ctheory.net/articles.aspx?id=590.

33. "The Institute for Infinitely Small Things," http://www.ikatun.org/institute/infinitelysmallthings/.

Blood

Jairus Grove

Blood may seem like either an obvious or a peculiar character for a book about things in international relations (IR). IR's geopolitical tradition marched through the Rhineland of the nineteenth century, tracking blood and soil through the circuitous pathways of geographers such as Alexander von Humboldt and nationalist historians such as Leopold von Ranke until their triumphant unification in Schmitt's triad of land, state, and people. Blood as identity and national or tribal continuity draws lines of enmity that constitute the political of international politics. Blood is a major player in the nation-state and in the world of nation-states and has been for at least as long as something like geopolitics have existed.

Certainly, blood politics is not restricted to Europe's heartland; in the nineteenth century, it was a defining feature of the gigantic emerging federal power across the ocean as well. After all, there is no object of American jurisprudence and legislative history more infamous than a single drop of black African blood. The "one-drop rule" refers to Virginia's antebellum hypodescent laws, which codified a long-standing mythology of blood and blood difference that hopscotched from biblical interpretation to phrenology to Nazi anthropology and back to American eugenics. The 1924 Racial Integrity Act and the subsequent eugenics policies restricting miscegenation and institutionalizing compulsory sterilization that continued late into the 1970s demonstrate the formative and pervasive horror of sanguine logic.

As this story unfolds, it is clear that this metaphorical blood was not the blood of circulation, oxygenation, and coagulation but an imaginary blood—sacred and profane—within the provincial world of human affairs. The real blood of platelets, hemoglobin, and lymphocytes has no use for this sordid history and in fact resists its own signification through its insistent indifference to racial difference. The metaphorical droplet of blood is no match for the pipeline of plasma that ran through Hawaii and then Europe and North Africa before returning to the Pacific during WWII and that continues to support global military operations today. This actual blood alters the course of conflicts, undermines long-held beliefs about racial evolution, and continually countermands the exuberant will-to-control of twentieth-century science.

Blood finds its material footing at precisely the moment biopolitics emerge as an organizing principle of global total war. The ecology of the global U.S. alliance system from World War II through the Cold War and beyond is not based solely on the civic republican ideals that were said to bind the Allies against the Axis. More than ideology holds the United States and its European and Asian allies together. In addition to treaties of mutual defense are the practical treaties of blood exchange. The symbolic pacts of loyalty are sealed with very real blood oaths that call upon vast infrastructures for the movement, preservation, and acquisition of blood products and even whole live blood from "walking blood banks."[1] This chapter affirms that, despite being the most overrun metaphorical dumping ground for intensive human drama and divisiveness, blood as a real thing has a place and a role in the formation and creative advance of the international-cum-global.

Human blood is both fugitive and indifferent as well as formative and insistent. The materiality of blood resists both the provincialism of human-manufactured racial difference and the hubris of a scientific mastery that believes itself capable of control via the breaking down of heterogeneous assemblages into their fundamental or component parts. The former greatly disadvantaged the Nazis during the Second World War, and the latter ensures that in many cases the U.S. military is able to use only 1.5 percent of its forward-deployed blood products before they rot on the shelf.[2] The great breakthroughs of blood-pressure supports such as the protein albumin and other blood products demonstrate the ubiquity and indifference of blood. However, the failure of such methods also performs the limits of and failures to master blood and demonstrates just how insufficient industrialized parts are for the sustaining of life. Blood is differentially generic and insistently univocal in its heterogeneity. To put it another way, blood is an assemblage that defies essence or formula while being predictable and consistent.

The complexity of blood is found in its being an assemblage of objects such as proteins, lipids, cells, water, and minerals. Further complicating matters, the object blood is also refracted through a series of technical and somatic connections that range from proximate to global. A natural history of the adoption of abstractable blood and blood products as a key component of modern warfare and their extraordinary waste under the guise of military readiness is meant to be more than a set of "cool facts" about the excesses and failures of global empire.[3] I find blood and its peculiar resistance to mastery and signification quietly heroic and worthy of their own exploration. Rather than playing some passive role in the ascendency of science as the lingua franca of the biopolitical state of affairs, blood sluggishly nags at the very ground of war.

Blood nobly refuses to submit entirely to humanity's petty squabbles. To try to capture blood's virtue, I will sketch out the emergence of the U.S. Armed Services Blood Program. Further, I will detail the role militarized blood procurement played in steering blood from a medium of medical intervention to a national strategic resource with all of its attendant and troubled global networks of acquisition, flow, and policing such that blood and blood products became a global commodity. On the way to the state of current practice, we will take a few necessary detours through Nazi Germany's failed blood program, as well as the demands for and success of the American-led Blood for Britain program despite its resonances with the racial logic of the Third Reich.

What I have in mind by *object* and *assemblage* is not much more than the commonsense definitions of these words, but there are a few attributes of these concepts that are not as common to sense. In casual conversation, an object is a thing. So far, so good. However, commonly things are grammatically and causally subordinated to subjects or, in some cases, to first causes such as gods. So, there are two necessary subtractions necessary before proceeding. First, everything is an object. And when I say "object," I do not mean to imply the opposite of a subject. Instead, I simply mean an actual thing that perdures or holds together against the grain of a universe that is winding down. Objects, as science fiction writer Stanislaw Lem calls them, are "islets of decreasing entropy" in a sea of noise.[4] Jane Bennett has similarly championed objects by channeling Spinoza's concept of *conatus* to describe the will or tendency of all things to hold on to existence.[5]

The second subtraction is somehow more invisible to our language and description. Objects are not passive receptacles or mechanical pieces in a Rube Goldberg device of subjective agency and causality. Objects have powers, capacities, and attributes that make them formative and collaborative. Nothing gets done without a crowded room of things working in relation with each other. Objects are real and continue to be so even when we are not looking. Trees fall in the forest and make a sound even when humans do not hear them or, more importantly, do not have the idea of hearing them. Objects are real and formative and continue to be so independent of human perception and cognition of them. So, there is not a divided world of formative conscious things (humans) and inert useable things (objects). Humans, like all other things in this story of blood, race, and conflict, equally have a role but not a lead role. All things are constrained and enabled by capacities and relations and so are ontologically equal. Put succinctly by computer programmer and alien-object advocate Ian Bogost, "[A]nything is thing enough to party."[6]

Assemblages are heterogeneous collections of objects whose relationships are differentially intense. As the intensity and organization of the collection or herd of objects changes, so does the expressive effect of the assemblage even if the population of things remains the same. Things do not dissolve into an assemblage, but neither can one have an atomistic explanation of an assemblage as "a strange irreductionist situation in which an object is reducible neither to its parts nor to its whole."[7]

Timothy Morton illustrates this seemingly contradictory position with the example of a coral reef. Such a complex thing is made up of coral, fish, sea anemones, sharks, dead coral, ocean water, microorganisms, sediment, sunshine, and so on. However, no one thing, even coral, embodies or is the essence of the reef. Neither does the absence of any one of these things deny the assemblage of a coral reef. Artificial reefs that are made of concrete or shipwrecks rather than coral are nonetheless teeming with life. An old tire or ubiquitous plastic bag is seemingly out of place but does not somehow eliminate what it is to be a coral reef. Despite an indefinite set of possible additions or subtractions, Morton points out, we still have little trouble distinguishing a coral reef from a parking lot.[8]

As with Morton's coral reef example, we can discern the difference between whole blood, plasma, red blood cells, hemoglobin, lipids, and white blood cells. Following Morton, this discernment is neither merely epistemological nor a subordinate relationship between part and whole. Despite the innovation of a technique called blood fractionation to discern and describe the parts of blood's distinct functions, the reassembly does not neatly add up to the collective we call blood. In fact, the body responds much better to fresh whole blood as compared to the defrosted cocktails of its reassembled parts. In other words, the body responds differently to whole blood versus plasma versus the volume-creating protein albumin versus the oxygen-carrying red blood cell or the protein hemoglobin that bonds with oxygen so that the red blood cell can hold on to oxygen.

Also, blood does not quite seem to not be blood simply because of a low occurrence of one of these components. We do not say that immunosuppressed individuals do not bleed simply because the fluid in their veins is missing substantial numbers of white blood cells. As will be seen in the fits and starts of fractionation research, the responses of the body to concentrated forms of blood components are whole responses; they are not part of a response that adds up to the whole assemblage blood. Thus, the difference between blood and plasma or plasma and albumin is a real difference that is experienced by our bodies as much as it is known or captured by our concepts, even though blood contains plasma and plasma contains albumin, among many other proteins.

I am saying that things can be different things depending on their relationship and that such differences are neither predictable nor fully knowable, even retrospectively. However, for a moment before dismissing such an outlandish claim for its logical contradiction, consider that it may not be the description that is contradictory or illogical but reality itself. In which case, why should we expect the world to live up to our standards of what we wish it were so that our logic could be neatly operative? What we are confronting is the difficult and irresolvable tension between atomism, form, and movement. Unfortunately, none of it exists at the instant or as we would like it. We cannot catch the becoming of objects or their collaboration as assemblages in the act, so to speak. Instead, we fumble around as provocatively as possible, in hopes of learning something from the world. This is what I think is meant by an object-oriented thinking or a speculative realism.

Raced Matters

> It shall hereafter be unlawful for any white person in this State to marry any save a white person, or a person with no other admixture of blood than white and American Indian.[9]

At the turn of the twentieth century, blood was suffuse with meaning. European and American humans in particular invested in their crimson fluid the legacy and essence of their civilizational difference. It is this concept of blood that is meant by phrases that equate tribalism or race wars, such as "blood feuds," "blood rivalries," "bloodlines," "blood is thicker than water," and "differences are in the blood." Blood was and often is a synecdoche for race, and so differences of blood are not merely differences; they have provided the distinctions for superior and inferior inheritance that were meant to justify or at least give grounds for war, colonialism, and slavery, and the subsequent radioactive fallout of racism.

The persistence of this racial story of blood is closely related to the sacred logic of blood. That something in blood was constitutive of one's essence is an affectively powerful belief. "Blood sacrifice," "blood ritual," "blood oath," "blood brothers"—the list goes on. To quote Spike from *Buffy the Vampire Slayer*, "It is never a 'lymph ritual'"; it is always about blood. Insofar as the symbolic tradition of blood has lost its sway on public policy and practice, this is due more (in my estimation) to the insistence and relative indifference of actual blood than to the attempts at demystifying race. In the case of World War II, the demand for blood—its biopolitical need—began to overshadow blood's resonance with earlier mythological technologies of sorting populations. The use of transfusions by some daring field medics in World War I had demonstrated the salubrious effect of additional blood in trauma cases. Despite limited knowledge of how to extract or store blood, it became immediately apparent to those treating the wounded that the ability to replace lost fluid substantially improved the chances of survival.[10] However, the use of blood was very limited because of a lack of knowledge and technology to carry out transfusions, and the homogeneous fighting populations of the First World War provided few encounters for dispelling the myths of "pure" and "impure" bloodlines.

Blood, for many reasons, is very difficult to utilize, even under perfect conditions, for exactly the same reason that it is such an asset. Blood's almost immediate inclination when removed from the body is to clot. Blood is also not entirely indifferent. Rather, it is quite attentive and observant of its surroundings and cohabitants, whether revival antigens or the presence of gases or injury.[11] Whereas the human concept of race is meaningless to blood, foreign cells and microorganisms provoke an almost immediate and often violent response. In some sense, blood is racist. The antigens present on red blood cells that give the fluid assemblage its color are, in some cases, very specific as to what other kinds of blood they will party with. Type A blood is accepting of other type A's and also type O's, but everyone likes type O's. Type B antigens play well with other Type B's and, of course, like O's as well. Type AB is the Andrew W. K. of blood and will party with anyone. Sadly, although type O is infinitely generous to others, its blood serum cannot tolerate anyone's antigens. So, from the standpoint of blood, there are more or less four races of humans, but that racial difference exists between parents and children, aunts and nieces, cousins, and other family relationships. However, in cases in which this difference cannot be tolerated, the consequences are lethal. Blood without antigen compatibility results in agglutination, a sudden and destructive clumping together of the unlike blood.

Two major breakthroughs emancipated blood from the dependency on one body's vessels, allowing for a serious application of transfusions and subsequently blood's global adventure. Sodium citrate was found to block the clotting agent in blood, so that usable viscosity could be maintained long enough to collect and then administer blood to a patient in need.[12] Before this discovery, arteries from the donor had to be directly sutured onto the recipient, often resulting in permanent damage to the donor. The second discovery was the ability to test for blood antigens so that type could be determined before a transfusion. These two innovations made blood portable and abstractable, and a new kind of war made portability and abstraction necessary.

The intense aerial bombing of England by the Germans in World War II, in particular the bombing of urban areas, led to serious injuries requiring transfusions but also made the ability to collect blood and store it extremely difficult. Even though sodium citrate can prevent clotting, blood quickly degenerates unless refrigerated. Frequent power outages and the need for medical resources to be on the move were in direct conflict with the demands of blood storage.[13] Across the Atlantic, in the United States, blood research had been progressing since World War I. The standardization of blood-typing and the ability to perform transfusions without permanent injury to the donor resulted in an immediate for-profit market in the United States. Under the watchful eye of the Blood Transfusion Betterment Association, professional donors were issued books for the recording and approving of all donations. To become a donor, one had to have a clean bill of health, refrain from drugs and alcohol, and, most importantly, have a telephone. As storage was nearly impossible, donors had to be reachable at all times.[14]

In 1937, Dr. Bernard Fantus coined the phrase "blood bank" and, along with many others around the world, began to investigate means for increasing the ability to accumulate and store blood. In Russia, cadaver blood initially appeared to be the solution. A massive hospital in the middle of Moscow known as the Sklif had thousands of beds and nearly constant trauma and emergency traffic filling it every day.[15] The centralization of care and the sheer number of bodies entering the door made it possible to acquire sufficient blood from the recently deceased, as well as discarded placentas and other parts.

News of the success in the U.S.S.R. inspired Fantus to begin collecting and storing blood at Chicago's Cook County Hospital.[16] Although he was not willing to harvest cadavers, the idea and method of storing blood was a conceptual breakthrough that inspired the possibility of blood's serious commodification, as it meant that donors were no longer needed to be present at the time of transfusion. Shortly thereafter, Charles Drew, a doctor from Howard University who had recently completed an advanced medical degree at Columbia University based on his blood research, attempted to push the capability of storage a little bit further. Drew made an exhaustive study of all available knowledge on blood and blood characteristics and ascertained that blood could be effectively broken down into its major components if allowed to settle.[17] He further refined the process of using the spinning of blood to increase the effectiveness and speed of its separation. One of the components, a syrupy yellow substance called plasma, remained stable for much longer periods of time without refrigeration. Although plasma lacked the oxygenating or immunity properties of red and white blood cells, Drew's research showed that it had the ability to raise and stabilize blood pressure, and that was found to substantially lower the death rate in victims for whom severe shock would normally cause a rapid decline.

Again, plasma was only a temporary solution for hemorrhaging patients, as it could not provide oxygen. But the temporary solution was all that was often necessary to stabilize patients so that more complex procedures could be completed. Furthermore, the breakthrough of plasma was not just in its storage life. In fact, what makes plasma so interesting is that it has no antigens. The degree to which blood distinguishes itself from other types of blood is not present in plasma. In this, Drew had found a thing that was completely generic and indifferent to human difference while still having the effect of stabilizing patients.

The finding was of immediate interest to the U.S. Department of Defense (DOD), which was trying desperately to supply England with blood. Plasma represented something that, if acquired in sufficient quantities, could be shipped to England. As might be expected, there were immediate concerns over purity on the part of both the DOD and the British government. Although the British were desperate for blood, the fear of killing people with tainted blood was quite high, as was the general reluctance to have a substance from another body pumped into one's own (transfusions were still quite novel). The process of producing plasma in significant quantities while assuring quality control also required extraordinary and unprecedented technical expertise, and so the concerns were reasonable. In response, the Blood for Britain program was created and given substantial resources to create the infrastructure necessary to provide England with safe and pure plasma. The DOD decided against medical advice that one component of the high standard for purity included collecting blood only from healthy white donors for plasma production.[18]

After substantial consultation and two national meetings of all available experts, a consensus emerged that only one man possessed the knowledge and capability to create and execute such a program: Dr. Charles Drew.[19] The only problem was that Drew was black. Despite the racial policy regarding donors, necessity overwhelmed the DOD's racism in choosing a program director. Drew was offered the job and, despite the race policy, accepted the position and honorably executed his duties. The program was successful despite the incredibly laborious task of collecting sufficient blood and producing tested and controlled plasma. The irony remains that Drew saved countless lives through his efforts but could not donate his own blood to the cause.

Drew's innovation had made it possible to produce a substance on an industrial scale that could be used in any body regardless of blood type, much less race. In 1942, the DOD, under pressure from black newspapers and activists, allowed the American Red Cross to begin collecting black blood for black soldiers.[20] Before this, black soldiers were told that they could receive white blood, as it did not contain the impurities that made black blood threatening to white bloodstreams. It is difficult to confirm how many African Americans lost their lives as a result of blood prioritization. However, there were definitive shortages in both the European and Pacific theaters that could have been offset by willing and vocal potential African American donors.[21] It is a matter of historical contingency and good luck that the population of African Americans in the U.S. armed forces and in the donor population was relatively small. Otherwise, many more U.S. lives would have been lost as a result of the DOD's refusal to listen to the racial indifference of blood.

The soldiers of Nazi Germany were not as lucky. Unlike the United States, which relied on the rumors of racial mythology and left the policing of that mythology more often to personal prejudice than to legislated penalty, Germany developed a robust scientific literature to support the empirical basis of racial difference. After substantial research, a less than 2 percent variance in the frequency of occurrence of type A blood among so called Aryan populations in contrast to type B blood in Slavic and Jewish populations came to form the basis of the 1935 Nuremburg blood-protection law. The first man punished under the law was a Jewish doctor who valiantly saved a patient by transfusing his own blood while preforming surgery. Unfortunately, the patient was not Jewish, and Dr. Hans Serelman was sent to a concentration camp for "polluting Aryan blood."[22]

An overwhelming majority of the best doctors and research scientists in Germany were sent to camps and murdered because of their so-called blood race. This had devastating consequences for the German military, as the Nuremberg Laws were enforced on the front lines as well. Legal blood transfusions were almost impossible because of the fear of tainting blood. The metaphorical one drop could, after all, be hiding in anyone. Further, research into transfusions and other lifesaving medical procedures was hobbled by the imprisonment and massacre of the German medical class. After the capture of Dr. Paul Schultze in 1942, it was revealed that battlefield injuries for the Germans were resulting in dramatically higher death rates than the Allied forces were encountering. What's more, the only method developed for coping with hemorrhagic trauma and severe wounds was a derivative of vinyl that was being injected into soldiers in an attempt to increase clotting, as additional blood was just not feasible to acquire.[23] Comparatively, in 1942, the United States successfully exported 31,250 gallons of blood and plasma.[24]

The United States made further improvements because of Dr. Edwin J. Cohn's development of a method called fractionation, which allowed for the isolation of the protein albumin, identified as responsible for plasma's ability to expand and support blood pressure. The stabilizing aspect of plasma could now be isolated and, importantly, dried into an easily transportable high-density powder. The breakthrough was immediately classified as a permanent Allied edge. Albumin, like plasma, was indifferent to blood type but, unlike plasma, could be easily transported in smaller quantities and reconstituted with minimal risk of contamination and could transmit almost no diseases, substantially lessening the demands of testing and quality control.

After the classification of albumin, blood and blood products were a permanent part of the American arsenal. Blood, like resources for war such as rubber, iron, and petroleum, was a strategic reserve that had to be maintained and sufficiently stockpiled to guarantee military readiness. By March 1945, more than 2,000 units of blood were being shipped per day to forces in the Pacific and Europe, totaling more than 500,000 units, or 62,500 gallons, of blood in thirteen months.[25]

By 1950, the extensive research on blood and the horror and embarrassment of Nazi blood laws led to a standoff between the American Red Cross and the Department of Defense.[26] The decision was made to end the segregation of blood, in part because of the political mobilization to end the practice; but that mobilization was significantly aided by blood's properties, which reflected empirically that race was superstitious animus, not reality.[27] The shaping of civilizational difference as biological and sanguine, represented by nineteenth-century concepts of species based on the coordination of phenotype and geography and further indexed by lingering and conflicting mythologies of multiple descent, was, in policy at least, coming to an end. In its place, another sense of the term *race* began to emerge, as described by Michel Foucault. Described as the petty normative, cultural and class differences that characterize the internal war of politics and externalized by modern war as a productive national body politic slowly displaced earlier notions of blood difference with a concept of difference that would pit the "race" of Americans against the "race" of Soviets. Foucault called this "state racism."[28] Race became a squabble over forms of life rather than strains of blood. Blood's material necessity for the nation eclipsed the value of archaic notions of blood race.

However, biopolitics are not sufficient to explain this process of deracialization. The capacity of blood and blood type to continually defy racial logic plays a significant part in the possibility of national security's winning the upper hand against archaic racial blood. Blood's indifference to the human superstitious investments in difference undermined attempts at reinventing racial mythology. Even in the case of Nazi Germany, where the resources of a whole nation were at the disposal of race thinking, manufacturing scientific racial knowledge was ultimately both impossible and catastrophic. The plasticity of meaning could not, in the final instance, hold up to the recalcitrance of blood. Further, blood's ubiquity and mysterious capacities inspired and brought together a global community of scholars. The research necessary to execute Dr. Drew's program for Britain, and its extrapolation into the industrial-scale production of blood for the Allied war effort, required that the circulation of information generated by blood's peculiar capacities and attributes crisscross the planet from Paris to Chicago to Moscow and, even more unlikely, cross the American color line.

The Assemblage Strikes Back: Fractionation and the Limits of Atomistic Science

If blood provided a challenge to racial thinking, it also continually asserted itself in the face of scientific control. Although in many respects the initial results of plasma and albumin were promising and ultimately made a huge difference in the war effort, it became clear well before D-day that neither was a substitute for whole blood. Field medics and doctors on the front lines were reporting cases of soldiers who, initially stabilized by plasma or albumin, would begin gasping for air and ultimately die.[29] These soldiers were suffering from a lack of oxygen, because they did not have sufficient red blood cells to bond with the oxygen being drawn into their lungs. They were suffocating from the inside out. In the North African theater, American doctors had noticed that French and British doctors less enamored with the technological breakthroughs of U.S. blood products were producing lower morbidity rates by using wholeblood transfusions. It is worth noting that the French were capable of this because they wholly embraced African and Arab donors as patriots fighting for the French cause. The French republic in exile did not recognize phenotypical racial difference at the level of blood, and blood in turn affirmed the republic's egalitarian practices of transfusion.[30]

Subsequently, the presumed causal object of blood-pressure stability was increasingly under scrutiny. Albumin and plasma would continue to be important tools for trauma doctors, but the full assemblage of blood was necessary to ensure higher survival rates among soldiers. The challenge was to similarly commodify and industrialize whole blood so that it could be brought to bear on the battlefield. This was not an easy task. In fact, to this day the ability to collect and store whole blood is severely limited.

Portable refrigeration and other techniques would be deployed, but it became common knowledge that the window for usable whole blood was about two weeks. This temporal fragility of blood is expressive in two ways. First, blood expresses the failure of its own disaggregation. Second, it is now clear to anyone paying attention that the stockpiling of blood means eminent invasion. The window for the successful use of blood, unlike stockpiled bullets or tanks, is very short, and as whole blood must be collected in a form that cannot be concentrated, the number of donors required necessitates the last-minute vast recruitment of the civilian population. As a result, the closely guarded secret of when D-day would commence was inadvertently announced by the scramble to gather sufficient blood.[31] The limitations and insistence of blood's integrity as an assemblage to produce the desired salutary effect made secrecy nearly impossible.

Despite even more significant advances in the storage of blood, it is still true that changes in the flow of blood express coming war and thus are closely monitored by intelligence agencies.[32] In part, this is because fractionation is not capable of atomizing blood so that it can be stored and reassembled. Fractionation has also not been capable of manufacturing a blood substitute. There is no "true blood," despite the fact that extensive chemical and physical analysis of each component of the assemblage has been completed repeatedly. The parts rendered and concentrated have measurable and important effects, such as stabilizing blood pressure, oxygenating, or improving clotting. However, it is not possible to somehow reassemble or synthesize these parts in a way that fulfills the body's demand for blood.

These limitations on blood's shelf life and the failure to devise a viable substitute have produced a vast global network for the U.S. armed forces. The Armed Forces Blood Program (AFBP) was institutionalized after a failure to restart the World War II ad hoc blood process during the Korean War. It became clear that blood had to be continuously on demand wherever U.S. forces might be. Further compounded by the post–World War II reconstruction, and because of the extraordinarily high casualty estimates and diverse geographic scenarios for potential war during the Cold War, the AFBP, like the U.S. military, is deployed worldwide. This places demands on the program that require not only a vast internal infrastructure for collection, screening, storage, and transport but also a number of treaties and procedures for the local acquisition of blood when the ability to acquire and deliver a sufficient supply from home is not possible.³³ The result of the treaties and regularization of blood flow creates a kind of allied blood supply, organized and codified through NATO. Blood's fragility, combined with the demands of U.S. empire, has created a new racial-geographic bloodline. Safe and secure blood comes from an assemblage of bonded countries and populations held together by mutual defense, a vast interoperable medical infrastructure and surveillance network, and the exchange of sacred fluids. Despite blood's resistance to archaic racism, its circulation as a commodity has been institutionalized such that blood's collection, storage, and transport are organized around lines of enmity not so distinct from the racial lines of the previous century.

So, in part, the creation of the AFBP was driven by the failure to master blood and the vulnerability created by that lack of mastery. There was a strategic and mutative interaction with this medical institutional failure and the material constructions of new identities being shaped by the terrain of violent geopolitics. Further amplified by months of very high casualties in Korea, the AFBP became a permanent organization whose job it is to oversee and coordinate blood collection and the assessment of blood needs for each of the military branches. Further, the practical necessities of blood governmentality developed in tandem with new investments in authority and security as a form of governance animated by future threats rather than existential needs. The AFBP empowers the surgeon general of the United States to set the level of the strategic blood supply based on the assessment of the domestic and global threat level.

The shift from fulfilling practical demand to the logic and organization of security is most visible in the immense waste produced to overcome the short shelf life of blood. The AFBP mandates that five days' worth of blood, estimated using theater conditions, be stockpiled and resupplied in theater at all times.³⁴ In the second Iraq war, of the 11,250 gallons of blood shipped in the first year of the conflict, fewer than 212 gallons were used. That is a wastage rate of roughly 98 percent.³⁵ Blood is an interesting material measure of the logic of security, as it demonstrates the degree to which security's construction is built using more than words and discourse.

During the Kosovo conflict, for every 100 units of blood provided to the battlefield, fewer than 2 units were able to be used before the blood had to be disposed of.³⁶ Given comparatively few casualties, this rate of waste was sustainable. However, at the height of the second Iraq war, the U.S. military was consuming 10,000 donors' worth of blood every six months. In many cases, not only was this beyond what could be shipped over, but it was beyond what could be collected or even purchased within the United States. This is all despite significant advances in storage for red blood cells and other blood products that are used successfully in civilian settings. As a result, transfusions of fresh whole blood (FWB) were common during Operation Iraqi Freedom despite the high risks of disease transmission. This practice runs contrary to every basic public-health requirement for transfusions dating back to the 1920s, and yet it demonstrates the power of blood as an assemblage. FWB has been demonstrated to interrupt cycles of coagulopathy that are common when high amounts of frozen red blood cells or other blood products are used. The transfusion of whole blood also reduces cases of hypothermia and rebleeding that occur in trauma patients receiving large volumes of frozen blood or blood products.[37] However, FWB can be used only with proximate donors, like those of the earliest blood banks. In the 1920s, on-demand donors were called "blood on the hoof," but in the military context, such individuals are now referred to as "walking blood banks."[38]

Security's unquenchable demand for blood in its encounter with the limitations for controlling and sustaining blood stockpiles has created unique problems for the AFBP. The result is the necessity to institutionalize policy for the collection of blood in theater beyond the circle of friends created through NATO. Blood and security's peculiar future-oriented assemblage has further deterritorialized the imposed boundaries of enmity. As stated in *Planning for Health Service Support*, an army field manual, frozen blood is meant to support soldiers only while self-sustaining blood programs are set up. The field manual is explicit that in mature theaters, blood supply is based on fresh liquid red blood cells and fresh frozen plasma from the donor base.[39]

To achieve this often-impossible goal, host countries are assessed for their level of cooperation and supply, which in the case of Iraq and Afghanistan includes a population that ranges from the financially desperate, to friends, to enemies. Civilian and World Health Organization studies of blood supply and infrastructure, reviewed ostensibly for development goals and humanitarian aid, are funded by the ASBP to gather sufficient intelligence about disease and infrastructure. For example, an article in one of the leading blood journals, *Transfusion*, written by Farooq Mansoor and colleagues and titled "A National Mapping Assessment of Blood Collection and Transfusion Service Facilities in Afghanistan," details the location, quality, and supply liabilities for the entire nation of Afghanistan, province by province, not unlike a geological survey for oil, listing statistical variance of infectious disease and capacity limitations. The last section of the article is titled "Conflict of Interest." Under this heading reads,

"The authors declare no conflicts of interest," despite the fact that the article's research was fully funded by the Military Infectious Disease Research Program and the Armed Services Blood Program and was further reviewed by the Walter Reed Army Institute of Research for any classified or objectionable material.[40]

In addition to the assessment of supply, the field manual goes further to clarify the legal status of enemy blood and its use. According to the manual's interpretation of the Geneva Conventions, there is nothing restricting enemy prisoners of war from "donating blood." Given this, the field manual recommends that the blood types of prisoners ought to be collected and this information ought to be included with other intelligence regarding available resources. Further, blood captured from the enemy ought to be turned over to blood-bank platoons to be used in medical-treatment facilities.

The imposed lines of friend and enemy or the biopolitical redefinition of state "race" produced around the securing and threatening of whole populations is institutionally defined through blood treaties, but in conflict zones, blood exceeds even those boundaries. The need for blood and its generic quality supersedes the vital difference of the enemy even as it is often racialized or signified as blood. The U.S. military cannot help but make a policy that demonstrates that enmity is arbitrary in the face of actual blood that is the real substance of the enemy.

Enmity can, of course, be said to be merely political and therefore to occupy a different realm from biological considerations, but the political is often a decision on a scale of war that can be found only in the essence of a mass enemy. Without essence and identity, the contemporary techniques of war cannot take extreme risks with whole populations. At some level, "they" must be somehow fundamentally like the enemy that hides among them. At least "they" are sufficiently different from "us," because if they were not, how could we take such risks? The archaic blood myth cannot help but sneak back onto the battlefield, as war requires not objectification of the enemy but racialization of the enemy in the making of a population. Such racialization may take the form of treating those whom we fight as objects, but that only raises a more fundamental question: Why do we treat objects so badly? If one takes for a minute the position of blood, if one becomes the object rather than objectifying, it is impossible not to see the irony in practices of modern biopolitical warfare that is waged in the name of a population difference (race) by human bodies that seemingly have no problem sharing blood before and after they spill it.

Notes

1. *Planning for Health Service Support,* Field Manual 8-55, Department of the Army, September 9, 1994, http://armypubs.army.mil/doctrine/DR_pubs/dr_a/pdf/fm8_55.pdf.

2. John Hedley-Whyte and Debra R. Milamed, "Blood and War," *Ulster Medical Journal* 70 (2010): 125–34.

3. Lauren Berlant argues that we have to resist "cool facts" becoming "hot weapons in arguments about agency." I hope that these facts resist such deployment in their disruption of concepts of agency. See Lauren Berlant, *Cruel Optimism* (Durham, N.C.: Duke University Press, 2011), 101.

4. Stanislaw Lem, *Summa Technologiae* (Minneapolis: University of Minnesota Press, 2013), 5.

5. Jane Bennett, *Vibrant Matter: A Political Ecology of Things* (Durham, N.C.: Duke University Press, 2010), 2.

6. Ian Bogost, *Alien Phenomenology; or, What It's Like to Be a Thing* (Minneapolis: University of Minnesota Press, 2012), 24.

7. Timothy Morton, *Realist Magic: Objects, Ontology, Causality* (Ann Arbor, Mich.: Open Humanities Press, 2013), 44.

8. Ibid.

9. "Racial Integrity Act of 1924," *Wikisource,* http://en.wikisource.org/wiki/Racial_Integrity_Act_of_1924.

10. Hedley-Whyte and Milamed, "Blood and War," 132.

11. Douglas Starr, *Blood: An Epic History of Medicine and Commerce* (New York: Knopf, 1999), 31–32.

12. Hedley-Whyte and Milamed, "Blood and War," 132.

13. J. R. Hess and M. J. G. Thomas, "Blood Use in War and Disaster: Lessons from the Past Century," *Transfusion* 43, no. 11 (2003): 1622–33.

14. Starr, *Blood,* 60.

15. Ibid., 65–66.

16. Ibid., 71.

17. Charles Wynes, *Charles Richard Drew: The Man and the Myth* (Chicago: University of Illinois Press, 1988), 32.

18. Ibid.; Starr, *Blood*; "Jim Crow Blood Policy Laid to War Department," *Afro-American,* May 5, 1945.

19. Starr, *Blood,* 96–97; Wynes, *Charles Richard Drew.*

20. "Jim Crow Blood Policy."

21. Starr, *Blood,* 96–97; John Hess, "Blood Use in War and Disaster: The U.S. Experience," *Scandinavian Journal of Trauma and Resuscitation and Emergency Medicine* 13 (2005): 74–81.

22. Starr, *Blood,* 72.

23. Ibid.

24. Hess and Thomas, "Blood Use in War and Disaster," 1623.

25. Ibid.

26. Wynes, Charles. *Charles Richard Drew: The Man and the Myth* (Chicago: University of Illinois Press, 1988).

27. Starr, *Blood,* 108.

28. Michel Foucault, *Society Must Be Defended,* trans. David Macey (New York: Picador, 2003), 237.

29. Armed Services Blood Program, *Strategic Plan, 2009–2012,* 2009, http://www.militaryblood.dod.mil/articlearchive/ASBP_Strategic_Plan.pdf.

30. I do not want to suggest in any way that the lack of blood racism meant that the French were not racist. Instead, my point is that blood provided a bond that militated against forms of civilizational and colonial superiority in ways that American and, particularly, Nazi racism did not.

31. Starr, *Blood,* 136–37.

32. Ibid., xi.

33. *Armed Services Blood Program Readiness,* Office of the Inspector General, Department of Defense, Report No. D-2001-059, February 23, 2001; Assistant Secretary of Defense for Health Affairs, *Armed Services Blood Program (ASBP) Operational Procedures,* DODI-6480.4, August 5, 1996; *Planning for Health Service Support.*

34. Assistant Secretary of Defense for Health Affairs. *Armed Services Blood Program (ASBP) Operational Procedures.*

35. David Kauvar et al., "Fresh Whole Blood Transfusion: A Controversial Military Practice," *Journal of Trauma: Injury, Infection, and Critical Care* 61 (2006): 181–84.

36. Hess and Thomas, "Blood Use in War and Disaster," 1626.

37. Kauvar et al., "Fresh Whole Blood Transfusion," 182.

38. Starr, *Blood,* 53; Olle Berseus, Tor Hervig, and Jerard Seghatchian, "Military Walking Blood Bank and the Civilian Blood Service," *Transfusion and Apheresis Science* 46 (2012): 341.

39. *Planning for Health Service Support,* 3–4.

40. Farooq Mansoor et al., "A National Mapping Assessment of Blood Collection and Transfusion Service Facilities in Afghanistan," *Transfusion* 53, no. 1 (2012): 1 and 6.

Bodies

Lauren Wilcox

RPAs [remote-piloted aircraft] are now part of our DNA.
—Maj. Bryan Callahan

Over the past decade, the use of drones to target and kill suspected terrorists, insurgents, and, incidentally, thousands of civilians by the United States, United Kingdom, and Israel (many other states are acquiring the technological capacities) has resulted in debates over the politics and ethics of robotic warfare. Drone warfare as an artifact of the international is discussed elsewhere in this volume; in this chapter, I discuss drone warfare as an exemplar of how human bodies are made international. This approach may seem counterintuitive to many, as the use of sophisticated technological systems to launch weapons at suspected terrorists or militants from great distances (even from different continents) can seem like an especially disembodied form of war (at least for those who are not at risk from its missiles). Drone warfare illustrates a way to think the internationalization of bodies not as individual units impacted by violence, or global economic processes, but as the subjects' experiences of being bodies. Drone warfare is symptomatic of the body in late capitalism; it is neither flesh nor information but a hybrid that is best understood as an intersection or border of the material and the informational, a mode that is an enabling condition of the death-worlds of targeted killings. The use of drones pushes our thinking about agency and embodiment from the human to the posthuman and, more specifically, to the human body as an assemblage of organic, technological, and cultural materials and forces that are contingent on different environments and (re)combinations. Drone warfare, I argue, pushes us to think about how our bodies and worlds are mutually imbricated and of the political effects of the bodies we are becoming.

Most people are used to assuming that human bodies end where their skin ends; such a body is best thought of as the normative body, a body image that corresponds to able-bodied (and male) norms of embodiment.[1] Drone warfare, by contrast, is best theorized as an embodied assemblage. The human/drone assemblage extends this body's capacities for vision and violence far beyond its unaided capacities. Theorizing drone warfare as a posthuman assemblage focuses our attention on bodies as the radical materiality of subjects, as they exist in relation to each other and to assemblages of global violence. As parts of an assemblage, objects have no essence but are formed in processes. In this view, bodies are not entities endowed with particular characteristics but, with other bodies (human and nonhuman), entities that form assemblages with certain capacities. Bodies are not objects with a political existence in and of themselves but are unstable elements in a configuration that changes and transforms.

Even one's own body is not singular, without other forms of matter and lifeforms necessary to sustain its life. Our bodies are rich ecosystems, assemblages of multiple bodies.[2] N. Katherine Hayles argues that our bodies are, and have always been, posthuman, a mode that captures the experience of bodily life in the digital age:

> The posthuman view thinks of the body as the original prosthesis we all learn to manipulate, so that extending or replacing the body with other prostheses becomes a continuation of a process that began before we were born.... [T]he posthuman view configures human being so that it can be seamlessly articulated with intelligent machines. In the posthuman, there are no essential differences or absolute demarcations between bodily existence and computer simulation, cybernetic mechanisms and biological organism, robot teleology and human goals.... [T]he posthuman subject is an amalgam, a collection of heterogeneous components, a material informational entity whose boundaries undergo continuous construction and reconstruction.[3]

The human/technology interface is about extending not only the capabilities of the human body but also what psychologists refer to as the body schema: the sensory and motor capacities of bodies. In this view, drone warfare is corporealizing; that is, it establishes a mode of embodiment. The posthuman condition of embodiment is not limited to the addition of technology to an already-existing body. Such an understanding would reify the distinction between the individualized body (bounded by the skin) and technology as external. The posthuman perspective on embodiment posits that being a body is always already to be in relation to technologies, to the environment, and to language and cultural practices. The body is, first of all, enculturated, shaped and molded by culture and training. The body is defined by its skill in melding with technology, one such iteration of militarized masculinity that is shaped not only by technology but also by military culture and training, as well as its broader cultural training.[4]

Drone warfare is, in many respects, similar to the use of manned airplanes, bombers, and helicopters; the difference is that the piloting crew is located not in the aircraft but on the ground, from an air base that can be either nearby (as in bases in Afghanistan or Djibouti) or very far away (such as in Creech Air Force Base, outside of Las Vegas, Nevada, or six other air bases around the United States). The CIA controls drone missions in Pakistan, Somalia, and Yemen from its headquarters in Langley, Virginia, and from a network of secret bases.[5] The drone assemblage involves a pilot and a sensor operator (almost a copilot) working as a team, as well as a ground team that manages the takeoff and landing of drones and their maintenance. The pilot and sensor operator are located in small buildings about the size of trailers, surrounded by multiple screens and keyboards. The screens, displaying information provided by the drones, are also monitored by military command centers, in a separate location. Drone assemblages use artificial-intelligence technology to take over from humans some of the tasks of flying and targeting. MQ-9 Reapers can loiter in the air for twenty-four hours while transmitting video to their controllers, and can launch laser-guided missiles and up to fourteen Hellfire missiles, whereas MQ-1 Predators can carry two Hellfire missiles. The U.S. Air Force's drones are flown by fully trained pilots. Drone operators are taught to feel as though they were in the Predator, or other drone, itself,[6] and the cockpits in the drone assemblages are designed to approximate the design of the military aircraft that the pilots who currently fly drones are used to, in order to make use of their training by employing a kind of noncognitive knowledge.[7] Overall, humans and drones are incorporated into an assemblage in which cognitive and physical tasks are networked and distributed.

Technology changes how our bodies gather information; the technologies associated with drone warfare are no exception. Drone warfare particularly transforms the field of the visual. As an advance from the technology that allows bombs to be aimed with lasers or using coordinates from GPS satellites, drones can hover over potential targets for hours or days, providing images to the teams of drone operators and sometimes to on-the-ground combat units as well. Surveillance video from drones has become crucial to the military campaigns in Iraq, Afghanistan, and Pakistan, and recently it has been used in the United States itself. Army brigadier general Kevin Mangum said, "It's like crack, and everyone wants more" drone-supplied video.[8] The U.S. military currently has too much information from surveillance drones that may be useful in future operations but is not needed immediately. To solve this problem, the military is working with the Defense Advanced Research Projects Agency to build "machine–machine" tools that can automate the cameras and help process data.[9]

The operations of the drone assemblage are dependent upon a satellite or radio link from the piloting station to the aircraft. Such links make drones at risk for attack, sabotage, or being taken over and redirected.[10] The U.S. Defense Department has set a goal for humans to be "on the loop" versus "in the loop," meaning that humans will monitor and override if necessary, rather than control certain aspects directly.[11] The goal of intelligent drones is that they be fully autonomous in cases of air-to-air combat; in other words, that they develop the computational abilities to adapt to complex situations involving the goal of self-preservation without human interference.

Artificial intelligence is not just a programming feature of drone assemblages; it represents a fundamental shift in the conceptualization of the human body to a posthuman form of embodiment that is described in terms of code. Artificial intelligence has its origins in Norbert Weiner's efforts to develop a machine capable of calculating the movements of aircraft in flight as well as the complexities introduced by a human pilot—and capable of learning and evolving using this information. The U.S. government was and is heavily involved in the development of the computing and artificial-intelligence industries as means of solving their own war-fighting dilemmas. Artificial-intelligence researchers use a framework that first interprets human behavior to inspire machine design and then reinterprets human behavior in relation to machines.[12] Artificial intelligence and cybernetics not only imagine displacing the embodiment of intelligence but also image the human body as a kind of computer itself, capable of sending, receiving, and processing information as code. Computing and cybernetics remove the meaningfulness of human communication and replace it with information that can be manipulated outside of human embodiment as abstract patterns of communication and control.[13] The functions of the brain, for example, are imagined as information flows through bodies as pulses in neurons that fire just like binary code. In short, cybernetics was developed as a way to build better killing machines that ended up redefining and reorganizing the boundaries of the human body.

Cognitive science and neuroscience reject Cartesian mind/body dualism, theorizing the workings of human consciousness and cognition as a bodily effect. Nor are these features only part of the brain, the bodily organ most closely associated with the mind. Our bodies contribute to our cognitive capacities through feelings.[14] Cognitive processes are, in part, constituted by bodily movements in conjunction with environmental structures.[15] Emotional states are a response to changes in bodily states. William James wrote, "[T]he bodily changes that follow directly from the perception of the exciting fact and that our feeling of the same changes as they occur IS the emotion."[16] Such a view of the immediacy of emotion is backed up by contemporary neuroscience.[17] The mind and the human subject that the drone operator or operators supply are both embodied, and embodied in a way that incorporates other elements of the drone assemblage, such as vision technologies, the control panels, and the hardware and software of the drone itself.

As assemblages, posthuman bodies are formed not only in relation to other bodies and materials but also by information, which here encompasses norms and affective relations. *Affect* refers to preindividual or preconscious capacities to affect and be affected, abilities that are theorized not just in terms of the human body but also in bodies as mediated by various technologies and cultural formations.[18] In affect theory, the unconscious is not repressed or suppressed but is a bodily and perceptual capacity that is in excess to what can be consciously articulated.[19] In other words, bodies also have affective capacities that are not driven by conscious thought—capacities to affect and be affected beneath the layer of cognition.

The "killability" of targets and civilians is not only a matter of the extension of executive privilege to the creation of kill lists or other sovereign privileges, it is also a product of the bodily world inhabited by (post)humans. The posthuman assemblage of the drone operation is a necessary condition for these bodies to become killable in a sense that is at the same time physical and political. The machine-intelligence and -imaging capacities of drone assemblages are necessary for the individualization of human bodies (rather than towns or buildings) as targets. The destruction of human bodies by various forms of aerial bombardment is not new; what is distinctive about drone warfare and its various ancestors and cousins of precision-targeting techniques is the integration of the human and the technological in a way that not only augments the human (as if this category had a prior essential meaning) but also highlights the extent to which drone warfare is a particular configuration of embodiment, of bodies-in-formation.

Just as the category of human is transformed into information processor, the targets of drones are transformed into data patterns. This takes place literally, in the viewing of the bodies of targeted people through optical technologies that allow screens to zoom in for close-up views, even when the drones are fifteen thousand feet away and are controlled by teams seven thousand miles away. Human bodies are represented as images on a screen and as heat signatures at night.[20] The individualization of bodies is a new visual capability of drone assemblages; previous precision-targeting systems located buildings or military installations rather than individual human bodies. Before drones, a person living in an ungoverned or ungovernable territory was illegible and inaccessible by state or other sovereign intervention. The visual capacities of the drone assemblages, combined with their lethal weapons technologies and hovering capacities, make every human in the world a potential target as an individual rather than as a member of a particular population that may be bombed. Artificial intelligence is also used to decide which people exhibit a "pattern of life" consistent with militant activity. "Signature strikes" target people in Pakistan, Yemen, and Somalia based on the computer-aided analysis of their activities.[21] Here, the bodies are not transformed into information processors but are produced as information patterns or code. Even when the drone assemblages are working as they are designed, "accidents" are built into the assemblage, just as inaccuracy is built into the targeting systems themselves, which are considered accurate within a particular range, called the "circular error probability."[22] Even when the component of human error is removed, the bombs are still constructed on the principle of inaccuracy. Furthermore, neither the bomb apparatus nor communications technologies are instantaneous, given the distance the signals must travel as well as the time between releasing the missiles and impact; even when the system is operating as it should, there is a temporal gap in which people besides those targeted may come into a space being bombed, leading to even more civilian casualties.[23] The same production of (some) human bodies as information processors in drone assemblage is a necessary condition for the production of other bodies as data patterns and error probabilities in which the dead are not murder victims but regrettable, if unavoidable, casualties.

Although some people fear that the use of drones, robots, and artificial intelligence more generally in war will result in warfare's becoming automated and therefore somehow less humane, Hayles is less optimistic that the dream or nightmare of divorcing human intelligence from human embodiment is likely to come true, or is even possible. Despite the lack of violent injury or death facing drone pilots and operators, their bodies are still mired in and constituted by affective relations that are not, and cannot, be reduced to their enculturation as information processors. The ability of drone assemblages to reach out and touch the other makes one's body open to touch as well. Here I am not just referring to the capacity of drones to be shot from the sky or whose satellite or radio-communications systems make them vulnerable to sabotage or redirection; the posthuman embodiment of the drone assemblage enables violent relations of death and destruction. However, bodies that touch are, by definition, touched in return.

Affect includes bodily responses that one is not consciously aware of, such as autonomic responses. Affect is precisely about the irreducible materiality of embodied life. Pilots admit to flinching when something comes toward the drone[24] and sometimes report feeling less present in their own bodies and more present as the drone itself. The autonomy of affect, which Brian Massumi defines precisely as its openness and its unpredictability, suggests that, despite the abstract desire to rid warfare of vulnerable bodies (on one side, at least), the transformation of the human into information and machine and machinelike intelligence (thereby disembodying war) is incomplete. This is evident by the stress and even trauma that drone operators experience, which has lasting effects on their bodily assemblages. Drone operators shoot weapons at people and may see their victims disintegrate, burn, bleed, and die. The limits of the disembodiment thesis regarding artificial intelligence in war are shown by the bodily affects of the drone operators, suggesting that their bodies are becoming international. Phenomenological accounts of embodiment contend that we experience the boundaries of bodies only through interaction with the outside world. Affect theory points toward the materiality of embodiment, which is in excess of the norms placed upon bodies, such as, in this case, the norms of a seamless integration in a technological system of information retrieval and processing and death dealing.

The violent worlds produced by drone assemblages ripple back through emotions as bodily affect. The touch of the drone folds back onto the pilots with a new form of the bodily experience of war. A recent study has found that drone pilots experience similar levels of health problems such as depression, anxiety, and posttraumatic stress as pilots of manned aircraft deployed to Iraq and Afghanistan, although information about the experiences of CIA pilots is not available.[25] One reason given for these conditions is that, compared to operators of manned aircraft, drone operators have clearer and more sustained visual contact with those they are killing. Col. Chris Chambliss, who commands the 432nd Wing at Creech Air Force Base, outside Las Vegas, Nevada, describes the effects of drone warfare on the pilots: "You have a pretty good optical picture of the individuals on the ground. The images can be pretty graphic, pretty vivid, and those are the things we try to offset. We know that some folks have, in some cases, problems."[26] Drone assemblages create proximity, with operators sometimes watching people for months as they go about their daily lives. Through this, war and killing become intimate and personal.[27] Despite being thousands of miles from their targets, psychologically drone operators are in combat. Their rates of posttraumatic stress disorder (PTSD) may be lower than those of military personnel who participate more directly in combat and are more likely to witness deaths of enemies, civilians, or their fellow soldiers up close, but they are still affected. The embodied experience of drone operators suggests that, rather than emotionless video-game warfare, war-by-drone assemblages produce affective relations through emotional responses. When human bodies are combined into drone assemblages, they are not just information processors but experiencing, feeling beings.

PTSD, a controversial diagnosis, refers to a condition with three sets of affective symptoms: hyperarousal, a constant expectation of danger; intrusion, the lingering imprint of trauma on one's body and mind; and constriction, or the numbing of feelings.[28] The hyperarousal is a conditioned response to combat, which suggests that even though drone operators may be safe from the reciprocal violence that once defined warfare,[29] their bodies are still affected by combat. Studies also show that close connections between comrades are essential to adapting to the trauma of warfare; such connections are often lacking for drone operators who are not serving overseas but who go home to their families every night.[30] This disjuncture between the affect of war and the affect of home is thought to contribute to the symptoms of PTSD experienced by drone operators. "This is a different kind of war, but it's still war," Air Force colonel Hernando Ortega has said. "And they [drone operators] do internally feel it."[31]

This discussion of the effects of drone warfare on the military and intelligence personnel involved in perpetrating its violence is not intended to pose some kind of equivalency of suffering between operators and victims. The internationalization of bodies does not stop with the violent disintegration of some bodies and the possible traumatization of others; the affective relations radiating from the violence of posthuman bodies compels affective states in those who have lost loved ones in drone attacks and have seen their villages destroyed. Life for those living under drones is transformed. Aside from being at real risk from bombings, suffering the loss of family and friends as well as their homes and livelihoods, people living in areas affected by drone attacks are reported to suffer anxiety. A person who lost both his legs in a drone attacked said, "Everyone is scared all the time. When we're sitting together to have a meeting, we're scared there might be a strike. When you can hear the drone circling in the sky, you think it might strike you. We're always scared. We always have this fear in our head."[32] People who live under drones, especially those who have survived or witnessed drone attacks, suffer symptoms of anticipatory anxiety, insomnia, and PTSD, a state of embodiment created by the posthuman bodies of drone assemblages.

The distributed cognition and dynamic partnership between humans and machines represented by drone warfare is evidence not of the "dehumanization" of war but of warfare's redefining the human and reforming the worlds in which bodies relate. More research needs to be done on the affective aspects of other examples involving digital technologies and other modes of the transmission of affects, including the transmission of violence. Drone warfare is an instructive site for thinking about how bodies are international in a way that poses a challenge to a certain narrative in Western thought that links technology either with progress and utopianism or, more pointedly, with fascist tendencies. Posthumans, or human–technology assemblages, may be produced by and implicated in what Donna Haraway has referred to as the "informatics of domination,"[33] but we need not see only bleak futures of destruction in our internationally posthuman bodies. Our bodies as international posthuman assemblages are sets of potentials that enable multiple potential worlds of violence but that can yet be (re)written. The materiality of bodies, as a relational process, entails possibilities that cannot be known or anticipated in advance, and that transcend definitions that fix the meaning of materiality to constriction or determination.

Notes

1. Margrit Shildrick, *Embodying the Monster: Encounters with the Vulnerable Self* (London: Sage, 2002), and "Beyond the Body of Bioethics: Challenging the Conventions," in *Ethics of the Body: Postconventional Challenges,* ed. Margrit Shildrick and Roxanne Mykitiuk (Cambridge, Mass.: MIT Press, 2005), 1–28.

2. Erin Manning, *Politics of Touch: Sense, Movement, Sovereignty* (Minneapolis: University of Minnesota Press, 2007), 95.

3. N. Katherine Hayles, *How We Became Posthuman: Virtual Bodies in Cybernetics, Literature, and Informatics* (Chicago: University of Chicago Press, 1999), 3.

4. Terrell Carver, "The Machine in the Man," in *Rethinking the Man Question: Sex, Gender, and Violence in International Relations,* ed. Jane L. Parpart and Marysia Zalewski (London: Zed Books, 2008), 70–86.

5. Nicola Abé, "Dreams in Infrared: The Woes of an American Drone Operator," *Spiegel Online International*, December 14, 2012, http://www.spiegel.de/international/world/pain-continues-after-war-for-americandrone-pilot-a-872726.html.

6. Matt J. Martin, with Charles W. Sasser, *Predator: The Remote-Control Air War over Iraq and Afghanistan* (Minneapolis: Zenith Press, 2010).

7. Raytheon, "UCS: Advanced Multi-unmanned Aerial System's Cockpit," *Defense Update,* 2006, no. 2, http://defense-update.com/products/u/UCS.htm.

8. Noah Shachtman, "Army Preps 'Unblinking Eye' Airship for Afghanistan," *Wired,* June 17, 2010, http://www.wired.com/dangerroom/2010/06/army-preps-unblinking-eye-airship-for-afghanistan/.

9. Spencer Ackerman, "Air Force Chief: It'll Be 'Years' before We Catch Up on Drone Data," *Wired*, April 5, 2012, http://www.wired.com/dangerroom/2012/04/air-force-drone-data/.

10. Noel Sharkey, "Saying 'No!' to Lethal Autonomous Targeting," *Journal of Military Ethics* 9, no. 4 (2010): 369–83.

11. Ibid., 378.

12. N. Katherine Hayles, "Computing the Human," *Theory, Culture, and Society* 22, no. 1 (2005): 131–51.

13. Antoine Bousquet, *The Scientific Way of Warfare: Order and Chaos on the Battlefields of Modernity* (New York: Columbia University Press, 2009), 104.

14. Antonio Damasio, *Descartes' Error: Emotion, Reason, and the Human Brain* (New York: Penguin, 1994), 245.

15. Raymond Gibbs, *Embodiment and Cognitive Science* (Cambridge: Cambridge University Press, 2006), 12.

16. William James, *The Principles of Psychology* (New York: Dover, 1890), 449.

17. Antonio Damasio, *The Feeling of What Happens: Body and Emotion in the Making of Consciousness* (New York: Harcourt), 1999.

18. Patricia T. Clough, ed., *The Affective Turn: Theorizing the Social* (Durham, N.C.: Duke University Press, 2007); and Gregory J. Seigworth and Melissa Gregg, "An Inventory of Shimmers," in *The Affect Theory Reader*, ed. Melissa Gregg and Gregory J. Seigworth (Durham, N.C.: Duke University Press, 2010), 1–28.

19. Manning, *Politics of Touch*; and Brian Massumi, *Parables of the Virtual: Movement, Affect, Sensation* (Durham, N.C.: Duke University Press, 2002).

20. Martin, *Predator*, 52.

21. Greg Miller, "Increased U.S. Drone Strikes in Pakistan Killing Few High-Value Militants," *Washington Post*, February 21, 2011, http://www.washingtonpost.com/wp-dyn/content/article/2011/02/20/AR2011022002975.html.

22. Maja Zehfuss, "Targeting: Precision and the Production of Ethics," *European Journal of International Relations* 17, no. 3 (2011): 543–66; and Stephen L. McFarland, *America's Pursuit of Precision Bombing, 1910–1945* (Washington, D.C.: Smithsonian Institution Press, 1995).

23. Abé, "Dreams in Infrared."

24. Rob Blackhurst, "The Air Force Men Who Fly Drones in Afghanistan by Remote Control," *Telegraph* (London), September 24 2012, http://www.telegraph.co.uk/news/uknews/defence/9552547/The-air-force-men-who-fly-drones-in-Afghanistan-by-remote-control.html.

25. James Dao, "Drone Pilots are Found to Get Stress Disorders Much as Those in Combat Do," *New York Times*, February 22, 2013, http://www.nytimes.com/2013/02/23/us/drone-pilots-found-to-get-stress-disorders-much-as-those-in-combat-do.html?_r=0&pagewanted=print.

26. Chris Chambliss, quoted in Scott Lindlaw, "Drone Operators Suffer War Stress," *Military Times*, August 7, 2008, http://www.militarytimes.com/news/2008/08/ap_remote_stress_080708/7.

27. Abé, "Dreams in Infrared."

28. On the controversies of the PTSD diagnosis, see Alison Howell, *Madness in International Relations: Psychology, Security, and the Global Governance of Mental Health* (New York: Routledge, 2011).

29. Elaine Scarry, *The Body in Pain: The Making and Unmaking of the World* (New York: Oxford University Press, 1985), 85.

30. Joshua S. Goldstein, *War and Gender: How Gender Shapes the War System and Vice Versa* (Cambridge: Cambridge University Press, 2001); and Judith Herman, *Trauma and Recovery*, rev. ed. (New York: Basic Books, 1997).

31. Hernando Ortega, quoted in David Zucchino, "Stress of Combat Reaches Drone Crews," *Los Angeles Times*, March 18, 2012, http://articles.latimes.com/2012/mar/18/nation/la-na-drone-stress-20120318.

32. Stanford International Human Rights and Conflict Resolution Clinic and Global Justice Clinic at NYU School of Law, *Living under Drones: Death, Injury, and Trauma to Civilians from U.S. Drone Practices in Pakistan*, 2012, http://livingunderdrones.org/.

33. Donna Haraway, "A Cyborg Manifesto: Science, Technology, and Socialist-Feminism in the Late Twentieth Century," in *Simians, Cyborgs, and Women* (New York: Routledge, 2001), 149–82.

Tanks

Michael J. Shapiro

The Tank Object

In this chapter, I analyze tanks as objects that have been most familiar as rolling weaponized steel behemoths, killing machines that became the major land-based strategic devices in industrialized warfare. Like many military vehicles—for example, amphibious sea–land craft, which dissolved the boundary between the sea and the land—the tank changed the spatiality of warfare. It was a weapon that "broke over the static geography of trench warfare with all the force of a freakish sea crashing over land."[1] However, there are many other ways to read the tank-as-object. For example, since its invention as a military device that provides "mobility, protection, and firepower,"[2] the tank has become part of a worldwide exchange of recognition, a sign of vitality and prestige. As for its vitality, in a critical reflection on metal, Jane Bennett provides a perspective that can address that aspect of the tank. Aligning herself with a philosophical tradition that sees objects as vital rather than inert, she favors what she calls a "vital materialism [asserting that] there is no point of pure stillness, no indivisible atom that is not itself aquiver with virtual force."[3] Bennett's vital materialism is implicitly articulated in the experience of a protagonist in the Lebanese writer Elias Khoury's novel *Little Mountain,* about an Arab-Israeli war, which also speaks to the tank's vitality and semiotic effect. At one moment in the text, the character Saqr Quraysh describes his reaction to seeing a tank. Observing a Lebanese "relief army" appear to liberate his town from an Israeli invasion, he sees a tank "up close" and puts his hand on the metal of it, exclaiming, "Metal is such a pleasure!" He then reflects on the recognition aspect of the value of the tank-as-object. Speaking of the tank's sign-function value, he remarks, "A tank makes you hold your head up high and more."[4] Of course, in contrast with the positive sign the tank emits for those who possess it, it has a negative valence for its victims. Viewing the tank in the context of particular historical moments, one can observe its negative sign-function value—a notable example of which I came to appreciate while visiting the city of Prague in 2011.

During my visit to the Czech Republic, after a brief trip to the countryside, I returned to Prague, where my visit had begun. Leaving the train station on foot, I was unable initially to orient myself on my city map to find my way back to my hotel. However, after wandering around in narrow streets, I became oriented on my map when I found a broader street that gave me a view of the Saint Wenceslas statue at the top of Prague's main avenue, Vaclavske nam. The statue of the Czech saint is associated with allegiance to an early founding period involved in the historical consolidation of the nation's people. However, the street also references the period when Soviet tanks invaded on August 21, 1968, to quell a national uprising. The Czechs have left the pockmarks in the National Museum just above the statue (which resulted from the shots fired from the Soviet tanks' cannons) in order to commemorate the uprising.

Ironically, the tanks the Soviets used to repress the Czechs' attempt at independence were assembled in their former consolidated nation of Czechoslovakia. Those tanks-as-objects were containers of a Slovak vitality: "[I]n Slovakia tanks . . . constituted the livings of tens of thousands of people. From the late forties, Stalin turned Slovakia into a logistical zone dedicated to the supply and equipment of his expanding war machine. Vast steel and aluminum works were developed, along with a massive heavy armaments industry that manufactured Soviet-designed tanks and other weapons, both for the Warsaw Pact and also for export to liberation struggles, terrorist groups and pro-Soviet forces throughout the world."[5]

Tanks as Commodities

The tanks used to repress independence movements in the nation-states of the Warsaw Pact (Hungary and Czechoslovakia) have not disappeared. They have become commodities in a global arms-trafficking network. The purchasers emerge in a modern arms market rather than a Cold War market, but they buy Cold War–era tanks. An image from the feature film *Lord of War* (2005) references part of the story. The array of idle tanks in the background in one scene are no longer the implements of repression that they were during the period when the Soviets were crushing rebellions in the Eastern Bloc. They are being acquired and sold by the film's protagonist, the entrepreneur Yuri Orlov, a man with ideological indifference who represents a real-life counterpart, the notorious Russian arms trafficker Viktor Bout (now serving a twenty-five-year sentence in a U.S. prison).

The toppled statue of Lenin in the foreground of the scene, juxtaposed with the calculator that Orlov is using, speaks to the displacement of the ideologically inflected Cold War basis of global weapons deployment by entrepreneurial motivation. To situate the economic geography and the personae involved in such transfers, I turn to a passage in Mathias Énard's novel *Zone*, which refers obliquely to the global arms traffic. In the novel, Énard's protagonist, Mirković, a reformed member of a Croatian militia during the Balkan wars, is on a train headed to Rome from Milan. He has collected an archive of atrocities in "the zone" (the Mediterranean area), which he plans to sell to the Vatican. At one moment in his journey, he observes a group of nearby passengers, which he describes this way: "Egyptian, Lebanese, and Saudi businessmen all educated in the best British and American prep schools, discretely elegant, far from the clichés of colorful, rowdy Levantines, they were neither fat nor dressed up as Bedouins, they spoke calmly of the security of their future investments, as they said, they spoke of our dealings, of the region they called "the area," the zone, and the word "oil." . . . [S]ome had sold weapons to Croats in Bosnia, others to Muslims."[6] Elsewhere, I have responded to this passage by noting, "The map of cultural difference is being supplanted as persons from diverse national cultures become assimilated as predatory entrepreneurs in a world in which global capitalism is redrawing the map as it indulges its various clienteles, profiting for example from ethno-national enmities. As a result, warring bodies are treated not in terms of national and cultural allegiances but as human capital, as a clientele, and fates are determined by a network of initiatives in which violence is a consequence rather than an intention."[7] As Énard continues, "[O]ur businessmen from the zone didn't see the threat behind the outstretched hand, the deadly games that would play out in the course of the years to come."[8]

The "Egyptian, Lebanese, and Saudi businessmen" on Mirković's train in Énard's *Zone* have careers akin to that of Viktor Bout, "distinguished [in the eyes of 'officials' in Washington] not by cruelty or ruthlessness but by cunning amorality," who was sentenced to twenty-five years for arms trafficking in a New York court in February 2012.[9] A Russian entrepreneur who began his arms-trafficking business in the early 1990s, Bout made a remark about one of his first places of business that is pregnant with implications for understanding the tank-as-commodity, whose sales are made possible through a relationship between arms trafficking and international juridical space. Early in his arms-trafficking career, Bout lived in and operated from Sharjah in the United Arab Emirates ("a kind of postmodern caravansary"), because, as he told a reporter, it was a place with "practically no law."[10]

Bout's strategy took advantage of an uneven juridical global cartography in a way strikingly parallel to the way HBO's *Deadwood* series (three seasons on television, 2004–2006) begins by observing the nineteenth-century juridical cartography in the American West. In the first episode of season 1, a main protagonist, Seth Bullock, a marshal in the Montana Territory, decides, after dealing with the difficulties of law enforcement, to move to the town of Deadwood in the Dakota Territory, where, he says, "there's no law at all," to open a hardware store. It turns out that in the Deadwood of post-1876, legal apparatuses had yet to be institutionalized. Neither violence nor business ventures were significantly inhibited by law enforcement.

Doubtless, the "wild West" model still fits much of the global scene. Juridical space contains many "gray zones" (one of Viktor Bout's favored kinds of venue). Legally speaking (the literal meaning of *jurisdiction*), "war crimes are limited to situations of international armed conflict," and according to the 1949 Geneva Conventions, international tribunals can prosecute only individuals who perpetrate a "grave breach committed against a person protected by the Convention."[11] That excludes many people from the population of juridical subjects and leaves many amoral commercial predators free to supply the weapons that wreak havoc. With respect to commerce, many places—such as Sharjah, where Bout launched the airfreight business he ultimately used to transfer arms—exist outside of global trade-regulation laws and their implementation apparatuses.

The arms-sales enterprise of Viktor Bout has a legacy that includes not only individual entrepreneurs—for example, the early twentieth-century arms dealer Basil Zaharoff, a Turk who provided "the template for those who followed him"[12]—but also many states: "The U.S., Russia, France, Germany, Sweden, Holland, Italy, Israel and China are regularly identified as the largest producers and traders of weapons and materiel."[13] However, the consolidating term *states* is misleading, because the weapons they sell are products of multiple-state commodity production. For example, the "American-made" F-16 Fighting Falcon fighter plane, a global best seller, contains "high-tech components from Germany, Israel, Japan, and Russia," and its assembly is outsourced to various global sites. In general, as Ann Markusen points out, "[a]rms manufacturers are following the lead of their commercial counterparts and going global, pursuing transnational mergers, and marketing operations abroad."[14] As a result of the mergers, which bring together entities with different maps of friends versus foes, the clientele for sales emerge from diverse, often conflicting security topographies (adding to the disjunctures between security- and commodity-sales cartographies). As Sam Opondo and I have noted elsewhere, "There is a multiplicity of cartographies in the world, each predicated on a particular political imaginary. For example, there is the 'Pentagon's New Map,' which differentiates states on the basis of their American capitalism-friendly susceptibilities."[15] Moreover, as a result of the arms-production mergers, the Pentagon, a client for arms, has decreasing control over where "U.S." weapons end up. As Andrew Feinstein puts it, "The arms industry and its powerful political friends have forged a parallel political universe that largely insulates itself against the influence or judgment of others by invoking national security. This is a shadow world."[16] And as was the case with Viktor Bout's shadow world, that universe's victims are as likely to perish because of the pursuit of economic advantage as they are from structured interstate antagonisms—or at times from a combination of the two.

The Tank as Vision Machine: Remote Killing

Apart from the tank's value as a commodity, lives made precarious by its use result not only from how it arrives but also from how it "sees." The Lebanese novelist Hanan al-Shaykh captures the implications of the tank-as-vision-machine in her novel *Beirut Blues* through her protagonist, Asmahan's, epiphany when she enters a tank: "Now I understand why when they're in a tank, soldiers feel they can crush cars and trees in their paths like brambles, because they are disconnected from everything, their own souls and bodies included and what's left is this instrument of steel rolling majestically forward. I feel as if I've entered another world. . . . There is no window where we are, and the feeble light comes from a bulb, or filters through from the small windows in the driver's area."[17] What Asmahan observes is dramatically enacted in the Israeli director Samuel Maoz's film *Lebanon* (2009), which features an Israeli tank crew during the First Lebanon War (1982). Almost all of the film's shots depict either the interior of the tank or what can be seen from the tank.

As the film drama proceeds and the dialogue within the tank develops (the camera explores the crew's faces in a way that highlights their anxieties and lack of confidence in themselves and each other), it becomes clear that the tank crew is a proletarian workforce highly alienated from the commands of its superiors. Crew members bicker among themselves in response to the commands of their inexperienced leader, Assi.

Given the Geneva Protocols, targets should be a function of the tank as an "ethical killing machine." However, in practice, what constitutes ethics is not based on a list of moral imperatives from which rules of engagement are derived (to guide either persons or robots). Rather, a series of fraught negotiations within the tank and between the tank crew and the outer command structure determine what they will shoot at, as the crew responds to the threatening environments that their gaze anticipates while at the same time their ability to see is radically compromised by the technologies of their weapon. As I have noted elsewhere, "[M]orality as traditionally understood is about deriving imperatives from fixed moral codes, while ethical imperatives are invitations to negotiate meaning and value, given situations of either competing and incommensurate value commitments and/or alternative perspectives on what is the case."[18] Ultimately, to the extent that the tank's mission in *Lebanon* enacts an ethical inhibition in the crew's targeting, it is a result of the resistance of one individual within (Shmulik) to commands that he fire at almost anything he and the rest of the crew see.

To appreciate the negotiations within the tank in Maoz's film, we have to heed the complex assemblage of the crew. The tank crew can be regarded as a proletarian workforce that contrasts significantly with the protagonists in the preindustrialized warfare of the nineteenth century. Earlier war technologies were "manned" by a different social stratum that was resistant to the devices involved in the industrialization of warfare: "Ever since the days of Frederick the Great the military establishments of Europe, with the temporary exception of France, during the First Republic, had known a remarkable continuity, with fathers and sons and grandsons passing through the same regiments, and absorbing the same orthodoxy about the unchanging nature of war. Even worse, the hierarchical organization of the army made it into a gerontocracy, which ruthlessly discriminated any initiative or originality from below." And this hierarchy favored a "conception of warfare . . . firmly rooted in the past, in an age when the musket, bayonet and horse, particularly the latter, had been the decisive weapons on the battlefield," which they saw as a space in which the "individual officer and gentlemen counted for something." Hence, they resisted the introduction of the machine gun, for it "represented the very antithesis of this desperate faith in individual endeavor and courage."[19]

As is evident in *Lebanon*, the tank's reluctant crew members, engaged in tense discussion about their responsibilities, mirror some of the social and ethnic cleavages in Israeli society. A microcosm of Israeli society, they indicate through their conversations that they hail from different nonprofessional occupational segments. And although the mission commander, Jamil, has an Arab Israeli name, the members of the tank crew all have European ancestry. Their names—Shmulik, Hertzel, Assi, and Yigal—place them as Ashkenazim. And although the contentious negotiations in the tank have a significant determining effect on the targeting and the resulting killing of many civilians, the inexperienced tank gunner Shmulik undergoes a dramatic change in demeanor and outlook during the ongoing assault. Throughout the film, he is a becoming subject affected by the looks of others who return his targeting look. Ultimately, his looks back at the military gaze free him from the demands of his superiors. To put it conceptually, through Shmulik's transformation, rendered in the sequence of close-up shots of his face and through the returned looks of the war's Others—Lebanese civilian victims and a captured Syrian soldier who is taken as a prisoner into the tank—the film enacts a contrast between the antagonistic military gaze (and the looks that the gaze engenders) and "empathic vision," where the latter is an ethical concern with victims that develops in subjects who undergo perspective-disrupting encounters and are affected by what they see.[20] (I will return to this contrast in the conclusion of the chapter.)

Thus, although the discourses of enmity that have accompanied the tank crew's incursion into Lebanon derealize the singularities of the Lebanese they encounter, it is a series of face-to-face encounters that afflict the raw recruit Shmulik, turning him into a becoming subject, one whose orientation to the officially engendered antagonism of the conflict becomes radically altered. At the outset of the film, Shmulik is ordered to fire the tank's cannon at an oncoming passenger vehicle, a BMW, which turns out to contain attackers. Seeing a close-up of the driver on his scope, and never having fired at actual people, Shmulik panics, and the resulting firefight mortally wounds one of the Israeli soldiers.

As the incursion into a Lebanese town proceeds, Shmulik, seeking to atone for his earlier hesitation, fires on a truck that turns out to be driven by a civilian delivering chickens (the man ends up being mortally wounded with an arm blown off). He subsequently overcomes his inhibition again and fires his cannon into a building, thereby participating in assault-force atrocities that maim and kill noncombatants, many of whom are seen as bloodied and partly fragmented corpses. Tellingly, in addition to many close-up shots of Shmulik's horrified expression when he sees the victims through his scope, there are three moments when surviving Lebanese walk up to the tank and return the tank's gaze, looking directly into Shmulik's scope.

Ultimately, Maoz's film has Shmulik pass from perception to action, but what is most significant with respect to how the film narrates the changes in Shmulik is Maoz's use of what Gilles Deleuze calls "affection images" (registered in Shmulik's changing facial expressions). In a passage on film images that fits the way this film manages Shmulik's transformation, Deleuze writes, "Affection is what occupies the interval [between perception and action], what occupies it without filling it in or filling it up. It surges in the center of indetermination, that is to say in the subject, between a perception, which is troubling in certain respects and a hesitant action. It is a coincidence of subject and object, or the way in which the subject perceives itself, or rather experiences itself, feels itself 'from the inside.'"[21]

The Tank Passes Over into Robotic Warfare

Speculating about how the rules of engagement can be observed when the observing subject is a weapon, Peter Singer imagines an internal software checklist for the rules of engagement built into autonomous robotic weapons that he suggests might ethicize the robots: "Is the target a Soviet made T-80 tank? Identification confirmed. Is the target located in an authorized free-fire zone? Location confirmed. Are there any friendly units within a 200-meter radius? No friendlies detected. Are there any civilians within a 200-meter radius? No civilians detected. Weapons release authorized. No human command authority required."[22]

What's wrong with Singer's "ethical killing machine" fantasy? Samuel Maoz's film *Lebanon* supplies an answer. If we return to the inside of the tank in the film, we discover ways of seeing targets that involve contentious commands and reactive negotiations among the tank crew and an officer who occasionally joins the crew to organize the mission and to impose targeting protocols. The protocols, even when followed, are shown to render the tank anything but an ethical killing machine—partly because of the "fog of war," partly because of inadequate training, and partly because some of the commanders involved have no motivation to discriminate among targets.

Where will inhibition be situated now that tanks have been situated in a partially robotic man-machine assemblage? Now with info-war and the correlative "digitalization of the battlefield,"[23] the tank, like aircraft and missiles, participates in the derealization of its targets. The battlefield for the contemporary tank is a cinematic location in which its "warriors" experience their antagonists in the same way they experienced them when they underwent a simulation training, that is, when their targets were wholly depersonalized and dematerialized virtual enemies seen on computer screens. No longer merely a weaponized "heavy metal object,"[24] the tank carries mobile firepower directed by computer systems inside. As a result, instead of directly looking out, its crew has informational and visioning prostheses, and much of the decision making that leads to the use of the tank's firepower comes not only from deliberations within the tank's interior command structure but also from interpretive agency built into the tank computer's history archive as well as from remote observers utilizing additional perceptual and informational technologies. Among the implications of the increasingly robotic tank's way of seeing is an ambiguity with respect to the responsibility for war crimes and atrocities that such "man-machine" assemblages as the modern tank represent.

Of course, that issue will soon fade with the attenuation of the terrestrial battlefield. The weaponized drone has become a major strategic weapon and is now attracting most of the debate about the ethics of targeting. Nevertheless, despite how remote the targets are from human perceptions that—either in real time or as programmed—direct drone targeting, the militarized gaze is still being returned, at least indirectly.

I have evoked the concept of empathic vision in the analysis thus far. Now let me elaborate. An episode of empathic vision is a way of seeing that derives from an encounter that yields "an openness to a mode of existence or experience beyond what is known by the self."[25] The protagonist in the text I have analyzed, Shmulik in the film *Lebanon,* evinces that kind of vision in a way that distinguishes him from others in their combat units. He has to be affected enough to extract himself from the coercive force of the gaze that has directed his unit toward its targets. In the case of Shmulik, his empathic vision is provoked, first, by seeing civilian casualties, which leads him to resist his tank commander's orders to keep firing indiscriminately, and second, by the presence of a battered Syrian prisoner, whom his fellow soldiers have abused before shackling him and bringing him into Shmulik's tank. No one but a changed Shmulik heeds the man's requests to let him relieve himself. At the end of the film, we see Shmulik unzipping the man's fly, taking out his penis, and holding a bucket for the man to urinate in, all the while making eye contact with him.

As the direct presence of such victims continues to attenuate, more responsibility will fall to resistant artistic interventions to supply what automatic vision is effacing. Speaking of the empathic vision that the arts can contribute, Jill Bennett states, "Art makes a particular contribution to thought, and to politics specifically: how certain conjunctions of affective and critical operations might constitute the basis for something we can call *empathic vision.*"[26] As the arts are increasingly deployed against what Foucault famously calls the "truth weapons" that deny the atrocities associated with their use, they perform politics of aesthetics by "undoing the formatting of reality produced by state-controlled media, by undoing the relations between the visible, the sayable, and the thinkable."[27]

Notes

1. The quotation is from Patrick Wright, *Tank: The Progress of a Monstrous War Machine* (New York: Viking, 2002), 54.

2. Ibid., 17.

3. Jane Bennett, *Vibrant Matter* (Durham, N.C.: Duke University Press, 2010), 57.

4. Elias Khoury, *Little Mountain,* trans. Maia Tabet (Minneapolis: University of Minnesota Press, 1989), 60.

5. Wright, *Tank*, 386.
6. Mathias Énard, *Zone,* trans. Charlotte Mandell (Rochester, N.Y.: Open Letter, 2010), 21–22.
7. Michael J. Shapiro, *Studies in Trans-disciplinary Method: After the Aesthetic Turn* (London: Routledge, 2012), 78.
8. Énard, *Zone,* 23.
9. Nicholas Schmidle, "Disarming Viktor Bout," *New Yorker,* March 5, 2012, http://www.newyorker.com/magazine/2012/03/05/disarming-viktor-bout.
10. Ibid.
11. Christopher Rudolph, "Constructing an Atrocities Regime: The Politics of War Crimes Tribunals," *International Organization* 55, no. 3 (Summer 2001): 663.
12. Andrew Feinstein, *The Shadow World: Inside the Global Arms Trade* (New York: Farrar, Straus & Giroux, 2011), 3.
13. Ibid., xxiii.
14. Ann Markusen, "The Rise of World Weapons," *Foreign Policy* 114 (Spring 1999): 40.
15. Samson Okoth Opondo and Michael J. Shapiro, introduction to *The New Violent Cartography: Geo-analysis after the Aesthetic Turn,* ed. Samson Okoth Opondo and Michael J. Shapiro (London: Routledge, 2012), 1. See Thomas P. M. Barnett, *The Pentagon's New Map: War and Peace in the Twenty-First Century* (New York: Putnam, 2004).
16. Feinstein, *Shadow World,* xxvii.
17. Hanan al-Shaykh, *Beirut Blues,* trans. Catherine Cobham (New York: Anchor, 1995), 67.
18. Michael J. Shapiro, "Slow Looking: The Ethics and Politics of Aesthetics" (review essay), *Millennium* 37, no. 1 (2008).
19. The quotations are from John Ellis, *The Social History of the Machine Gun* (Baltimore: Johns Hopkins University Press, 1975), 175.
20. The expression "empathic vision" belongs to Jill Bennett; see her *Empathic Vision: Affect, Trauma, and Contemporary Art* (Stanford, Calif.: Stanford University Press, 2005).
21. Gilles Deleuze, *Cinema 1: The Movement-Image,* trans. Hugh Tomlinson and Barbara Habberjam (Minneapolis: University of Minnesota Press, 1986), 65.
22. Peter W. Singer, *Wired for War* (New York: Penguin, 2009), Kindle edition, location 7528.
23. Wright, *Tank*, 418.
24. Ibid., 431.
25. Bennett, *Empathic Vision*, 9.
26. Ibid., 21.
27. The quotation is from Jacques Rancière, *The Politics of Aesthetics,* trans. Gabriel Rockhill (New York: Continuum, 2004), 13.

Drones

Joseph Pugliese

When the *Washington Post* broke the story in 2012 that the "Obama administration ha[d] been secretly developing a new blueprint for pursuing terrorists, a next-generation targeting list called the 'disposition matrix,'"[1] there was barely a raised eyebrow, as this latest development was largely consumed as just another aspect of the inexorable ascendency of drones and their institutionalization within the United States' military arsenal. The disposition matrix is a massive database that expands the United States' interventionist role in the international arena well beyond the current drone kill lists managed by the U.S. government, the military, and the CIA. In this chapter, I examine how the United States' use of drones is redefining the concept of the international through the lens of this disposition matrix. If the Internet's conditions of historical emergence were firmly embedded within developments inextricably tied to the "military-industrial-media-entertainment network,"[2] then this relation is being further entrenched through the establishment of interoperable databases that work effectively to construct and enable killing assemblages predicated on unmanned military machines (drones) that kill at a distance.

In fleshing out the material resonances of the disposition matrix, I focus on how killing at a distance through drones has now congealed, precisely as disposition, into a habitual tendency undergirded by a military-industrial infrastructure or matrix. Pierre Bourdieu's concept of "durable dispositions" will be mobilized here in order to articulate the configuration of a "drone habitus"—as constituted by a series of habitual practices—at both local and international levels. In his theory of the habitus, Bourdieu brings into focus the manner in which sociocultural practices, values, and habits are essentially generated by a discursive matrix that, because it is immanent, is invisible to the very subjects of the habitus, even as it constitutes their dispositions to the world in which they live. In the course of this chapter, I draw upon Bourdieu's concept of the habitus precisely because it will enable me to map the otherwise invisible, because immanent, ensemble of elements that effectively constitute and reshape our understandings of the international as it is being crucially reshaped by the disposition matrix of drones. In the context of the United States, the drone habitus enables the affective disassociation of killer from killed. In the Pakistani context, for example, it radically redefines and circumscribes the very practices of everyday life, including performing nightly prayers during the holy month of Ramadan or attending a funeral. As what I have elsewhere termed the "prosthetics of empire,"[3] drones emerge as technologies of extraterritorialization that override international sovereignties, borders, and human-rights protocols and conventions. As objects enmeshed within the lived specificities of local spaces, drones at once reconfigure the practices of daily life, even as they establish new conceptualizations of the relation between the local and the international. I close the chapter by remarking on the unintended effects that the U.S. administration's use of drones is generating in the regions of drone strikes.

The use of unmanned aerial vehicles, or drones, for military purposes can be traced back to the First World War. It was during the period of the Vietnam War, however, that drones began to be deployed in an intensive manner for purposes of surveillance and reconnaissance. The war on terror that was unleashed by the United States soon after 9/11 marked both the arming of drones with missiles and the positioning of drones as the weapon of military choice in the conduct of this war. Military drones are unmanned planes equipped with sensors, cameras, and radar that can identify targets through smoke, fog, haze, and clouds. They are also equipped with laser-guidance technology and missiles; the laser designator in the nose of the plane locks onto a target and guides the trajectory of the missile once it has been fired. The drones are usually launched from a military base close to the theater of war, but they can actually be controlled from a ground-control station thousands of miles away. Many of the drones deployed in the war in Afghanistan are controlled, via satellite links, from U.S. Air Force bases located more than seven thousand miles away, in Nevada, New Mexico, or Virginia. From these ground-control stations, drones are controlled by a pilot, two sensor operators, and screeners—personnel with video-analysis expertise. The pilot navigates the plane while the two sensor operators control the plane's cameras and sensors, firing the drone's missiles when it locks onto a target. The decision as to who is designated as a legitimate target that can be killed by a drone strike hinges on a still-secretive, and thus controversial, process.[4] The approval of a target's name on a so-called "kill list" places the individual within the crosshairs of a drone.

The positioning of drones as the U.S. administration's weapon of choice in the conduct of the war on terror and the extraordinary expansion of the U.S. drone program across the globe underscore the manner in which this military object has assumed a normative status in the contemporary arsenal of war. In geopolitical terms, the United States is encompassing the globe with a constellation of what I have elsewhere termed "drone archipelagos."[5] The United States' imperial drone archipelago encompasses a range of key geopolitical sites that include Sicily, Afghanistan, Pakistan, Iraq, the Arabian Peninsula, Djibouti, Ethiopia, the Seychelles, Turkey, Uzbekistan, the Philippines, and Hawaii. This drone archipelago stretches from the Mediterranean across the Indian and Pacific Oceans to the Atlantic, with clusters of drone bases dotting the entire breadth of the U.S. mainland from Alaska to Florida. From its drone bases across the expanse of this militarized archipelago, the United States is conducting globalized shadow wars in central and western Asia, the Arabian Peninsula, and Africa. Situated in this geopolitical arena, drones emerge as the key weapon of choice for the globalized maintenance and extension of U.S. empire. In the words of the U.S. military, drones "provide global vigilance, global reach, and global power."[6] In this duplicative summation of the power of drones, the global is framed as coextensive with the U.S. state: the inside (the U.S. state) has encompassed its outside (the rest of the world). The United States speaks in the name of the global precisely because it now enfolds it as an extension of its sovereign domain. In one sense, we can now say that the material assemblage of drones has enabled the arrogation of international space by the U.S. state and its consequent enfolding within a sovereign domestic sphere that now overrides geographical limits and national sovereignties. Even as the U.S. administration has argued that the conduct of its various drone wars is in keeping with international law, Ben Emmerson, who is heading a United Nations investigation into deaths from U.S. drone strikes, states that the U.S. drone campaign "involves the use of force on the territory of another state without its consent and is therefore a violation of Pakistan's sovereignty."[7]

Situated in this globalized context, the U.S. government's development of a disposition matrix of drones signals the enmeshment of drones within a complex assemblage of military power, information networks, geopolitical spaces, and target subjects of the global South. Before I proceed to flesh out the manner in which the disposition matrix of drones effectively interpellates, constructs, and resignifies the international, I want briefly to focus on the Bourdieuian dimensions of this governing metaphor. A matrix signifies the conceptual and material infrastructure that constitutes an object's conditions of emergence and possibility. The matrix of drones is constituted by that ensemble of power, politics, economy, science, and technology otherwise known as the military-industrial-media-entertainment complex. The disposition matrix of drones is amplifying the killing capacity of military technologies by enmeshing them within networked relations of informatic power enabled by massive, interoperable databases. The disposition matrix of drones, as I will demonstrate, now enables a reconfiguring of the everyday practices of life of the subjects of the global South by placing them at risk of drone-driven injury and annihilation.

The manner in which these killing-at-a-distance practices have now been normalized by the U.S. state in the everyday conduct of its unending war on terror is succinctly captured by the qualifying term *disposition*. The terms *disposition* and *matrix* have profound Bourdieuian resonances; specifically, they evoke his foundational concept of the habitus. The disposition matrix of drones, as habitus, refers to an integrated assemblage of institutions of power, technologies, material sites, and embodied agents that enables and orients an ensemble of identifiable and reproducible practices. The habitus, precisely as matrix, signifies for Bourdieu "the durably installed generative principle" that enables the production and reproduction of a range of practices immanent within the logic and rules of the matrix.[8] The power of the habitus is defined, Bourdieu underscores, by the manner in which the practices that it engenders, both collectively and individually, are perdurable: the immanent laws, codes, and conventions that engender a series of definable practices assume the "form of durable dispositions" across different subjects, sites, and institutions. A habitus, Bourdieu writes, must be understood "as a system of lasting, transposable dispositions which, integrating past experiences, functions at every moment as a matrix of perceptions, appreciations, and actions and makes possible the achievement of infinitely diversified tasks."[9] Positioned in this Bourdieuian context, I want to term the geopolitical configuration of the drone habitus a "dronescape"— a material assemblage of technologies, subjects, spaces, and relations of power. In the context of dronescapes, I will now examine the manner in which the disposition matrix of drones generates a series of perceptions and actions that are critically resignifying the international.

In the heat of a summer's night, an Afghan couple make love on the roof of their house. Unbeknownst to them, this intimate, private, and domestic act is being captured by the surveillance cameras of a U.S. drone hovering high in the night sky. Uplinked via satellite and relayed to video screens situated at a drone ground-control station in the United States, their private act of love becomes an international event that will be scrutinized and analyzed by drone screeners who will adjudicate the intelligence value of this intimate moment. Captured by infrared cameras, their lovemaking is objectified and transmuted into a screen spectacle to be consumed by U.S. military personnel located on the other side of the globe. What should have been a private moment now becomes an international event enmeshed within telemediated relations of militarized power. "I saw them having sex with their wives," said Brandon

Bryant, a drone pilot located in a ground-control station in New Mexico. "It's two infrared spots becoming one."[10] Two living human beings become, once they are networked in the disposition matrix of the drone assemblage, disembodied digital spots. Reconstituted as nothing more than a pixelated cluster, they are officially classified as "patterns of life"[11] that, once they are inserted into the database of the disposition matrix of drones, will be assessed to determine whether they are "objects" of interest that need to be further surveilled or threats that might need to be liquidated via a drone strike. From the safety of their U.S. ground-control stations, drone pilots omnisciently surveil the lives, loves, and deaths of their targets. Bryant describes a post-drone-strike scene: "And after the smoke clears, there's a crater there. You can see body parts of the people. But the guy who was running from rear to front, his left leg had been taken off above the knee, and I watched him bleed out. The blood rapidly cooled to become the same color as the ground, because we were watching in infrared. Then I eventually watched the guy become the same color as the ground that he died on."[12] Everything here is described as if in cinematic terms: a slow-motion unfolding of dismemberment, trauma, and eventual death as telemediated spectacle. A human body is transmuted into nothing more than the infrared pattern of radiant heat; under the gaze of an objectifying lens, the body entropically declines into an inert object that is indistinguishable from the ground upon which the drone target dies. In the lethal context of the drone matrix, neither the wives nor children of suspects are safe. The wives of terrorists, indeed, have been termed by one U.S. academic as legitimate "drone bait" and their children as "terror spawn."[13] The CIA Counterterrorism Center's chief has boasted that, thanks to their drone-automated execution program, "We are killing these sons of bitches faster than they can grow them now."[14] In the words of a U.S. military officer, "Our major role is to sanitize the battlefield."[15] Sanitizing the battlefield via drone strikes includes the killing of the innocent children of suspects. The Bureau of Investigative Journalism has documented the drone killing of at least 204 children across Pakistan, Afghanistan, and Yemen.[16] Facilitating this sanitizing of the battlefield is the use, by drone pilots, of the term "bugsplat" to describe the victims of drone strikes: bugsplat effectively renders the human subjects who are liquidated by Hellfire missiles into nothing more than entomological waste.[17] In order to contest this reduction of civilian subjects by drone operators into inconsequential detritus, a Pakistani artists collective has installed a gigantic photo of a child who lost her family in a drone strike to face the drone surveillance cameras in the "heavily bombed Khyber Pukhtoonkwha region of Pakistan, where drone attacks regularly occur. Now, when viewed by a drone camera, what an operator sees on his screen is not an anonymous dot on the landscape, but an innocent child victim's face."[18] Constituted by thousands of pixels, this huge photo of a child has now been captured by both drone cameras and satellite imaging, thereby entering the surveillance circuits and databases of the U.S. military. Situated within these flows of information, this child's face speaks back to the systematic rendering of drone targets into faceless and disposable patterns of life.

Abdulrahman al-Awlaki was an innocent child killed by a drone strike. He was the son of the Islamist cleric Anwar al-Awlaki. Both father and son were U.S. citizens. Regardless of this status, Anwar al-Awlaki was never given the right to prove his innocence or to argue his case via due process of law. He was assassinated by a drone in Yemen. Two weeks after his father was killed, sixteen-year-old Abdulrahman, who had set off to find his father, was also killed by a drone, together with his cousin. Tom Junod, of *Esquire*, described the drone killing of Abdulrahman: "He was a boy who was still searching for his father when his father was killed. . . . He was a boy among boys eating dinner by an open fire along the side of a road when an American drone came out of the sky and fired the missiles that killed them all."[19] Once he was situated within the fraught locus of a dronescape, Abdulrahman was precluded from being merely "a boy among boys." Once he was caught within the crosshairs of the disposition matrix of drones, it was impossible for him to occupy any of those demographic categories that confer protection from military attack according to international laws of war: a minor, a civilian, a citizen, and so on. The disposition matrix of drones structurally disqualified him from embodying any of these noncombatant, nonmilitary descriptors. The disposition matrix of drones, and its taxonomies of life and death, inscribed Abdulrahman with the lethal descriptor of a MAM. In the U.S. drone lexicon, a MAM refers to a military-age male who has not necessarily proved to be military personnel. MAMs, as such, can be killed despite the fact that their identities are actually unknown and that there is no proven evidence of them having committed any terrorist acts. In their deployment of the MAM target descriptor, drone pilots are effectively killing acronyms rather than human targets. In the international landscapes of drone warfare—Afghanistan, Pakistan, and Yemen—the civilians of the global South are, once they have been processed by the drone matrix and recoded by its violently objectifying language, transmuted into "spots," "lumps," "squirters," "dismounts," and "patterns of life."[20]

"Patterns of life" that are deemed to be drone-killable are classified under "signature strikes" (in contradistinction to "personality strikes," where the names of the targets are known). The devastating effects on civilians of anonymous signature strikes have been well documented. A report called *The*

Civilian Impact of Drones draws attention to one particularly disturbing case: "US experiences in Afghanistan illustrate the risks of targeting with limited cultural and contextual awareness. On February 21, 2010, a large group of men set out to travel in convoy. They had various destinations, but as they had to pass through the insurgent stronghold of Uruzgan province, they decided to travel together so that if one vehicle broke down, the others could help. From the surveillance of a Predator, US forces came to believe that the group was Taliban."[21] The group was subsequently branded with the imprimatur MAMs, with its attendant right to kill:

> As described by an Army officer who was involved: "We all had it in our head, 'Hey, why do you have 20 military age males [MAMs] at 5 a.m. collecting each other?' . . . There can only be one reason, and that's because we've put [US troops] in the area." The US forces proceeded to interpret the unfolding events in accordance with their belief that the convoy was full of insurgents. Evidence of the presence of children became evidence of "adolescents," unconfirmed suspicions of the presence of weapons turned into an assumption of their presence. The US fired on the convoy, killing 23 people.[22]

The category of civilian, sanctioned by international laws of war to protect nonmilitary subjects from attack, is rendered redundant by the drone matrix. Civilian homes, once they are situated in the crosshairs of drones, are stripped of their civilian status: they become noncivilian "structures" or "buildings" rather than "houses."[23] Civilian women, children, and men are precluded from inhabiting the potentially lifesaving category that they in fact embody. The disposition matrix of drones suspends this category. Afghan, Pakistani, and Yemeni civilians consequently live and die under the sign of its violent elision. They are enabled to embody the category of civilians only retrospectively; after they become the victims of militarized violence, the military will term them CIVCAS (civilian casualties): "It [the Obama administration] in effect counts all military-age males in a strike zone as combatants, according to several administration officials, unless there is explicit intelligence posthumously proving them innocent."[24] This is the prerogative of empire: to kill first and only afterward, if at all, to bother to find out whether those killed were innocent. These are the disposable deaths of empire that fail to figure in the U.S. administration's modest and sanitized body counts, as the identities of the dead often remain unknown. "They count the dead," says one Obama official, "and they're not really sure who they are."[25]

Drone pilot Brandon Bryant recalls the drone killing of a child whose identity remains unknown:

> There was a flat-roofed house made of mud, with a shed to hold goats in the [drone's] crosshairs. . . . When he [the soldier controlling the drone] received the order to fire, he pressed a button with his left hand and marked a laser. The pilot sitting next to him pressed the trigger on a joystick, causing the drone to launch a Hellfire missile. There were 16 seconds until impact. . . . Suddenly a child walked around the corner. . . . Bryant saw a flash on the screen: the explosion. Parts of the building collapsed. The child had disappeared. Bryant had a sick feeling in his stomach.
> "Did we just kill a kid?" he asked the man sitting next to him.
> "Yeah, I guess that was a kid," the pilot replied.
> "Was that a kid?" they wrote in a chat window monitor.
> Then, someone they didn't know answered, someone sitting in a military command center somewhere in the world who had observed the attack. "No, That was a dog," the person wrote.
> They reviewed the scene on video. A dog on two legs?[26]

Inserted within the drone matrix, the death of an innocent child is internationally networked across multiple screens, only to be designated an inconsequential "information" event equivalent to the death of a dog.

In October 2006, a drone strike on a madrassa (Muslim seminary) in Bajaur Agency of Pakistan's federally administered tribal areas (FATAs) killed sixty-nine children between the ages of seven and seventeen.[27] An examination of the list of names and ages of children killed by U.S. drone strikes includes infants who are only one or two years old.[28] One of the key principles of international laws of war is discrimination: "Discrimination has always been the principle of the law of armed conflict that has been most respected by belligerents in war because it is the least ambiguous principle. The Geneva Convention calls for the clear division of all people and targets into two main categories: soldiers/combatants and civilians/non-combatants, or targets and non-targets."[29] If some of the drone-killing of children can be categorized under the problematic (because it is objectifying and sanitizing) category of collateral damage, the killing of sixty-nine children by a drone strike on a madrassa cannot be so easily relegated to accident or mishap. The drone killing of sixty-nine children in a school suggests that the principles of discrimination and proportionality can be violated with impunity by the U.S. military and that drone technologies are effectively rendering key aspects of international law irrelevant.

The objectifying effects of drone-screen technologies are further amplified by the drone operators' use of terms that render the human victims of drones into nonhuman animals: "dogs," "bugsplat," "bees," and so on. Drone operators jokingly call survivors who flee a drone strike "squirters," because the visceral terror of the experience may cause them involuntarily to urinate. The disposition matrix of drones is predicated on a hierarchy of life that precludes the subjects of the global South from inhabiting the position of human-rights-bearing subject. Caught in the crosshairs of drones, they become nonhuman biological matter that is at once disposable and killable.

Drones, in effect, can be seen to reconfigure the international along unequal relations of power that divide along global North and South axes: observer/observed, human/animal, sovereign/subjugated. The drone matrix renders the lives of the targets of the global South open to unbounded surveillance. The private lives and everyday practices of the subjects of the global South's dronescapes are televisually captured by drone cameras and relayed across multiple U.S. screens and databases. "We watch people for months," says Col. William Tart. "We see them playing with their dogs or doing their laundry. We know their patterns like we know our neighbors' patterns. We even go to their funerals."[30] This knowing of "their patterns like we know our neighbors' patterns" cuts only one way. The configuration of militarized power enabled by drones is marked precisely by its asymmetry: for the drone subjects of the global South, there is no equal right of reply to being the observed target; there is no due process of law available to them to clear their suspect status; and there is no procedural recourse to law once their lives have been deemed killable. In the context of a recent report by Stanford University and New York University, *Living under Drones,* Col. Tart's observation "We even go to their funerals" assumes particularly chilling dimensions. The report documents the manner in which "[i]nterviewees stated that the US drone campaign has undermined the cultural and religious practices . . . related to burial, and made family members afraid to attend funerals" because of the repeated strikes by drones of funeral processions.[31]

In his mediation on the life of things, Bruno Latour writes, "Objects become things . . . when matters of fact give way to their complicated entanglements and become matters of concern."[32] Drones, once they are situated and inextricably entangled in the everyday lives of the subjects of the global South, must be viewed as matters of concern. The entanglement of drones in the spaces and lives of the subjects of both the global South and the global North is effectively producing a new international formation that, as mentioned earlier, I have termed "dronescapes." Dronescapes are profoundly reshaping civilian and military lives, practices, and rituals. In what follows, I want to examine the attributes of dronescapes as they shape the lives of the subjects of the global South and North. Viewed within dronescape formations, as entangled ensembles of spaces, technologies, and human subjects, drones emerge not as mere technological effect of human cause but as actors instrumental in the very processes of shaping, conditioning, and producing new local and international spatial relations, subjectivities, and cultural practices. I use *actor* here in Latour's extended sense: "[T]hings might authorize, allow, afford, encourage, permit, suggest, influence, block, render possible, forbid and so on, in addition to 'determining' and serving as a backdrop to human action."[33] Complicating and attenuating reductive conceptualizations of cause and effect, Latour locates agency in nonhuman objects and things: "[A]nything that modifies a state of affairs by making a difference is an actor."[34] Taking my cue from Latour, I want to map the various ways in which drones, in their respective North and South dronescapes, are authorizing, blocking, rendering possible, and forbidding, and are thereby constituting a series of significant transformations of the local once it is situated within international relations of drone-enabled power.

In the global South, the habitus of a dronescape is one shaped by a disposition to fear, injury, threat, and imminent death. In the context of a dronescape, over which drones hover seven days a week, twenty-four hours a day, the civilian subject of the global South is afraid to attend a funeral, go to school, or attend the mosque and offer prayers: "Gul Wazir Dawar, a resident of Ghulam Khan town [Pakistan]" recounted in 2012 how he had stopped performing prayers outdoors because of his fear of drone attacks. "'I feel ashamed that I am not going to offer prayers just because of death fear. . . . Drones cannot figure out whether it's a militant's meeting or a religious congregation. They just fire missiles,' he said, referring to various misguided attacks."[35] A farmer from Dattakhel, Pakistan, exclaimed, "We are not safe anywhere. . . . Death can catch us anytime anywhere."[36] A young Afghan detailed the effects of drones on his family life: "When I'm home and study at night, my father and mother are very worried and tell me not to stay up too late because they may make a mistake and bomb the house. When my younger brother knows of a drone incident, he says he won't go to school or get out of bed early because the drones may come."[37] The children's fear that their school might be struck by a drone missile is driven by the fact that at least ten drone bombings of schools or former schools have thus far been documented.[38] In the locus of the dronescape, the material threat posed by a hovering drone, or the anticipation of a strike by a drone, produces what might be best called a generalized climate of fear that inflects, shapes, and disables the civilian practices of everyday life that are elsewhere taken for granted. UNICEF, for example, "reports that in the FATA region, where many drone strikes occur, 450 community health centers were closed by the government in 2010 due to the unwillingness of personnel to work in the region."[39] In North Waziristan, a college student described how "[b]ecause of these drones, people have stopped coming or going to the bazaars.... [I]t has affected trade to Afghanistan."[40]

The disabling of civilian practices and rituals within the dronescapes of the global South raises a number of other questions that pertain not only to the reconfiguring of daily lives but also to the violent, asymmetrical relations of power that determine what constitutes an act of terrorism within the international domain. On March 17, 2011, a U.S. drone fired missiles into a jirga, a traditional gathering of elders and community leaders that is viewed as a "principal institution for decision-making and dispute resolution in FATA."[41] Contrary to claims by the U.S. administration that only insurgents were killed by this drone strike, the "evidence suggests that at least 42 were killed, mostly civilians, and another 14 were injured." Eyewitnesses described the post-drone-strike site: "[E]verything was devastated. There were pieces—body pieces—lying around. There was lots of flesh and blood." A survivor explained that the "funerals for the victims of the March 17 strike were 'odd and different than before.' The community had to collect [the victims'] 'body pieces and bones and then bury them like that.'"[42] These drone-strike survivor accounts closely parallel the descriptions of the devastation left by terrorists and suicide bombers who deploy planes as missiles or detonate bombs within civilian sites in such Western cities as New York and London. What remains after a drone or suicide-bomber strike is a field littered with virtually unrecognizable human remains. The question arises, then, how does a drone strike of a civilian gathering such as a jirga, and the resultant horror of shredded flesh and mutilated bodies, ethically differ from the terrorist action of a suicide bomber?

Haunting the U.S. drone killing of civilians in the trammeled dronescapes of the global South is the specter of the very thing that is driving the killings in the first instance: terrorism, a practice either formally legitimated by law and recoded as legitimate violence or entirely beyond the law but, in either case, devoid of ethics. What I discern in this double scene of killing violence is the twin exercise of retributive justice devoid of ethics: one, drone-driven and enabled by the legitimating imprimatur of law; the other, suicide-bomber-enacted and categorically outlawed in its status. Both acts of killing violence, despite the fact that they are differently motivated by the concept of retributive justice—respectively, for the 9/11 attacks and for the killing of civilians in the Iraq and Afghan wars—undermine the very notion of justice, precisely when justice is understood in its profound ethicophilosophical conceptualization as indissociable from ethics. The inequality of law and justice in the killing violence of drones underscores the manner in which law, once it has been emptied of its ethicality, may, in Derridean terms, "retain its rights (which . . . I distinguish from justice), the right to its rights, but it loses justice. Along with the right to speak of justice in any credible way."[43]

If the dronescapes of the global South are characterized precisely by the difficulty of embodying the subject position of civilian and the radical foreclosure of the range of civilian practices and rituals that can be conducted because of the arbitrary risk of death and mutilation, then a significantly different dronescape emerges in the context of the United States and its drone ground-control stations. In the U.S. context, the dronescape is tied to U.S. soil even as it is geographically unfettered because of the networked use of screen technologies and satellite links. The mediating effects of drone-screen technologies transmute killing into the stuff of video games, generating a type of causal disconnect between, and consequent disavowal of, the human operators' relation to the killing that transpires on the ground in remote Afghanistan or Pakistan. The following remarks by Predator drone operators located at a ground-control station in Nevada exemplify this spatiotemporal disconnect: "It's antiseptic. It's not as potent an emotion as being on the battlefield"; "It's like a video game. It can get a little bloodthirsty. But it's fucking cool"; "Most of the time, I get to fight the war, and go home and see the wife and kids at night."[44] "You have some guy sitting at Nellis [Air Force Base, near Las Vegas] and he's taking his kid to soccer. It's a strange dichotomy of war."[45] This strange dichotomy of war is enabled by a parenthetical logic that brackets off causal relations through a series of teletechnomediations that, in turn, transmute the real into simulacrum.

The ensconcing of war operations and the everyday deployment of lethal drone attacks within U.S. cities such as Langley, Virginia, gesture to a mutation in the conduct of war, a transformation of the figure of the military combatant, and a reconfiguration of the international due to the blurring of the lines between the space of the battlefield and the civilian locus of home. The manner in which drone operators can exterminate human targets during their assigned combat sessions, via their ensemble of telemediating technologies and military hardware, and then go home to take the kids to soccer or have a drink at their local bar normalizes war as something that is effectively part of the civilian continuum of everyday life practices. This continuum of practices is facilitated by the euphemisms of war: the screen medium that displays the atomization and incineration of bodies by drone missiles is called by the military "kill TV";

the formal guide to targeted drone killing is described by U.S. officials as a "counterterrorism 'playbook'";[46] the material violence inflicted on human targets becomes merely "kinetic activity," as though killing were just another form of gym exercise; and the human targets of drones are reduced to yet another abstract kinetic term: dismounts. The term *dismounts* effectively abstracts the human drone targets from the embodied materiality of their subjecthood: they are objectified and reduced to mere trackable movements without flesh-and-blood bodies. The U.S. military's videos of drone kills have been released to the public for consumption as a type of civilian kill TV. Drone-attack videos have, in fact, become YouTube hits. Dislocated from the harrowed ground of an actual drone strike, disembodied from the bloody victims of such a strike, and producing a strike that is rendered generically interchangeable with a conventional video-game sequence because of the telemediated nature of the medium, a drone becomes little more than yet another commodity circulating within international networks of visual consumption. Situated in this telemediated drone context, the singular death by drone of a subject in Pakistan becomes a screen spectacle networked across international monitors, an event of televisual spectrality that ensures, through endless reiterations, that a singular death becomes a type of revenant emptied of the material finality of death as such: it becomes a death that, through endless rescreenings, will not die.

The parenthetical logic that underpins the video-game dimensions of drone killings helps facilitate the transition from exterminatory combat operations to civilian sites and practices. In the words of an air force colonel of a Predator drone squadron, "It teaches you how to compartmentalize it [the reality of war]."[47] The everyday returns to civilian locations of home after a series of technologically mediated killings in another country can be seen to be inscribed by the forces of technological (dis)location that drive the operations of drones. Drone operators have remarked on how the trip home from their ground-control stations enables them to transition from battlefield to civilian mode; the hour's drive back to their homes gives them "that whole amount of time to leave it behind. They get in their bus or car and go into a zone—they say, 'For the next hour I'm decompressing, I'm getting re-engaged into what's it's like to be a civilian.'"[48] This return to a safe home is the privilege and prerogative of the drone-enabled resident-soldier of the global North. In the dronescapes of the global South, the at-home is open to the anomic violence of drones and the ever-present risk of obliteration of home, friends, and family. The militarization of civil space in the dronescapes of the global South establishes the impossibility of living the civil—the safe space for the unfolding of quotidian lives. Seemingly banal civil practices such as using a mobile phone (which a drone detects and can use as a tracking device) or assembling for a community gathering (such as a funeral) place their subjects at risk of being killed. In the fraught spaces of the dronescapes of the global South, even basic humanitarian practices such as helping to rescue the survivors of drone strikes have been rendered impossible, as the U.S. military is now deploying what it calls "double-tap" or "follow-up" strikes: after the initial drone strike, a drone returns after a brief respite in order to fire on the rescuers who have come to the aid of the victims of the first strike. Journalist Noor Behram, who has been covering the drone strikes, related that, as a result of double-tap strikes, "what the tribals do, they don't want many people going to the strike areas. Only three or four willing people who know that if they go, they are going to die, only they go in. . . . [O]nce there has been a drone attack, people have gone in for rescue missions, and five or ten minutes after the drone attack, they attack the rescuers."[49] Hayatullah Ayoub Khan recounted how he was driving between Dossali and Tal in North Waziristan when he witnessed a drone strike three hundred meters in front of his own car:

> Hayatullah stopped, got out of his car, and slowly approached the wreckage, debating whether he should help the injured and risk being the victim of a follow-up strike. He stated that he got close enough to see an arm moving inside the wrecked vehicle, someone inside yelled that he should leave immediately because another missile would likely strike. He started to return to his car and a second missile hit the damaged car and killed whoever was still moving inside. He told us that nearby villagers waited another twenty minutes before removing the bodies, which included the body of a teacher from Hayatullah's village.[50]

One humanitarian organization has consequently developed a "policy to not go immediately [to a reported drone strike] because of follow-up strikes. There is a six hour mandatory delay."[51] The practice of killing rescuers violates international humanitarian law. In the words of Christof Heyns, United Nations special rapporteur on extrajudicial, summary, or arbitrary executions, "[T]argeting rescuers is a war crime."[52] In the volatile context of dronescapes, a lethal climate of arbitrary and indiscriminate violence, unbounded fear, and paranoia has impacted the ability of humanitarian workers to deliver various forms of aid to drone-affected areas.[53] This climate of fear and paranoia has been amplified by the United States' issuing of bounties for high-value drone targets and the payment of local accomplices to install transmitter chips in Taliban and al-Qaeda houses "signalling specific targets for CIA strikes." One captured accomplice has "'confessed' . . . 'to throwing chips all over' because the money was so good. The story bred fear and suspicion throughout Waziristan, where residents are 'gripped by rumours that paid CIA informants have been planting tiny silicon-chip homing devices' that attract the drones."[54]

In keeping with the complex ways in which the effects that objects produce once they are deployed in the social field can never be entirely predicted or controlled, even as drones are being used by the U.S. administration in order to liquidate terror suspects and networks, drones appear to be fomenting the growth of new terror cells and recruits. One of the unintended effects of the American drone campaign has been, because of the killing of innocent civilians, the radicalization of communities that are drone targets, which has led to the easier recruitment of anti-American militants and their supporters. Extremist websites operating from the drone-strike regions, such as Yemen, post horrific "pictures of the attack's aftermath, with bodies tossed like rag dolls on the road . . . coupled with condemnations of the government and the United States. In Sabool and Radda, youths have vowed to join al-Qaeda to fight the United States."[55] "The drone war is failing," explains Abdul Rahman Berman, executive director of Yemen's National Organization for Defending Rights and Freedoms. "If the Americans kill 10, al-Qaeda will recruit 100."[56]

In his conceptualization of the material object as charged actor that "gathers around itself a different assembly of relevant parties" and that thereby "trigger[s] new occasions to passionately differ and dispute," Latour brings into sharp focus the agentic power of the object once it is situated within social space.[57] In the contemporary context, drones have mobilized two radically different assemblies of relevant parties caught in a passionate dispute over the legitimacy of the use of this technology of war. On the one hand, the U.S. administration celebrates the drone's surgical precision, cost-effectiveness, and the manner in which it enables the saving of the lives of U.S. soldiers because they no longer need to be deployed in the field of war. "Call it 'pain free' military action," remarked one U.S. drone pilot. "Except, of course, not for the enemy."[58] On the other hand, the regions of the global South that are experiencing the relentless and escalating unfolding of U.S. drone strikes view drones as war machines that violate international law in their indiscriminate killing of civilians and that, in their around-the-clock hovering over the towns and villages of targeted areas, are destroying the very fabric of civilian lives in their fomenting of fear and terror.

The two very different dronescapes that I have outlined in this chapter are critically reshaping the international. The seemingly most remote site—a tribal village in the FATA of Pakistan—is, through drones, satellites, and screen technologies, catapulted by telemediated networks of relations to U.S. metropolitan centers. The most private and intimate of acts—for example, lovemaking—are caught by the cameras of a hovering drone and enmeshed within a military database in order to become an international event to be scrutinized by drone screeners for any intelligence value. What is evidenced here is a technologically enabled reconfiguration of international relations that enmeshes drone subjects within a disposition matrix that topologically conjoins the everyday lives of subjects living in Afghanistan, Yemen, or Pakistan to military drone installations that are firmly rooted within U.S. urban and civilian sites. Strikingly different in the types of lives that can be lived within the contours of global North and South dronescapes, these new spaces bring into focus the manner in which drone-enabled military power is critically reshaping civilian lives so that the borders between civilian and military life have now become blurred and attenuated. In the contemporary context, the drone must be viewed as a key actor in the resignifying of the political within the larger context of international relations. The technology of the drone is critically reshaping international configurations of civilian and military lives precisely as it is overriding long-standing international laws of war and human rights and conventions.

Notes

I am profoundly grateful to Constance Owen for her brilliant research assistance.

1. Greg Miller, "Plan for Hunting Terrorists Signals U.S. Intends to Keep Adding Names to Kill Lists," *Washington Post*, October 24, 2012, http://www.washingtonpost.com/world/national-security/plan-for-hunting/lists/2012/10/23/html.

2. James Der Derian, *Virtuous War: Mapping the Military-Industrial-Media-Entertainment Network* (Boulder, Colo.: Westview, 2001), xi.

3. Joseph Pugliese, *State Violence and the Execution of Law: Biopolitical Caesurae of Torture, Black Sites, Drones* (Abingdon, Eng.: Routledge, 2013), 184–85.

4. See Dana Priest and William M. Arkin, *Top Secret America: The Rise of the New American Security State* (New York: Little, Brown, 2011), 204–8.

5. Ibid., 217.

6. United States Air Force Scientific Advisory Board, *Report on Operating Next-Generation Remotely Piloted Aircraft for Irregular Warfare*, SAB-TR-10-03, April 2011, http://info.publicintelligence.net/USAF-RemoteIrregularWarfare.pdf.

7. Ben Emmerson, quoted in Associated Press, "UN Says US Drones Violate Pakistan's Sovereignty," March 15, 2013, http://www.google.com/hostednews/ap/article/ALeqM5hOhDYlNogHG4Dtl8ZoXsfoCIL3sQ?docId=57c08a3afc9147c8945f33f145058184.

8. Pierre Bourdieu, *Outline of a Theory of Practice* (Cambridge: Cambridge University Press, 2011), 78, 81.

9. Ibid., 82–83.

10. Brandon Bryant, quoted in Nicola Abé, "Dreams in Infrared: The Woes of an

American Drone Operator," *Spiegel Online International*, December 14, 2012, http://www.spiegel.de/international/world/pain-continues-after-war-for-american-drone-pilot-a-872726.html.

11. David Cloud, "CIA Allowed to Kill Terror Suspects without Identification," *Sydney Morning Herald*, May 7, 2010.

12. David Greene and Kelly McEvers, "Former Air Force Pilot Has Cautionary Tales about Drones," National Public Radio, May 10, 2013, http://www.npr.org/2013/05/10/182800293/former-air-force-pilot-shines-light-on-drone-program.

13. Christine Fair, in "Challenging the US Drone Program," The Stream, Aljazeera, accessed June 10, 2012, http://stream.aljazeera.com/story/challenging-us-drone-program-0022195.

14. Daniel Bates, "Post 9/11 CIA Has Become a Killing Machine Focused on Hunting Down Terrorists 'Faster than They Can Grow Them,'" *Daily Mail Online* (London), September 2, 2011, http://www.dailymail.co.uk/news/article-2033052/Post-9-11-CIA-killing-focused-hunting-terrorists-faster-grow-them.html.

15. Quoted in Peter W. Singer, *Wired for War: The Robotics Revolution and Conflict in the 21st Century* (New York: Penguin, 2009), 34.

16. Bureau of Investigative Journalism, "Covert War on Terror: The Datasets," January 3, 2013, http://www.thebureauinvestigates.com/2013/01/03/yemen-reported-us-covert-actions-2013/.

17. See Pugliese, *State Violence*, 210.

18. #NotABugsplat, "A Giant Art Installation Targets Predator Drone Operators," http://www.notabugsplat.com/.

19. Tom Junod, quoted in Ryan Grim, "Robert Gibbs Says Anwar al-Awlaki's Son, Killed by Drone Strike, Needs 'Far More Responsible Father,'" *Huffington Post*, October 24, 2012, http://www.huffingtonpost.com/2012/10/24/robert-gibbs-anwar-al-awlaki_n_2012438.html.

20. See Pugliese, *State Violence*, 190–98.

21. Center for Civilians in Conflict, Columbia Law School, *The Civilian Impact of Drones: Unexamined Costs, Unanswered Questions*, September 29, 2012, http://civiliansinconflict.org/resources/pub/the-civilian-impact-of-drones.

22. Ibid.

23. Michel Chossudovsky, "The Children Killed by America's Drones. 'Crimes against Humanity' Committed by Barack H. Obama," GlobalResearch, Centre for Research on Globalization, January 26, 2013, http://www.globalresearch.ca/the-children-killed-by-americas-drones-crimes-against-humanity-committed-by-barack-h-obama/5320570.

24. Jo Becker and Scott Shane, "Secret 'Kill List' Proves a Test of Obama's Principles and Will," *New York Times*, May 29, 2012, http://www.nytimes.com/2012/05/29/world/obamas-leadership-in-war-on-al-qaeda.html?pagewanted=all&_r=0.

25. Quoted in ibid.

26. Abé, "Dreams in Infrared."

27. Brave New Foundation, *Youth Disrupted: Effects of U.S. Drone Strikes on Children in Targeted Areas*, Scribd, December 2012, http://www.scribd.com/doc/115147268/Youth-Disrupted-Effects-of-U-S-Drone-Strikes-on-Children-in-Targeted-Areas.

28. Chossudovsky, "Children Killed by America's Drones."

29. Armin Krishnan, *Killer Robots: Legality and Ethicality of Autonomous Weapons* (Farnham, Eng.: Ashgate, 2009), 93.

30. Abé, "Dreams in Infrared."

31. Stanford International Human Rights and Conflict Resolution Clinic and Global Justice Clinic at NYU School of Law, *Living under Drones: Death, Injury, and Trauma to Civilians from US Drone Practices in Pakistan*, September 2012, p. 92, http://www.livingunderdrones.org/.

32. Bruno Latour, "From Realpolitik to Dingpolitik; or, How to Make Things Public," in *Making Things Public: Atmospheres of Democracy,* ed. Bruno Latour and Peter Weibel (Cambridge, Mass.: MIT Press; Karlsruhe, Ger.: ZKM Center for Art and Media, 2005), 41.

33. Bruno Latour, "Nonhumans," in *Patterned Ground: Entanglements of Nature and Culture,* ed. Stephen Harrison, Steve Pile, and Nigel Thrift (London: Reaktion Books, 2004), 226.

34. Ibid.

35. Gul Wazir Dawar, quoted in "Drones Scare Pakistanis Away from Offering Prayers in Ramadan," Pakistan Defence, July 31, 2012, http://www.defence.pk/forums/social-issues-current-events/199361-drones-scare-pakistanis-away-offering-prayers-ramadan.html.

36. Quoted in "Drones Scare Pakistanis Away."

37. Quoted in Hakim and the Afghan Peace Volunteers, "Afghan Peace Volunteer Says Drones Bury Beautiful Lives," War Is a Crime.org, January 13, 2013, http://warisacrime.org/content/afghan-peace-volunteer-says-drones-bury-beautiful-lives.

38. Bureau of Investigative Journalism, "Over 160 Children Reported among Drone Deaths," August 11, 2011, http://www.thebureauinvestigates.com/2011/08/11/more-than-160-children-killed-in-us-%20strikes/.

39. Reprieve, *Drones: No Safe Place for Children,* March 2013, http://www.reprieve.org.uk/media/downloads/2013_04_04_PUB_drones_no_safe_place_for_children.pdf, p. 6.

40. Quoted in Stanford International Human Rights and Conflict Resolution Clinic and Global Justice Clinic, *Living under Drones*, 97.

41. Stanford International Human Rights and Conflict Resolution Clinic and Global Justice Clinic, *Living under Drones,* 58.

42. Quoted in Stanford International Human Rights and Conflict Resolution Clinic and Global Justice Clinic, *Living under Drones,* 59.

43. Jacques Derrida, *Negotiations* (Stanford: Stanford University Press, 2002), 101.

44. P. W. Singer, "Images," http://wiredforwar.pwsinger.com/multimedia/images.

45. Quoted in Singer, *Wired for War,* 331.

46. Greg Miller, Ellen Nakashima, and Karen DeYoung, "CIA Drone Strikes Will Get Pass in Counterterrorism 'Playbook,' Officials Say," *Washington Post*, January 20, 2013, http://articles.washingtonpost.com/2013-01-19/world/36474007_1_drone-strikes-cia-director-playbook.

47. Quoted in Singer, *Wired for War,* 367.

48. Associated Press, "Remote Control Warriors Suffer War Stress," NBC News.com, July 8, 2008, http://www.msnbc.com/id/26078087/.

49. Noor Behram, quoted in Kevin Gosztola, "Rescuers Reportedly Killed in US

Drone Strike in Pakistan," *The Dissenter,* January 3, 2013, http://dissenter.firedoglake.com/2013/01/03/rescuers-killed-in-us-drone-strike-in-pakistan/.

50. Hayatullah Ayoub Khan, quoted in Gosztola, "Rescuers Reportedly Killed."

51. Quoted in Gosztola, "Rescuers Reportedly Killed."

52. Christof Heyns, quoted in Gosztola, "Rescuers Reportedly Killed."

53. Heba Aly, "Analysis: The View from the Ground; How Drone Strikes Hamper Aid," *IRIN: Humanitarian News and Analysis,* March 20, 2013, http://www.irinnews.org/report/97690/the-view-from-the-ground-how-drone-strikes-hamper-aid.

54. Stanford International Human Rights and Conflict Resolution Clinic and Global Justice Clinic, *Living under Drones,* 27.

55. Sudarsan Raghavan, "When U.S. Drones Kill Civilians, Yemen's Government Tries to Conceal It," *Washington Post,* December 25, 2012, http://www.washingtonpost.com/world/middle_east/when-us-drones-kill-civilians-yemens-government-tries-to-conceal-it/2012/12/24/bd4dac2-486d-11e2-8af9-9b50cb4605a7-html.

56. Abdul Rahman Berman, quoted in ibid.

57. Latour, "From Realpolitik to Dingpolitik," 15.

58. Matt J. Martin with Charles W. Sasser, *Predator: The Remote-Control Air War over Iraq and Afghanistan: A Pilot's Story* (Minneapolis: Zenith, 2010), 147.

PART III
Things in Motion

MemeLife

Kathleen P. J. Brennan

Word-of-mouth meme thing. We don't really know what it does, yet. Whether it does anything, really. Where did you hear about it?
—William Gibson, *Pattern Recognition*

In his 2005 novel *Pattern Recognition*, William Gibson explores the phenomenon of memes spread on the Internet through the eyes of his protagonist, Cayce Pollard. In the book, the meme in question is an ever-expanding body of film clips that come to be known simply as "the footage."[1] On her own time, Cayce becomes curious about the footage after being randomly exposed to a film clip on the street and then finding an online forum, "Fetish:Footage:Forum" (F:F:F), filled with people interested in the footage: "The forum has become one of the most consistent places in her life, like a familiar café that exists somehow outside of geography and beyond time zones."[2] The F:F:F is an assemblage that developed alongside the footage meme, but it also took an active role in shaping the understanding and spread of the meme.

Like Cayce, I have found that my interest in memes cannot be ignored. In this chapter, I will explore Internet memes as generative of a particular version of the international. Internet memes are often jokes, usually based on a visual gag. These memes spread rapidly, some say virally, both in their original form and in user-generated variations. Memes develop in and in turn help create human-nonhuman assemblages on the Internet. This chapter will explain the way that people, ideas, interfaces, software, and protocols (to name a few of the players involved) come together to generate shared meanings in these memes. It will also examine the particular international that these memes then shape: an international that seems oblivious to geographic boundaries, focused on visual representation and centered on an overlapping set of popular culture texts in which people connect through emotion as much as, if not more than, through information, and through which people get interested and debate current events and issues. In my discussion, I draw upon a number of memes to demonstrate the continuing influence of early examples, such as "LOLcats," and to show the direct role that Internet memes have played in important events.

Briefly, I will discuss the material aspects of the Internet assemblages that support meme generation and propagation through the work of Jane Bennett and Alexander Galloway. I will survey the history of memes with a focus on the term's origin in the 1970s work of Richard Dawkins, which not only explains the genetic legacy of the term *meme* but also highlights some key attributes of memes that continue to shape the concept in its online iterations. I will then examine how John Protevi's understanding of cognition expands the potential and effectiveness of memes in the international through the role of emotion.

Internet-as-Assemblage: Materiality and Protocol

The Internet itself can be seen as a global assemblage. The Internet assemblage can include any, and at times includes all, of the following: devices, computers, cell phones, and tablets; people; various telecommunications platforms and their component parts; telephone lines, fiber-optic cables, cellular networks, and satellites; protocols governing the connections between computers and the general functioning of the Internet; massive server farms; the various companies and governments involved; the power grids and their component parts; the software on individual devices and in the servers, operating systems, browsers, and translators. This list could go on. It is clearly a mix of human and nonhuman parts, and, like any assemblage, the Internet is more than the sum of its many parts.

It is more useful, however, to think of the Internet, with its collection of smaller overlapping assemblages, as always interfacing with other global assemblages, such as the global economy. In this chapter, I first focus on the smaller overlapping assemblages within the Internet, and later consider memes as actors on the Internet and in global affairs. Examples of these smaller assemblages can be found in existing websites and forums, like Cayce's F:F:F, that have established user bases and thus a type of community. A slightly larger assemblage can often be found in the interactions between established websites that have contiguous user bases. When looking at these assemblages, it is important to take note of their existing limitations and capabilities: Can the platform handle images and videos? Can the servers that host the website handle massive spikes in traffic? Is the website organized in a way that is approachable to new visitors, or is it decipherable or accessible only to established users? And so on.

In *Vibrant Matter*, Jane Bennett asserts that one unique characteristic of assemblages is that they distribute agency broadly throughout their component parts and the interactions thereof.[3] The Internet-as-assemblage creates, and I think thrives by, such complications of agency. The Internet I refer to here is a conglomeration of human and nonhuman parts that all have a say in the role that the Internet plays in politics, with no one part dominating all of the others.

One smaller assemblage within the larger Internet assemblage is the set of ever-evolving protocols that allow the Internet to function. Internet memes usually have a visual aspect: they may involve a still image, a GIF (graphics interchange format), a video, or perhaps an ASCII-based drawing. The visual focus is logical, because the protocols of the Internet make it a visual medium. For example, you do not see the source code when you visit a website, you see the visual representation that the source code generates. In *Protocol*, Alexander Galloway describes the way source code generates the websites you see by the protocols that dictate the process: "During the process of developing the Web, [Tim] Berners-Lee wrote both HTTP [Hypertext Transfer Protocol] and HTML [HyperText Markup Language], which form the core suite of protocols used broadly today by servers and browsers to transmit and display Web pages."[4] These protocols are the result of early human and nonhuman assemblages that developed the Internet as it is known today, but they also continually shape these assemblages. Through the process of creation and adoption, established protocols are able to exert control, because if people and machines do not follow them, they cannot participate in the network.

Thus, these protocols, which emerged from a human and nonhuman assemblage, generated and continue to generate the Internet as it is today: a visual realm in which people from around the world interact, buy things, consume and publish information, and so on. The specific protocols that are in place, in conjunction with the increased sophistication of devices and the telecommunication platforms that connect them, result in an Internet that is uniquely capable of and geared toward visual memes that make use of high-resolution imagery and streaming video, which both involve moving considerable amounts of data.

History of the Meme

Richard Dawkins coined the term *meme* in his 1976 book *The Selfish Gene*. In his study of genes and evolution, Dawkins became curious about the idea that there may be units of transmission other than genes:

> We need a name for the new replicator, a noun that conveys the idea of a unit of cultural transmission, or a unit of *imitation*. . . . I hope my

> classicist friends will forgive me if I abbreviate mimeme to *meme*....
> Examples of memes are tunes, ideas, catch-phrases, clothes fashions, ways of making pots or of building arches. Just as genes propagate themselves in the gene pool by leaping from body to body via sperms or eggs, so memes propagate themselves in the meme pool by leaping from brain to brain via a process which, in the broad sense, can be called imitation.[5]

A tune or melody, for example, is played on an instrument or sung, which you then hear either directly or through some medium (CD, record, radio, television, or the like). The tune leaps into your brain, and you imitate the tune perhaps by humming or singing along, or performing your own version. For the sound to transmit at all, the materiality of the air through which it travels must be noted; add in the bodily organs of both performer and listener, plus any instruments or transmission media involved, and you begin to see both the formation of a meme and its accompanying assemblage.

The tune is a good example to use to explore two other aspects of memes. First, Dawkins gives memes a certain level of agency to parallel that given to genes: "Just as we have found it convenient to think of genes as active agents, working purposefully for their own survival, perhaps it might be convenient to think of memes in the same way."[6] Consider the type of tune that gets stuck in your head, an earworm, and it becomes easier to imagine memes as agentic. Once the meme-tune is in my mind, I might find myself humming or singing it at random throughout the day. The existence of the meme in your mind, your memory of it, exerts a certain self-propagating force that leads to you imitate the meme and thus assist its propagation.

The other aspect of memes that I want to discuss here is the way multiple memes come together in "co-adapted meme-complexes."[7] As in genetics, those memes which are able to "exploit their . . . environment to their own advantage" have a higher likelihood of surviving, and this exploitation produces meme-complexes in which individual memes use each other to improve their odds.[8] In terms of tunes, consider how a genre of music develops and becomes dominant, which makes it difficult for other types of music to become popular, to become memes, in the minds of listeners who are already conditioned by the dominant genre. You can also see this process occurring on a global scale in the way that Western, and in particular American, popular culture texts and references proliferate in the international. This pool of references dominates and shapes the international but can also act as a source for adaptation or co-option by anyone, or any meme, in that international.

In the 1989 re-release of *The Selfish Gene,* Dawkins made some additions to the original text through endnotes. The endnotes are critical here because of the rise and evolution of the Internet since the original publication date. Although still not quite what it is today, the Internet in 1989 is much more familiar to us than the computer setup Dawkins describes in the 1976 edition. In the endnotes he states, "It was obviously predictable that manufactured electronic computers, too, would eventually play host to self-replicating patterns of information—memes. Computers are increasingly tied together in intricate networks of shared information. Many of them are literally wired up together in electronic mail exchange. Others share information when their owners pass floppy discs around. It is a perfect milieu for self-replicating programs to flourish and spread."[9] Dawkins thus introduces the possibility of Internet memes of any kind, but he goes on to discuss only the dangers of viruses and worms while excluding the potential for other memes to flourish "in intricate networks of shared information."[10] Why limit his concept in this way? Although the world certainly still faces the danger of such insidious memes, people do not restrict the term in the same way Dawkins does here. Today, the term *meme* still refers to units of cultural transmission. However, in popular discourse it tends to refer solely to Internet memes. In these Internet memes, one can see key aspects that Dawkins laid out in the 1970s and 1980s: they spread through imitation and take up residence in our brains and hard drives as materially encoded patterns, they are agentic in a self-serving or self-propagating manner, and they work together in ways that make them more potent and thus more likely to survive.

The Role of Emotions in Internet Memes

John Protevi, in *Political Affect*, offers the opportunity to expand on Dawkins's explanation of memes through his use of the embodied-embedded approach to cognition, and in particular the role of emotion in cognition. In the preface, he mentions that in "the embodied mind school, you see the basis for a careful discourse on human nature grounded in such basic emotional patterns as rage, fear, and protoempathic identification," and he goes on to explain the impact of being "embedded in social practices that inform the thresholds and triggers at which those basic emotions come into play."[11] When I think about Internet memes, and in particular when I try to understand why memes are effective (meaning why memes are able to both make an impression on my own brain and also maintain some level of effectiveness over time), it is clear that the answer comes down to basic emotional patterns. Internet memes do not work just because they are logically sound or rationally legible (in fact, that is not even a requirement); they are effective because they have an emotional impact. The memes that have the most initial impact and the most staying power have to bring in emotion: maybe a given meme makes you laugh and so incites joy to some degree, or it makes you angry or upset, and so on. This emotional aspect of Internet memes that is integral to their function is generalizable to the broader category of memes.

In his chapter in *Political Affect* on Hurricane Katrina, Protevi discusses the existence and role of "prototypes by which we judge whether objects belong to a category by seeing how close or how far an object is to our prototype."[12] These prototypes interact with our "virtually available response repertoire."[13] Protevi points to the political physiology of the continuing fear of slave revolt or of blacks getting revenge for slavery as a "bogey" that is instilled in children.[14] The content of the rumors spread in the aftermath of Katrina ("mass rape and carnage")[15] was already available in the existing cultural bank of memes, in memes that people in the society are primed to believe or identify with. These memes, or prototypes, are based in a reality that does not exist now, and may have never existed in the form that was crystallized in the memes, and yet they are still efficacious. In fact, the historical roots of this slave-revolt meme, for most people, have to be forgotten for the meme to still be effective, because no one likes to think of him- or herself as a racist. This meme continues to be reinforced and to reinforce people's beliefs and behaviors synchronically: rather than citing the danger of slave revolt or black-on-white violence as revenge for slavery, people point to the dangers of gang violence and the drug war. The content of the meme changes, but the racialized form remains. The affective cognition remains as well, as the meme continues to evoke fear and panic.

One of the ways that Internet memes attain staying power is through their plasticity, and in particular the way that they are sometimes emptied of their content, particularly through coadaptation. One dominant Internet meme is the "LOLcat."[16] LOLcats have a particular form: an actual photo of a cat or cats with text in a bold white font superimposed on the photo, and the text often takes the form of LOLcat speech (which has a unique grammar and vocabulary).[17] Some people may argue that the LOLcat is a passing fancy, but it has been surprisingly long lasting and robust (robust in that the form and content have stayed quite consistent over time). One of the ways that this consistency over time can be explained is that the LOLcat form has also been adapted to many other memes. When I see any image on the web superimposed with a bold white font, I am primed for it to be an instance of an Internet meme that is likely to make me laugh. The content of a meme can be emptied while the form remains, and the form itself, because of our established responses to the earlier example of a meme, will still evoke the same emotions, or at least will prime people to respond in a similar way. The conditioning effects of LOLcats thus shape the available language on the Internet and the collective "response repertoire."[18]

Beyond the point that such plasticity is possible in Internet memes, and thus memes as a whole, the dominance of form over content also opens up the possibility that a meme can still be effective when the knowledge of its origins is lacking. The "condescending Wonka" meme (whose form is drawn from the LOLcat) can be effective without any awareness of LOLcats, meaning that the form of bold white font superimposed on an image can be used to communicate any content.[19] Thus, one possible characteristic of memes in general is that the longevity of a meme is dependent on the extent to which that meme can continue to be effective while being detached from its own history or context. Further, I think that plasticity, and specifically the ability to adapt to other existing and emerging memes, is the secret to longevity rather than a need to stay precisely the same. LOLcats may have maintained their form, but people constantly adapt the pictures used and the content of the text. This model of longevity opposes some of our commonly held assumptions.

There are meme-generating assemblages that depend on the visual nature of the Internet, which I described earlier in my discussion of protocols. These meme-generating assemblages are often centered on particular websites such as Reddit[20] and the 4Chan /b/ board,[21] which have overlapping, although not completely coinciding, communities of users. A website like Reddit involves a range of activities and discussions surrounding these memes (although it is important to note that Reddit is not limited to the creation and spread of the visual memes I am focusing on here). Community members generate new memes, generate new iterations of existing memes, combine memes, comment on which iterations they like and do not like, discuss the context of the memes, and, importantly, spread the memes outside of Reddit. Also, Reddit employs a unique system of upvoting and downvoting in which individual users vote on whether a post should be moved up (and thus become more easily seen) or down, and thus the community of more-established users determines what outsiders and casual users see on the main page of Reddit. In Gibson's *Pattern Recognition,* Cayce describes the F:F:F as "a fishbowl: it felt like a friend's living room, but it was a sort of text-based broadcast, available in its entirety to anyone who cared to access it."[22] The main page of Reddit is a great illustration of the fishbowl effect that Gibson describes. The Reddit fishbowl is key in understanding the mechanisms through which memes play a role in the international, because Reddit often acts as a source, or at least a clearinghouse, for new and newly coadapted memes, which Redditors (Reddit members or users) then spread more broadly via other social media platforms.

Memes and the International

Combining Dawkins's original understanding of memes as units of cultural transmission that propagate through imitation, Bennett's vibrant material assemblages, Galloway's protological Internet, and Protevi's affective cognition, I finally have the necessary tools to examine memes both in and of themselves and as they generate a particular international. To some extent, Internet memes merely participate in an already-existing international, which is both a shared and a variable space. Although the Internet is a global space, the extent to which people can access it freely, if at all, varies by geographic location. The variance is partially socioeconomic (accessing the Internet requires an interface and a communications infrastructure), but there are ways that people get around economic barriers: leapfrogging traditional communications networks by going straight to cell connectivity, using businesses such as Internet cafés, and so on. The variance also stems from state censorship and control of content, although this, too, can be subverted to a certain degree. With these variances taken into account, it is still useful to think of the Internet as a shared international space. As the Internet is the native land for the type of memes I focus on here, there is an in-built potential for memes to operate globally. When a meme is initially coming to be in a particular forum—say, the 4chan /b/ board—it can quickly become an epidemic, spreading rapidly within that limited community and then branching out to other websites with overlapping members. Like any good virus, once the meme begins to spread quickly in these limited regions, it may then become a pandemic as it stretches first across the Internet through common vectors like Twitter, Tumblr, and Facebook and then jumps to the offline world via major media outlets like CNN. Interestingly, as in the case of LOLcats, after the initial spreading of a meme, it will often become endemic in certain spaces online.

But why do memes spread this way? What makes them viral or contagious in the way that a biological parasite or bacteria is? First, memes, as I have mentioned, are inherently visual online. The repetition of common forms combined with the use of compelling or emotive images allows memes to transcend language barriers. For example, there are French LOLcats that mimic the form of English LOLcats (consisting of a picture of a cat with text in a bold white font) and even match the idiosyncratic LOLcat grammar but do so in French. The visual nature is also important because it taps into a shared visual reservoir of references. Much of Internet culture references Western popular culture texts, but the references are often legible around the world because of the export of these texts. Also, part of the acculturation process in general online, and specifically in the websites where these memes are generated, is learning these visual references. Memes do not limit their references to popular culture, though; they also play with current events. For example, when Israeli prime minister Benjamin Netanyahu went in front of the United Nations in fall 2012 to discuss the Iranian nuclear situation, images of him with his ACME-style cartoon-bomb poster (think Wile E. Coyote and the Road Runner) almost instantly became visual meme fodder.[23] The issue of a possible Israeli preemptive strike in response to Iran's enriching a certain amount of nuclear material is captured in this visual, and the image takes on a life of its own in meme form.[24] This is not an unprecedented process. I see Internet meme-ification as operating in a similar way to the legacy of important photographs in the twentieth century, such as the iconic "Napalm Girl" image from the Vietnam War. The difference comes in the way the images are manipulated, are crossed with existing memes, spread, and become haunting not only in their content but also in their form. When a meme's form can function without regard to its content, then it is ripe for being coadapted to another existing meme or to an emerging one.

One of the consequences of the visual nature of memes, which is also observable in auditory memes, is that people connect with them in an emotional or intensive way. Protevi's affective cognition is useful in this regard. The emotional aspect of the connection is another way in which memes transcend the language barrier. This aspect captures the way that memes can still be effective when they fail to communicate the details of an event or issue, meaning that they can continue to propagate because people still connect to them. A meme can visually pique your interest through humor, sadness, or outrage, which can then make what was an isolated issue a shared issue (or what was an isolated reference a shared reference) as you go out of your way to understand the meme. Emotional content also adds to the haunting or persistent quality that many of these memes have.

Through their visual nature and their ability to connect with people affectively, memes thus contribute to the increasingly shared international that can be found online. Memes, however, can also be used to mark out territorial (not geographical) boundaries within the international. When particular memes or meme-complexes territorialize parts of the Internet through demarcation, they also alter the entire shared space. For example, there are entire websites dedicated to cataloging and showcasing Internet memes, like www.knowyourmeme.com. On knowyourmeme.com, anyone can find information on the memes that he or she encounters elsewhere, and thus may become connected to the meme itself and to the events and texts it references. Such resources are particularly useful when trying to become part of an established community online that uses what can be an opaque set of endemic memes. Interestingly, meme images and references can also be found on protest signs now, which have the effect of marking a protest as Internet-adjacent. Examples of such signs were especially prevalent in the 2008–2009 international "Anonymous"-backed protests against the Church of Scientology.[25] These protest signs functioned as a territorial marker: the protests they accompanied were seen as an extension of the online action against the Church of Scientology, and of the communities in which both these protests and memes were spawned. Using memes in your icon or profile picture in a forum or on a social-networking site also identifies you as part of the shared international.

I Can Has Internet Memez, Plz?

Memes are most successful at participating in and shaping the shared but variable international that is present primarily on the Internet when they connect with people on an emotional level. Once the connection is made emotionally, not only is information spread, but its power can be augmented as well. Whether a meme makes you curious enough to search out more information about the texts and events that it is referencing because you want to make your own version of the meme or because you want to get involved with the issues it is raising, it is a meme that shapes the pool of available information and images and can add to the response repertoire.

Current-event-based memes can get people interested and involved in political issues that they were either unaware of or completely ambivalent to previously. Netanyahu's cartoon bomb and the rapid Internet meme response made Israel's position look ridiculous and had a lasting impact because of the inherent humor of the image. A more recent Internet meme is the "Harlem Shake."[26] You can find examples of Harlem Shake videos from all over the world (along with the lamentations of those who despise them), and for the most part you connect with them through either humor or annoyance. However, Harlem Shake videos from countries such as Tunisia and Egypt were met with state-led crackdowns and the possibility of violence and arrests.[27] The response to the videos produced in these Arab Spring countries renewed the focus on the regimes that arose from these protests, which began in 2010. The example from Tunisia seems to be a typical example, motivated initially by humor, but the examples from Egypt were more explicitly motivated by politics from the start.[28] As with other memes, though, the intent of the creators is less important than the impact of the meme, of which fact the students in Tunisia have now been reminded.[29] The Harlem Shake videos produced in Egypt and Tunisia made the post–Arab Spring struggles in these countries visible and engaging before the body and arrest counts again began to mount. The Harlem Shake craze came along at the right time to be adapted first for humor in Tunisia and then for politics in both Tunisia and Egypt. Although this craze came and went quickly, its role in turning eyes back to these countries became part of its legacy. The national boundaries, and the power struggles that go on inside of them, did not disrupt the spread of the Harlem Shake meme, but the meme was able to bring global attention to the continuing protests within those boundaries. Importantly, it never mattered that the dance associated with the meme was not the actual Harlem Shake; as with so many of the memes I have discussed in this chapter, the cultural origins of the content are less meaningful and impactful than the form. The legacy of the Harlem Shake meme not only exemplifies the part that Internet memes play in shaping and participating in a particular international, but also serves as a reminder of key attributes of this international: its ambivalence to geographic boundaries, its focus on the visual, its ability to draw on and bring attention to popular culture and current events, and its reliance on emotion as a connector.

Notes

1. William Gibson, *Pattern Recognition* (New York: Putnam, 2003), 3.
2. Ibid., 4.
3. Jane Bennett, *Vibrant Matter: A Political Ecology of Things* (Durham, N.C.: Duke University Press, 2010), 23.
4. Alexander R. Galloway, *Protocol: How Control Exists after Decentralization* (Cambridge, Mass.: MIT Press, 2004), 137.
5. Richard Dawkins, *The Selfish Gene,* 2nd ed. (New York: Oxford University Press, 1989), 192 (original emphasis).
6. Ibid., 196.
7. Ibid., 199.
8. Ibid.
9. Ibid., 329.
10. Ibid.
11. John Protevi, *Political Affect: Connecting the Social and the Somatic* (Minneapolis: University of Minnesota Press, 2009), xiv.
12. Ibid., 173.
13. Ibid., 17.
14. Ibid., 173.

15. Ibid., 178.

16. "LOLcats," KnowYourMeme.com, last modified February 2013, http://knowyourmeme.com/memes/lolcats.

17. A sample image may be found at KnowYourMeme.com, http://knowyourmeme.com/photos/199476-lolcats.

18. Protevi, *Political Affect*, 17.

19. "Condescending Wonka/Creepy Wonka," KnowYourMeme.com, last modified June 2012, http://knowyourmeme.com/memes/condescending-wonka-creepy-wonka.

20. Reddit, accessed March 1, 2013, http://www.reddit.com.

21. "/b/," 4Chan.org, accessed March 1, 2013, http://boards.4chan.org/b/.

22. Gibson, *Pattern Recognition*, 65.

23. "Netanyahu's Cartoon Bomb," KnowYourMeme.com, last modified September 2012, http://knowyourmeme.com/memes/events/netanyahus-cartoon-bomb.

24. Rick Gladstone, "Iran Ridicules Netanyahu's Bomb Theatrics, and Calls on Israel to Renounce Nukes," *New York Times*, September 28, 2012, http://www.nytimes.com/2012/09/29/world/middleeast/iran-ridicules-netanyahus-bomb-theatrics.html?_r=0.

25. Quinn Norton, "Anonymous 101: Introduction to the LULz," *Wired*, November 8, 2011, http://www.wired.com/threatlevel/2011/11/anonymous-101/all/.

26. "Harlem Shake," KnowYourMeme.com, last modified February 25, 2013, http://knowyourmeme.com/memes/harlem-shake.

27. Issandr El Amrani, "Arab Spring Blues?," Latitude, *New York Times*, February 27, 2013, http://latitude.blogs.nytimes.com/2013/02/27/in-the-arab-world-the-harlem-shake-turns-political/.

28. Mark Hogan, "'Harlem Shake' Like an Egyptian: Viral Craze Ruffles Authorities Worldwide," *Spin*, March 1, 2013, http://www.spin.com/articles/harlem-shake-egypt-student-protest-faa-investigation/.

29. Charlene Gubash, "How the Harlem Shake Is Being Used to Push for Change in Egypt," NBCNews.com, March 1, 2013, http://worldnews.nbcnews.com/_news/2013/03/01/17144225-how-the-harlem-shake-is-being-used-to-push-for-change-in-egypt?lite.

Videos

Rune Saugmann Andersen

Two decades ago, Michael J. Shapiro saw the first signs that international politics would take new forms in a new video age that would place spectators in the position of referees vis-à-vis international conflicts.[1] Yet the way in which developments related to video technology interact with the international (rather than the social as such) remains largely unexplored, at least beyond introductory references to increasingly global camera omnipresence in analyses of surveillance.[2] Video is described as a part of the larger communications environment in Ron Deibert's excellent study of media convergence and world order, is implicated in James Der Derian's work on the "virtual continuation of war by other means," and acts as an important background to Jean Baudrillard's reflections on how images and their referents collapse, rendering war indistinguishable from TV.[3] This chapter charts how video has emerged as one of the key technologies making the international on an everyday basis, albeit in ways different from those observed by Shapiro.

Video is doubly implicated in making things international. First, the already-international technology has been reconfigured from a technology that was employed mainly in domestic forms of documentation and spectatorship of mass-produced content to perhaps *the* international form of representation and spectatorship (for example, from home movies and Blockbuster videos to a vital tool for Occupy Wall Street and protesting students at UC Davis or Tahrir Square). Second, video documentation is today essential in making local events international (for example, the 2007 uprising in Burma, the 2009 Iranian postelection crisis, the later Arab Spring, and Taksim Square). Recent protests and uprisings in the Middle East and North Africa highlight how the digitalization of video technology has changed both the way that events come into being as international and the actors involved in making events international. These protests are represented through constant reappropriation and remediation between user-produced images, especially videos, social media, and mass media.

This chapter explores how the international character of video is changing, as well as how video is changing the international. It starts with a brief outline of the genealogy of video, charting how video is made international in new ways. Thereafter, it turns to the question of how video is implicated in politics, and finally how video makes the international today.

Making the Private Medium Public

Until the late 1990s, *video* referred mainly to analog video recording and VHS-cassette playback, both well integrated in a Western consumer culture and intimately related to the private sphere. Video recording was done with analog technology, and many households would have a video camera present at holidays and family events such as birthday parties. The videotapes were also seen primarily in relatively closed settings, thus constituting the medium as one of primarily private forms of representation and spectatorship.[4] When TV shows broadcasting funny episodes from citizens' videos began to appear in many countries in the 1990s, the titles would suggest the private character of the recordings (for example, *America's Funniest Home Videos*). Video recorders captured the cherished moments of the private sphere, marking it as authentic and familiar rather than international.

At the same time, VHS standards were used to distribute mainstream movies via video-rental chains. This second use of video technology was important especially to the worldwide spread of Hollywood films, and thus generated an important international political economy in and of itself, being a key ingredient in the cultural changes viewed by critics as the Americanization of the world.[5] These early incarnations of video were already international.

Even when the home video and the video machine in the home theater were the dominant ways in which video permeated Western culture, a few events pointed to the possibility that video might play an altogether different role, a public one rather than the private role of home entertainment. Abraham Zapruder's video showing the assassination of President Kennedy in 1963 marked the first time that a privately held video camera brought an event of world politics to the screens of TV spectators around the world.[6] Now, it is common that a citizen-journalist rather than a traditional media professional will be at the scene to capture an event of historical significance.

The 1991 video filming of the LAPD beating of Rodney King marked another step in the integration of video into international politics. Here again, a private, citizen-recorded video was spread by major news networks and thus enabled King to speak with a voice that could overcome institutional barriers and reconstitute a routine case of police violence as a political event of international significance, an event the whole world was watching.[7] Video, the Rodney King case showed, not only documents events that already fit the criteria for international significance but also has a transformative potential, changing events from local occurrences to topics of international politics.

The Zapruder and King videos reconstituted video recording not only as a private activity for domestic production and enjoyment but also as a capturing of reality that, potentially, the whole world is watching. The next section deals with how the political performativity of video has changed as video was transformed from a private technology of home entertainment to a ubiquitous communications technology crucial in making events international.

Video as International Speech and Spectatorship (Arm Yourself with a Camera)

Digital video became omnipresent in the late 1990s, integrated into everything from weapons systems to mobile phones.[8] Video-camera surveillance has permeated social spaces from mass transportation to kindergartens.[9] Mobile-phone video has been seen to bring about revolutionary changes in the public sphere.[10]

Traditionally, the representational regimes of world political debate have placed great emphasis on institutional authority in granting speaking rights to subjects, and have always excluded the majority of citizens from international politics. The spread of video carries a transformation of political authority; and, crucially, it influences who can speak effectively in political debate, because video presents itself as a faithful witness of real events.[11] Although the verbal testimony of a subaltern would seldom be enough to speak truth to powerful institutions, subaltern-produced video can. The videos and images from Abu Ghraib are exemplary in this regard. Verbal accounts of abuses in the Iraqi prison were known when the images surfaced, in 2003–2004, but the videos and photographs of the scenes propelled Abu Ghraib to the center of world politics.[12]

However, the Abu Ghraib images also tell a tale of the limitations of video as political speech. Even if video did empower subalterns of the U.S. armed forces to expose occurrences that powerful institutions would prefer to keep hidden, these subordinates remained largely powerless in controlling the political interpretation of the images they produced. The soldiers seen in the videos and photographs were ultimately framed as "bad apples" and disciplined for the prisoner abuse exposed in the Abu Ghraib images, whereas those responsible for the war and its conduct—the political and military elite—were able to disconnect themselves from the images. Errol Morris's film *Standard Operating Procedure* (2008) shines a rare light on the subordinate military officers and tells of their interpretation of the situations depicted.[13] Although unknown individuals do not have means to control the interpretation and remediation of the images they produce, and are therefore largely powerless when the media and political elite turn the blame on them, the image producers, Morris was able to use his status as a respected documentary filmmaker to destabilize the official interpretations and suggest that the individual soldiers saw (and filmed) what they did as a part of doing their job. Even if video makes it somewhat easier for ordinary people to gain a voice in debates on international politics and security, political debate is still very much influenced by the status and institutional position of those speaking; thus, the dominant political interpretation of videos is largely beyond the control of the nonelite videographer.

An important element in the authority of citizen-made video is its character of nonprofessionalism: blurry, shaky, and pixelated videos are signs of authenticity, linking video to two important genres: to the genre of home video that preceded today's omnipresent video cameras; and to the earlier documentation of important political events. First, the chaotic and unedited appearance of amateur protest-video shots resembles footage from the genre of home videos, such as children's birthday parties. Home videos, apart from being familiar, are perhaps the only visual genre in which the spectators are simultaneously the routine subjects of the representation. This collapsing of the exposed and the voyeur allows the spectator to subjectively establish the accuracy of the video mode of representation through her or his own having-been-there. The genre of amateur video carries this epistemological authority even when what it depicts is not familiar.

In 2005, when the launch of YouTube provided the first major platform dedicated to the publication of user-made digital video, its slogan, "Broadcast yourself," paid homage to the private origins of the authority of video. The slogan also points to the blurry pictures, unedited clips, and other visibly homemade (yet sometimes carefully choreographed) traits of videos as markers of authenticity in the emerging YouTube genres.[14] The very banality and unprofessionalism of the images became an asset, reminding spectators of the home video. In a spectacular world where slick, carefully choreographed, and professionally edited visual-media representations have long been decried as lying at "the heart of society's real unreality,"[15] the lack of professional elaboration powerfully reinforces the reality of the visuals, bolstering their claim to authentically "capture" something that was really there.[16]

Second, videos featuring blurry and unedited images of violence and death connect to the genre of citizen documentary, especially to past documentation of political violence and assassinations. Zapruder's film of the assassination of Kennedy in 1963 defines this genre. The political importance of past films in this genre legitimizes the voyeurism inherent in willfully and repeatedly being a spectator of the violent death or suffering of fellow human beings, alleviating what Frank Möller terms the looking/not-looking dilemma inherent in witnessing violence.[17] The political importance attached to videos of violence enables both the spectators and the media companies that profit from the circulation of such videos to push the boundaries of what can legitimately be shown and seen.

Videos enable visually represented political violence to become part of making up world politics in a way that is radically different from the role political violence played before. Whereas the Hobbesian version of international politics is concerned first and foremost with the effective monopoly on political violence, and not primarily with avoiding such violence, mediated representations of political violence are challenging this conception. In her work on mediated suffering, Lilie Chouliaraki argues powerfully that such suffering and the pity it provokes have become an important driving force in structuring international relations.[18] Political violence is thus reconfigured as a liability, and the monopoly over *visual representations* of political violence has become an important addendum to the simple monopoly over brute force.

Weapons systems, to an extent, also have a voice and speak, because the integration of video into weapons systems means that many modern weapons produce visualizations of the violence they inflict. This technology has been harnessed to battle in the two recent wars in Iraq, for example,[19] but it has also been used to make weapons into tools that speak against their own violence. In 2010, WikiLeaks was able to publish a leaked video from a U.S. attack helicopter in Iraq as a testimony to the brutality of the U.S.-led war there.[20] Such representation of the brutality of how the war is conducted would be impossible without the integration of video into the weapons that are used in the war, and the extensive visual archives of war that they produce.

The spread of digital video has been associated with a "democratization of representation,"[21] because it enables everyday people to make assertions about what count as facts in global politics, such as the fact that torture was taking place in Abu Ghraib or that U.S. troops were not taking great care to distinguish combatants from civilians when shooting at a group of people assembled at a street corner in Baghdad in 2007. Similarly, as seen in uprisings from Burma and Iran across the countries of the Arab Spring to UC Davis, video is a powerful tool for arguing about who is being subjected to political violence. Viewed in the communicative grammar of securitization theory, video can therefore be said to speak powerfully about referent objects of security, rather than in itself desecuritizing or securitizing such referent objects.[22] It is thus partially misleading to speak of a democratization of representation, because video does not enable citizens to speak about all topics with the same chances of successfully getting their message across. At present, it seems that although video is powerful in its constitution of referent objects of international politics, the dominant political interpretation of such video is still mostly elite-driven.

Video as Making the International

The effectiveness of video in producing the referent objects of global political debate suggests that video itself becomes an important condition of making local events international. Strikingly pronounced in images from around the world since the uprising in Burma in 2007,[23] video has become more than just another tool of representing events. In another example, the steady flow of videos recirculated in global news media as undeniable evidence that revolutions were under way became a hallmark of the Arab Spring uprisings. The existence of video has in itself become an important prerequisite for transforming local political violence into global political event. Besides the fact that it is frequently hard to see such videos clearly, they all carry the imprint of their mode of production—grainy and blurry images that suggest they were taken in conditions of urgency and danger.

Revolutionary images from unknown producers in Arab locations typically carry prominent signs of nonprofessionalism. Blurry and out of focus, the visual representation displays obvious traces of filming done using cheap and low-resolution handheld video-recording devices such as camera phones. When these images migrate from online media to mass media, the low quality acts as a signifier of authenticity. The stigma of authentic amateurism embeds the video and photographs in triumphant news-media narratives in which modern, technologically sophisticated protesters—those who are (most) like us, with Western media habits and employing freedom of (online) expression as means in a fight for democracy and civil rights—are facing some brutal regime characterized, either explicitly or by implication, by the opposite traits.[24] *What* is depicted is trivial compared to the meaning carried by *how* it is depicted; by the sheer existence of videos containing fire, fighting, and protest, the validity of the narrative equating digital imagery with resistance and a quest for freedom promises real action and concretizes the view that video cameras and Internet media are the new weapons of the weak.[25]

In the Western news media, these videos are not chiefly about representing something taking place. In their very being, these videos act as a material confirmation that video resistance is taking place—that is, their meaning is carried not so much by what they depict but mostly by their very existence. The video therefore becomes an object deriving its political significance simply from its existence as an object, not an object showing something else (other than itself) taking place.

Such revolutionary videos somehow become a negation of the argument, famously made by René Magritte in his artwork *La trahison des images,* that images are not reality but simply a depiction of a reality. In Magritte's work, a carefully painted reproduction of a pipe appearing above the text *Ceci n'est pas une pipe* calls attention to the difference between reality and representation by highlighting the image as merely a representation of a pipe, not something one can fill with tobacco and use for smoking. The revolutionary digital video, by contrast, seems to deny this dichotomy and assumes a double ontological status, both *representing* protest through new visual media and *being* the material embodiment of protest through new visual media.

When the video is no longer a mere representation of something else, it becomes difficult to argue that it distorts or is staged, that is, that the connection between the image and reality is not credible. When the existence of the video as a digital thing becomes the signifier, it becomes a sign of itself, not a representation of something else, and its role in representing the international changes. Often images from recent uprisings are taken simultaneously to represent a social practice (depicting some group as using modern visual communication tools for political purposes) and to embody that social practice (being an artifact produced by someone's practice of using modern visual communication tools for political purposes),[26] and they derive their power from the difficulty of denying the practice in the face of an artifact apparently produced by that practice.

The double ontological status makes it excessively difficult for an opponent—whether a skeptical spectator or one disadvantaged by the image in question—to disrupt the chain of signifieds linking the image to the use of modern visual technology, and to the revolutionary claims (favoring civil liberties, having modern attitudes) that are attached to struggling by modern technological means to achieve political change. At the same time, the status of the video as an object inscribed with progressiveness and revolutionary intents or ideals renders these inscribed values as commodities. This has consequences for the role of video in making the international, because it, first, enables the scrutiny of the ideas of different movements to be replaced by the spectatorship of the means they use—what is called "metacoverage" in media scholarship.[27] This trend toward metacoverage is not limited to the international but can be observed in most Western political systems. Second, the backgrounding of content enables or furthers the weaponization digital video. When the content of videos becomes less important, video becomes easier to use as a means of controlling the interpretation and characterization of different actors in the conflict—and such dichotomization makes it easier for outside actors to latch onto the part attributed to the progressive characteristics.[28]

After the prominent role of video in the Arab Spring, mediation quickly became a primary battlefront in the conflict in Syria, where provision of video-communication equipment became a way for the United States to intervene when international disagreement precluded lethal support. Providing satellite-communications equipment to help Syrian opposition forces was undertaken to help them "stay in contact with the outside world."[29] Such provision could be counted on to ensure that digital video images of the repression of the (at first nonviolent) protests would be forthcoming, thus keeping the conflict on the international agenda.

Such weaponization of video relies on the willingness of media to interpret protest videos according to the narrative I have described, as evidence of sophisticated protesters repressed by brutal regimes. Even if media such as the Associated Press and the *New York Times* recognize that arming actors with satellite video-communication equipment would "stretch the definition of humanitarian assistance and blur the line between so-called lethal and nonlethal support,"[30] the same media companies remain an essential cornerstone of making such tactics work.

As I have noted, digital video has greatly stretched the limits of media companies' determining which representations of political violence to distribute, further enabling the weaponization of video. The *New York Times*, for example, is dedicating a special project website, "Watching Syria's War," to redistribute mostly rebel-produced digital video from the conflict, even as the newspaper recognizes the militarization of the video.[31] Keeping in mind the point of view of the widening and deepening debate that has been going on in the discipline of international relations about the concept of security, it is interesting to note that the "Watching Syria's War" site banner highlights "online video that has allowed a *widening* war to be documented like no other" (emphasis added). The widening of the war to encompass digital video of decontextualized political violence, I argue, is one of the most interesting aspects of how video is changing the international—and is being changed in the process, from a tool of representation to a digital material artifact that is indeed employed as a weapon by the weak and the strong alike.

Video: From an International Technology to a Technology of the International

This chapter has looked at the intersections of video and the international. Whereas video recording and spectatorship were imagined in the twentieth century primarily as relating to leisure activities enjoyed in the private sphere, in the first decade of the twenty-first century the imagined viewing and recording communities have become global. Video recording has emerged as one of the prime markers branding conflicts as international, especially through the constituting of video-filmed political violence as a sign of weakness of the perpetrators of the violence and of the sophistication of the recorders of violence. As video has become a marker of technologically advanced protest against repressive regimes, what is visible in videos of distant political violence seems to have become of lesser importance than the sheer existence of video as a digital material artifact of sophisticated forms of dissent and the repression thereof.

For protest movements aspiring to speak in the international political debate, therefore, video of repression is an essential tool. Likewise, sophisticated video-communication equipment resistant to state restrictions on civil-communication channels is a weapon of intervening in distant conflicts. Early use of video in public debate about violent encounters led Shapiro to observe that "the media moves the public increasingly into a critical juridical space," because "spectators of [a violent] event were placed in a position to evaluate... official discourse on the basis of what they saw."[32] Yet even if the high priority given to video as a quasi weapon today testifies to the importance of public perceptions for international politics, the replacement of emphasis on the content of the video by the metacoverage of decontextualized video as a digital material artifact of sophisticated dissent suggests that the role of video in the public sphere may not be in refereeing events. Rather, video becomes important in testifying to a loss of the monopoly of representing political violence, evoking in Hobbesian thinking the loss of legitimacy of the state in question. Video is crucial in representing local events involving political violence as international events, as problems that are rightfully the referent objects of international political debate.

Notes

1. Michael J. Shapiro, "Strategic Discourse / Discursive Strategy: The Representation of 'Security Policy' in the Video Age," *International Studies Quarterly* 34, no. 3 (1990): 327–40.

2. Didier Bigo, "Security, Exception, Ban, and Surveillance," in *Theorizing Surveillance: The Panopticon and Beyond,* ed. David Lyon (Collumpton, Eng.: Willan, 2006); Jonathan Finn, "Seeing Surveillantly: Surveillance as Social Practice," in *Eyes Everywhere: The Global Growth of Camera Surveillance,* ed. Aaron Doyle, Randy K. Lippert, and David Lyon (Abingdon, Eng.: Routledge, 2012); and other contributions in *Eyes Everywhere.*

3. Ron Deibert, *Parchment, Printing, and Hypermedia: Communication in World Order Transformation* (New York: Columbia University Press, 1998); James Der Derian, "Virtuous War / Virtual Theory," *International Affairs* 76, no. 4 (2000): 771; Jean Baudrillard, "The Violence of the Image" (2004), http://www.egs.edu/faculty/jean-baudrillard/articles/the-violence-of-the-image; Jean Baudrillard, *The Gulf War Did Not Take Place,* trans. Paul Patton (Bloomington: Indiana University Press, 1995).

4. Nicholan Mirzoeff, *Watching Babylon: The War in Iraq and Global Visual Culture* (New York: Routledge, 2005), 19.

5. David Croteau and William Hoynes, *Media/Society : Industries, Images, and Audiences* (Thousand Oaks, Calif.: Pine Forge Press, 2003).

6. André Gunthert, "Digital Imaging Goes to War," *Photographies* 1, no. 1 (2008): 103–12.

7. Deibert, *Parchment, Printing, and Hypermedia,* 157.

8. Ibid., 117.

9. Doyle, Lippert, and Lyon, *Eyes Everywhere*; Torin Monahan, "Counter-surveillance as Political Intervention?," *Social Semiotics* 16, no. 4 (2006): 515–34; Kevin D. Haggerty and Richard V. Ericson, eds., *The New Politics of Surveillance and Visibility* (Toronto: University of Toronto Press, 2006); David Lyon, *The Electronic Eye: The Rise of Surveillance Society* (Minneapolis: University of Minnesota Press, 2004).

10. Celia Hannon, Peter Bradwell, and Charlie Tims, *Video Republic* (London: Demos, 2008).

11. Rune S. Andersen, "Remediating #IranElection: Journalistic Strategies for Positioning Citizen-Made Snapshots and Text Bites from the 2009 Iranian Post-election Conflict," *Journalism Practice* 6, no. 3 (2012).

12. Susan Sontag, *Regarding the Pain of Others* (New York: Picador, 2004); Gunthert, "Digital Imaging Goes to War."

13. *Standard Operating Procedure,* directed by Errol Morris (Sony Pictures Home Entertainment, 2008), DVD and Blu-ray.

14. Jean Burgess and Joshua Green, *YouTube: Online Video and Participatory Culture* (Cambridge, Eng.: Polity, 2009).

15. Guy Debord, *The Society of the Spectacle,* trans. Donald Nicholson-Smith (New York: Zone Books, 1994), para. 6.

16. Sontag, *Regarding the Pain of Others.*

17. Frank Möller, "The Looking/Not-Looking Dilemma," *Review of International Studies* 35, no. 4 (2009): 781–94.

18. Lilie Chouliaraki, *Spectactorship of Suffering* (London: Sage, 2006); Lilie Chouliaraki, *The Ironic Spectator: Solidarity in the Age of Post-humanitarianism* (Cambridge, Eng.: Polity, 2012).

19. Baudrillard, *Gulf War Did Not Take Place*; David Campbell, *Politics without Principle: Sovereignty, Ethics, and the Narratives of the Gulf War* (Boulder, Colo.: Lynne Rienner, 1993).

20. Mette Mortensen, "Metacoverage Taking the Place of Coverage: WikiLeaks as a Source for Production of News in the Digital Age," *Northern Lights* 10 (2012): 91–106.

21. Matt Ratto and Megan Boler, eds., *DIY Citizenship: Critical Making and Social Media* (Cambridge, Mass.: MIT Press, 2014).

22. Barry Buzan, Ole Wæver, and Jaap de Wilde, *Security: A New Framework for Analysis* (Boulder, Colo.: Lynne Rienner, 1998); Lene Hansen, *Security as Practice: Discourse Analysis and the Bosnian War* (London: Routledge, 2006); Axel Heck and Gabi Schlag, "Securitizing Images: The Female Body and the War in Afghanistan," *European Journal of International Relations* 19 (April 27, 2012): 891–913, doi:10.1177/1354066111433896.

23. For a Sundance-winning filmic account of this crisis that employs the narrative of sophisticated visual dissent, see *Burma VJ: Reporting from a Closed Country*, directed by Anders Østergaard (Dogwoof Pictures / Oscilloscope Laboratories, 2008), 35 mm.

24. Andersen, "Remediating #IranElection."

25. Ratto and Boler, *DIY Citizenship*.

26. Andersen, "Remediating #IranElection," 9.

27. Mortensen, "Metacoverage Taking the Place of Coverage."

28. For an analysis of such weaponization in relation to the 2009 postelection conflict in Iran, see Rune S. Andersen, "Citizen 'Micro-journalism': How #IranElection Was Exploited in Politics and Newspaper Stories," in *Social Media Go to War: Rage, Rebellion, and Revolution in the Age of Twitter*, ed. Ralph D. Berenger (Spokane, Wash.: Marquette Books, 2014), 344.

29. Steven Lee Myers, "U.S. Joins Effort to Equip and Pay Rebels in Syria," *New York Times,* April 1, 2012.

30. Ibid.

31. "Watching Syria's War," New York Times.com, http://projects.nytimes.com/watching-syrias-war.

32. Shapiro, "Strategic Discourse / Discursive Strategy," 339.

Garbage

Michele Acuto

Garbage is one of the most pervasive elements of our society. The world currently generates about four billion tons of waste per year, sustaining a $433 billion industry that discards, transforms, and moves garbage across cities, states, and continents. Waste management might account for as high as 6 percent of employment globally, and it has been proved to have a substantial impact on climate change both via greenhouse-gas (GhG) emissions, such as carbon dioxide (CO_2), and by influencing the near-term production of short-lived climate pollutants, such as methane and black carbon. A long-lived feature of civilization, garbage is now, more then ever, intertwined with global environmental questions. Municipal waste alone now represents almost 5 percent of total global greenhouse-gas emissions, and city landfills are today the third-largest human-made source of methane—a GhG more than twenty times more potent than CO_2. As has been recognized in several multilateral statements, such as the UN General Assembly declaration "The Future We Want," issued after the 2012 Rio+20 United Nations Conference on Sustainable Development, waste now plays a vital role in the development of sustainable futures and in the deployment of effective environmental actions worldwide. Yet garbage remains largely relegated to marginal positions in policy and scholarly analysis, when not black-boxed altogether out of everyone's sight. Seeking to reconnect this international influence with the mundane realities of urban life, this chapter takes garbage as a point of entry to illustrate the links, disjunctures, and translations between the everyday practices of discarding and the international dynamics of global environmental governance. Drawing on assemblage thinking and relying on material from an empirical study of waste management in Sydney, Australia, I attempt to deploy an object-oriented analysis to contrast simplistic views of the mundane as local and prepolitical. In presenting a snapshot of the connection between everyday gestures and global challenges, I illustrate that, if things can be made international, the opposite is also true.

More Than an Object: Waste as Assemblage

Several recent works in the social sciences have been advocating for an understanding of politics as located within that "hidden geography of objects" which has for long unproblematically characterized political analysis.[1] Scholars in science and technology studies (STS) have in the past few years called for a better understanding of the material participation of things in public life, recently spurring a renewed interest in "cosmopolitics."[2] This approach is understood not as the ethical cosmopolitan-politics stance on global citizenship well established in international theory but more holistically as an appreciation of a "politics of the cosmos" that focuses on the dynamics of forging viable cohabitation among all participants in society, whether human, object, or other.[3] In cosmopolitics, things are allowed a greater room for action and interaction with society: the "matter" (read as both "issue" and "materiality") of politics moves from a largely fictional neutrality as "matter of fact" into a participatory position where it takes part in society as "matter of concern."[4] In this sense, then, we may ask, how can the matter of waste be made international as a matter of concern to political scientists?

Students in international relations, urban studies, and political economy have tended for the most part to be "shy of the *stuff* of waste."[5] When it comes to theorizing and rationalizing its societal impact, the materiality of waste seems elusive at best: garbage is produced in innumerable forms and shapes, composed of endless blends of elements and marked by high degrees of local specificity. For instance, the World Bank estimates that the composition of waste varies substantially from low- to middle- and high-income countries, with organic matter accounting for as much as 64 percent in the first and as low as 28 percent in the richer nations, and paper remaining a marginal factor in poorer countries (as low as 5 percent of the total) whereas it represents a larger percentage than organic matter in high-income contexts, where it totals up to one-third of the overall waste mass.[6] Main drivers of this disparity are, of course, economic conditions as well as the substantial variation in management systems available in these places, from something as simple as waste-collection recipients to more refined complexes for sorting, promoting specific discard practices, and preventing wastage in the first place.

Wealth and infrastructures are not, however, the only drivers of this divergence: social, legal, political, and religious norms are all embedded in what ends up (and what does not) in the bins of the world. What makes waste particularly complex and yet easily appreciable is its close relationship with our identities:

we are what we discard, and waste is at least partly what we are. If we stop for a second to consider the contents of our kitchen bins, we will be met by the ghosts of our culinary choices, furniture failures, and private medical matters. Agglomerates of fish bones, shattered lightbulbs, and Band-Aids speak of what we have been in a way that few social-scientific papers can. The same could also be argued for larger-scale containers such as municipal Dumpsters, construction- and demolition-debris trucks, and large shipping tankers cruising off our ports. It might, then, be useful to think of waste as an assemblage of elements whose heterogeneous (that is, inextricably social *and* technical) connections are continuously recast. As assemblage, waste is the emergent effect of, but cannot be limited to, these heterogeneous connections between things, peoples, and places.[7] Garbage, then, takes us a step further in the cosmopolitical agenda, presenting us with networked views of objects and people as inherently intertwined in sociotechnical relations, and offering a reminder of both the precarious nature of political orders and the mundane basis of their stabilization.

Waste represents the conjugation of consumption, circulation, and metamorphosis, and cannot be restricted to a simple set of qualities and a specific detail of geographical coordinates. This unfinished and pervasive quality might seem, at least on the surface, to be an unnecessary theoretical somersault when trying to investigate the international relevance of mundane things such as waste. Yet, although epistemologically challenging, an assemblage view of waste probes the ontological and normative limits of international thinking. Scrutinizing waste, and its international ramifications, through the figure of the assemblage can help us to unpack the reciprocal agency of people and materials in connecting disparate geographical contexts, situated practices, and globalized processes.[8] As Nicky Gregson points out, the assemblages of waste are inherently "fragile achievements," both because their extensive presence in society is continuously recast by millions of people and because of their multiscalar nature: as a mangle of things, norms, and institutions, waste is assembled in certain locations, then moved, reassembled, and mutated in other places, traveling not just materially across geographical extents but also discursively across political-economic processes.[9] From this viewpoint, waste prompts us to challenge the stability of those preconceived political orders that social theorists tend to deal with, whether in their international or national form, but at the same time offers us a practical entry point from which to investigate the creation of such orders in the everyday lives of those who are embedded into them.

Making Garbage International

One of the most common myths about waste is that garbage is dealt with locally and simply ends up buried somewhere. Although it is certainly true that, as the World Bank estimates, dumps and landfills constitute almost three-quarters of the waste-management options in both low- and high-income countries, the story of trash does not end at the dump and often even bypasses it directly.[10] Most waste quite quickly stops being accounted for as waste by authorities and private business. It mutates into recycled materials for domestic and international export, or dissolves into energy via incineration plants, or even becomes a tradable asset precisely because of its waste quality, ending up disposed of in places and countries other than its production sites.

Waste is made international in two often dissociated ways. On the one hand, garbage is rematerialized in something other than waste and dispersed globally via global markets and commercial flows. On the other hand, waste has more recently begun to be internationalized within the global politics of the environment by being rediscussed as an issue of direct relevance to the success of national and international sustainable-development plans. Let me start from the latter point.

The global impact of waste flows has been, for a few decades now, the focus of some international environmental institutions. This is not a surprising fact if we consider the sheer dimension of these flows. A number of treaties have been put in place to deal with this mass of waste. The most well-known case is that of the Basel Convention on the Control of Transboundary Movements of Hazardous Waste and Their Disposal, signed in 1989. Now effective in more than 179 countries, it was established to prevent unregulated export of hazardous waste from developed to developing countries and to ensure minimum international standards of environmental management of toxic wastes. International cooperative action on cross-national movements of waste continues to the present day both in stand-alone forms, such as the Basel Convention, and in more complex environmental instruments, such as the 2001 Stockholm Convention on Persistent Organic Pollutants or the 1998 Rotterdam Convention.

These conventions have thus far had a relatively effective role in signaling several malpractices of waste exporting. However, the actual international impact of these instruments remains questionable, and regional politics have often sought to provide alternative solutions or complementary refinements to widely perceived shortcomings of (near-)global treaties such as the 1991

Bamako Convention. As with many other areas of global environmental politics, nongovernmental organizations (NGOs) have been particularly proactive in pushing for more accurate international oversight and in advocating against perceived disparities and abuses in the commerce of waste. The most-cited case is that of the Basel Action Network, an NGO named after the homonymous treaty, which works as a watchdog over the convention's ratification and the fair implementation of regulations regarding toxic and hazardous waste. Yet, overall, the global outlook of waste production and management is far from comforting.[11] The landscape of global environmental action on waste is still characterized, as Kate O'Neill noted in 2001, by delays in implementation, weak enforcement, and a continuous possibility of regulatory capture in international negotiation.[12] Trends that O'Neill identified as particular to the control of hazardous waste (such as pesticides and radioactive waste) have now spread across the whole waste-management landscape: in particular, substantial changes in the structure of the waste-disposal industry have seen the emergence of major multinational corporations such as Veolia, which are now shaping both localized and transboundary movements of waste.[13]

Though encouraging, initiatives such as the Basel Convention signal the limits of the international attention for waste. One could argue that waste is generally made international in the same way as several other transnational environmental problems, such as ozone-depleting substances in the 1987 Montreal Protocol process. Yet there remains a selective attention by diplomats to particular waste and to specific materials, such as nuclear refuse, that, although certainly worthy of scrupulous consideration, tend to catalyze the already-marginal attention garbage gets on the international stage. If, in a classic international theory fashion, we were to tell the story of how waste is made international by focusing on these regulatory and transnational-advocacy dynamics only, we would be missing the vast majority of those four billion tons and of the $433 billion industry giant characterized by the yearly complex of waste generation and disposal. We would also be missing the scalar depth of the (global) cosmopolitics of waste, which unravel concurrently across a variety of sociopolitical layers while not overlapping neatly. A "hidden mountain" of everyday waste is black-boxed out of our sights, produced deep inside the state casing of most of these processes, characterized by less hazardous connotations, and moved across political channels, commercial pathways, and governmental layers that tend to stand off the radar of world politics.[14] For the most part, waste management and policy remain sidelined matters seen as the purview of cities and local governments. We need to go "deep," beyond the flatness of international regimes, to understand the visceral pervasiveness of waste.

Garbage has become, in the past few years, an increasing presence in discussions other than those of municipal policy, legal international trade, or local sustainability regimes. Networks of local governments and nongovernmental actors, in particular, have been developing alternative solutions beyond the transnational movement of waste, which seems deadlocked in the stalling negotiations of international processes aimed at tackling the localized management and environmental impact of waste at its origins. This is the case, for example, of the Climate Leadership Group (or C40), a network of fifty-eight major metropolises gathered to confront GhG emissions and climate impacts through information exchange and coordinated municipal activities. The C40 has been proactive in the past few years in promoting cross-boundary discussions on the potential for cities to project international influence "from below," via direct action at the city level, which can be an effective response to global challenges. On garbage, C40 has promoted the establishment of a solid-waste partnership among several of its member cities and has aimed at better waste management for GhGs and a methane-emissions abatement. This initiative has been recently expanded by linking with the newly inaugurated Climate and Clean Air Coalition (CCAC), coordinated by the UN Environmental Program in partnership with six major international players to manage the impact of short-lived climate pollutants.[15] The C40-CCAC connection, as I have argued elsewhere on the C40 in particular, illustrates how cities are increasingly intertwined with more-than-local responses to transnational challenges.[16] Connecting waste management to coordinated global action by city governments, these initiatives further promote the multilevel reciprocity of world and urban politics.

From waste-specific discussions such as those of the Basel Convention, waste is being reassembled into dominant discourses of global environmental governance such as climate change or green growth. Waste is made a matter of international concern in itself, while it is also being weaved into more complex global environmental discourses. The UN Rio+20 resolution "The Future We Want" embodied much of this trend. It called not only for the customary increase in cooperation among covenants and institutional instruments already in place, but also for a greater role of public-private partnership in waste management, including in prevention and in the promotion of a view of waste as resource.[17] Whether in the shape of the C40's Solid Waste Partnership or as part of larger statements such as those of Rio+20, the boundaries of what garbage means internationally are currently being recast—and cities and multinational companies have a key stake in these shifts. The materiality of waste flows that circulate across the seas, then, is only part of the story of the internationalization of garbage. As processes like the Basel Convention testify, waste is also assembled discursively into the wider domain of global environmental governance. Overall, we could then argue that waste is made international by moving physically across boundaries as much as by emerging as a relevant element of global environmental discussions and processes, not only in the international arena but across the whole spectrum of governance, urban and home politics included.

Making the International into Garbage: A Climate Issue

How do these macropolitical processes translate deep into the everyday mundanity of waste? In order to problematize how this happens across a variety of scales, from the international system to the kitchens of central Sydney, it might be useful to rely here on Noortje Marres's frame for "issuefication" as a process through which everyday objects come to "resonate with particular matters of concern" by being assembled, or "charged," with environmental, economic, and (ultimately) political issues.[18] An outlook into the issuefication of global environmental politics into mundane urban practices could help us reconnect international and localized city processes as components of a cosmopolitical whole far more complex and multitiered than an international regime.

As is illustrated earlier, the reassembling of waste into international processes has been progressively sustained by the action of local governments, which appear to bear the greatest responsibilities for this problem, and their private partners, which have the most pervasive reach and implementation capacities in the current landscape of environmental management. In Australia, garbage has featured among other commonly cited urban outputs as a central concern in the City of Sydney's long-term environmental plans. Waste has in this sense become a cornerstone participant in the city council's goal of reducing GhG emissions by 70 percent by 2030.[19] Garbage was already featured in the vision of the original *Sustainable Sydney 2030* document, issued in 2007, as a substantial component of the council's attempt to curb global warming. Under the spotlight are more than the improvement of existing waste-management systems and the direct environmental impact of landfilled garbage. Garbage is consistently framed as a core area where Sydney could offer environmental leadership both nationally and globally, and the plans for Sydney's waste are directly linked to wider environmental solutions. Rather than discarded as trash, garbage is being reassembled in the management of Sydney's resources as a powerful asset for energy generation and climate action. In 2007, for instance, waste was envisioned to play an important part in the council's decision to build a network of "green transformers" that would cut the carbon content of electricity and provide low greenhouse hot water, heating, and cooling for both new and existing buildings. The green-transformers approach would allow the city of Sydney to "digest" half of its waste with the aim of returning it as electricity into the local distribution network.[20] Waste management has since become progressively embedded in the city's efforts to curb climate change.

This reassemblage of waste into broader environmental plans is not, of course, just the child of an isolated plan like Sydney 2030. State frames set up by the New South Wales (NSW) government via its Waste Avoidance and Resource Recovery Act of 2001 and Strategy of 2003 have also been influential, with roots in NSW's legislation back to the Protection of the Environment and Operations Act of 1997. Confirming the growing centrality of private actors, substantial policy lobbying and strategic advocacy have also been developed by the business sector through bodies such as Environment Business Australia and the Australian Business Roundtable on Climate Change. (Notably, the impact of the federal authorities remained relatively marginal throughout.) Yet in 2007 the city council itself asserted its primacy in shaping the politics of Sydney's environment and began to push more extensively beyond efficient planning for local problems such as residential waste management by explicitly linking these problems into integrated responses to wider (where not international) challenges, such as climate change and economic austerity. Improving waste management has become as much a matter of local sustainability as a cornerstone of a strategic approach aimed at projecting environmental leadership globally and at attracting tourists and investors via Sydney's prime green lifestyle.[21] In connection with Sydney 2030, the City of Sydney's 2007 Environmental Management Plan made specific target provisions aimed at 66 percent resource recovery of residential and commercial waste by 2014, and classified as a high priority the development of "a comprehensive strategy" to meet reduction targets while formalizing and streamlining many of the aforementioned efforts into a single policy document that complemented the 2030 vision by looking at the same objectives and time frame.[22] This goal does not simply evidence a centralization of waste management in the hands of local government; as many documents leading to, emerging from, and relating to Sydney 2030 can testify, garbage as an environmental issue has been tightly interwoven with a more comprehensive understanding of Sydney's long-term sustainability—a process that has been promoting "integrated management" of waste into other sectors, including energy, transportation, and even food consumption. This move embeds waste more holistically within Sydney's overall environmental outlook. The state-owned company WSN Environmental Solutions, for instance, focused its 2007 campaign to promote public sensibility toward the issue of climate change.[23] Providing people and local governments with guides such as *Waste: It's a Climate Change Issue*, WSN targeted communities and authorities in Sydney to illustrate how decisions made about waste management can reduce the impact of climate change and how local councils hold a crucial role in "leading the community forward in recycling and waste reduction," for both sustainability and climate purposes.[24] In documents like these, garbage is being charged with normative capacities, making waste (and wast*ing*) resonate with a complex of sociopolitical issues ranging from good citizenship to long-term responsibility and sustainable urbanism in a time of economic crisis.

Initiatives like WSN's have a key role in issuifying climate change into waste not only as framing devices, depicting, for instance, how waste accounts for at least 3 percent of Australia's net GhG emissions, but also as connective tissue between waste management and broader sustainability concerns. In *Waste: It's a Climate Change Issue*, WSN coupled council-oriented suggestions such as performing GhG inventories with an analysis of the broader urban ecological footprint and of the direct benefits of recycling for avoiding tons of CO_2-equivalent emissions, saving gigawatts of electricity and megaliters of water each year. As in Sydney 2030, waste intersected with a wide spectrum of problems, from environmental concerns over sustainability and the long-term quality of life in Sydney to economic reasoning over responsible use of natural resources and more affordable lifestyles in a time of financial downturn. This has had the capacity to turn a matter of fact such as garbage into a matter of concern to a variety of key challenges for both municipal policy and individual practices, even if only for a brief moment of reflection in the mundane routines of the urban dweller.

So, how are these complex environmental processes issuified into domestic dynamics? Take sorting trash as an example. Coercion and continuous law enforcement are both extremely expensive alternatives in a large system such as that of waste management. The waste flow is typically regulated instead through the co-option of individuals into community-oriented agencies. Because garbage is produced by and initially managed across the complex threshold of private property and individual domestic practices, and because the mounting limitations to public financing have been coupled with a growing mass of urban dwellers, waste has been progressively controlled by local authorities such as the City of Sydney Council via complex instructions to its citizenry. This has usually been promoted through advocacy and incentive schemes. One such campaign, for instance, has been that against the usage of plastic bags to dispose of waste in the "yellow bins" (for recyclable materials) available to Sydney residents. The council has gone to some substantial length to avoid plastic bags' clogging machinery and contaminating recycling at materials-recovery facilities in the city, both deploying advertisements to alert dwellers of the issue and to provide information as to proper disposal facilities, and teaming up with major retailers and providing free carrier bags made of alternative materials at its own one-stop shop.

Now, as the council has been formulating more explicit plans for climate change and long-term sustainability, these commonplace practices of green campaigning and proper-disposal behaviors have been injected with additional environmental concerns. Conspicuous in the present Sydney landscape is the council's effort to publicize the idea of a "zero-waste" Sydney. Zero Waste is a programmatic approach to garbage that is broadly geared toward reorganizing resource life cycles so as to avoid the disposal of materials in landfills, promoting instead reusing, recycling, and lowering the need for raw materials. With roots that go back to the 1970s, this approach has been taken up by many cities to emphasize the possibility of "closing the loop" against limited conceptions that identified waste management in a linear way wherein the dweller's output is controlled via various technical fixes that ultimately dispose of waste in a safe manner. Zero Waste seeks to provide a more holistic information set to urban dwellers as to the positioning of waste management (both domestically and by the council) within the broader spectrum of sustainability. The city has aimed at recasting not only the dweller's relation with waste disposal and appropriate recycling practices, as through educational online tools and information leaflets, but also the role of the wider waste-management system in Sydney as a key participant in the city's effort to maintaining Sydney as what it calls "one of the most beautiful cities in the world."

Forging connections between waste and sustainability, as well as more specifically between waste and climate change, is important. Yet, precisely because of the variable nature and mundane origins of sustainable development, its promotional shift from problem to asset needs to be necessarily reconnected to the everyday. The new management of waste beyond linear "end-of-pipe" models, which saw garbage being simply removed and disposed of, and the contemporary imagery of waste as a potential resource for energy, material recovery, and ultimately new sustainable futures have been spearheaded by "loop" models of management, such as Zero Waste. Sydney has sought to illustrate not simply how garbage is to be removed out of our sight but how the benefits of its responsible management return to our very households in the form of utilities such as energy and heat, as well as renewed materials such as recycled paper, not to mention even wider (and certainly less tangible) environmental benefits, such as reduced air pollution and mitigated climate-change effects. Zero

Waste campaigning has also helped the council reconnect disparate instruction sets that were generally given as self-standing suggestions to the city's dwellers. For instance, Zero Waste has opened a gateway for information on environmentally sound disposal practices, has promoted interactive tools to instruct the constituency about appropriate sorting in recycling bins, and has functioned as a streamlining channel to advertise council initiatives such as free e-waste drop-off days that ensure recycling rates of problematic waste of up to 95 percent. Injecting broader environmental considerations into the very mundane routines of waste management requires, then, a reassemblage of the everyday into alternative practices and redefined habits. This often means, as in WSN's *Waste: It's a Climate Change Issue*, illustrating the more-than-local impact of garbage via further clarifications and campaigning on the mundane practices of urban dwellers. Changing habits also entails the creation of alternative pathways to the projected new circular networks of waste management, as well as the emergence of new conceptions of waste. An example of this is the council's promotion of composting and community farming. The City of Sydney has put in place a sustainability-oriented Community Gardens Policy since 2009 as part of the broader Sydney 2030 effort, and has progressively been advocating with its residents the twin local-global advantages of this practice, currently hosting eighteen gardens and three footpath pilot projects. Gardens have been used to promote composting of organic waste as a means to reduce the percentage of landfilled materials as well as to further promote a circular view of waste as directly intertwined with food. The council has in this sense been encouraging dwellers with a joint economic-environmental message of multiscalar reach: "Reduce your carbon footprint, save on your groceries." This initiative also illustrates that the role of the city does not stop at pure campaigning but extends to helping prospective community gardeners with grants (both cash and in-kind support, such as hiring a public venue) and helping to kick-start gardens and to encourage volunteering and community action.

Making the International into Garbage: An Everyday Practice

The issuefication of climate change into the mundane practices of waste disposal is not just about new initiatives. It also implies tinkering with the ordinary assembling of garbage. This demands a redisposition of the elements that make up that assemblage, not just the recasting of habits. The enactment of broader environmental concerns into a mundane practice such as deciding what needs to be discarded and where it has to happen requires redesigning the fixed passage points through which the highly mobile materiality of waste is controlled. Particular objects become necessary "interfaces" connecting disparate issues and governance networks.[25]

Bins are the most common example of this process in the domain of waste management. The City of Sydney operates using two major containers, a red-lid bin for general waste and a yellow-lid bin for mixed recycling materials, along with an additional green-lid bin for garden organics. Yet waste cannot be scripted with political and environmental prescriptions. Rather, as an infinitely variable assemblage of individual choices, consumer profiles, and changing materials, garbage requires authorities (whether local or national) to mold the mundane dynamics that lead to waste and that emanate from it. A lot of efforts from municipal authorities have thus far gone into illustrating the correct practices of waste disposal, both in the form of sorting instructions (what goes in which bin) and in highlighting alternative and responsible recycling avenues (what does not and should not go in bins). For the most part, these efforts have taken the form of public-awareness campaigns, such as one campaign that addresses taking special care of problematic garbage (e-waste, used batteries, and the like), all of which have generally reached urban dwellers as leaflets, billboards, public-awareness town-hall meetings, and more recently through new-media campaigns and community events.

These efforts are now also recalibrated toward embedding those broader environmental concerns emerging in Sydney 2030. Interfaces that allow for the mobility of waste, like bins, become key collaborators for the "mundanization" of international issues such as climate change into the everyday. Yet purely technological fixes cannot be entrusted to solve the translation of (new) concerns into domestic sorting practices. Interfaces often require additional support to prompt a change in habits and a reassemblage of waste. Bins, for instance, are equipped with instruction sets (as sticky labels on the bins' lids) illustrating what can and cannot be recycled, but these need to be read, translated, and made habitual by the city dweller. This process allows us to notice how different political spheres intersect with the everyday habits of urban dwellers to make and remake the complex system governing waste. The task of inserting environmental issues into everyday dwelling experiences is far less that of a mission creep that can be carried out by local and central governments through top-down scripts and regulations, and more that of a (partly) open negotiation between the individual and his or her political allegiances, which ultimately creates the very texture of governance realms like home and city. As Kersty Hobson noted in surveying one of the Sydney-based sustainable-living programs, bins are not simply the "conduits for normalization," imposing sustainable living onto urban dwellers and their garbage: they can also "solicit practices that forge specific socio-material relations"—but these relations still rest on the dialectic of individual behavior and governmental imposition.[26] Crucial for the translation of environmental concerns into mundane practices are not only the council's direction but also the response and initiative of dwellers who themselves play a key role in charting novel geographies of garbage. For as powerful as the sustainability framing of Sydney 2030 might be, dwellers still have to play their part in discarding, and discarding according to rules. Yet individuals might also opt for unexpected paths. For example, Sydney has often pushed for the resale and reuse of potential waste materials. In 2010, a small group of residents from the famed beachfront neighborhood of Bondi promoted a "garage-sale trail" that extended among 126 households in a single day, moving an estimated fifteen shipping containers' worth of unwanted items and secondhand goods, and profiting an average $750 per sale.[27] Inspired by this initiative, the city council sought to match the pop-up spirit of Sydneysiders with its environmental commitments to double the amount of municipal waste diverted from landfill toward recovery, following suit in 2011 by promoting an even larger garage-sale trail sprawling across the suburbs of Surry Hills and Glebe. Programs such as Garage Trail and Community Gardens, then, have had the effect of illustrating alternative paths to environmental management, not just on the macro scale of the city but also in negotiation with the habits, pop-up initiatives, and everyday gestures of dwellers, embedding possibilities and paving the way (at least rationally, but often also legally and physically) to the development of new waste rationalities that move from a view of waste that "just is" to one of waste that "can be." Garbage has undergone a progressive transformation from a common management problem to an essential asset to environmental strategies and long-term green visions. These shifts cannot just be embedded into people's lifestyles via new interfaces or coercive new practices; the reassemblage of waste management needs to be negotiated into preexistent and extremely complex systems of habits, governance networks, and material flows. Bins, of course, are but one of the type of actants that can be reassembled in the intricate connections of waste management. As incentives toward recycling and waste-to-energy conversion are put in place in strategies such as Sydney 2030, the multiscalar assemblages of waste management from home to the international market see a shift in actants' positioning and responsibilities. Examples abound: new materials-recovery facilities become nodes in the connections of waste mobility; alternative shipping routes and transport vehicles make their appearance on the scene; new jobs emerge; and new policy or management officers move into alternative positions at city hall as much as in the state and federal government. Garbage becomes analytically crucial in this changing multiscalar context, then, because it offers an analytical window into the stabilization of environmental-governance processes through mundane practices. As Anna R. Davies notes, waste also "marks the beginning of relocation and re-materialisation processes which are conducted at varying scales, from the molecular to the international over different time periods."[28] Looking into the dynamics that embed environmental governance into the practices of waste management also provides us with a vantage point from which to unpack how the politics of the environment are charted, from the everyday realm of kitchens and backyards to the big-picture contexts of treaties, conventions, and global trade.

Conclusions

Garbage offers a crucial test for the critical capacity of a cosmopolitical line of reasoning to open up the analytical horizons of international theory. However, as I have illustrated, it is necessary to think of waste more as an assembled entity whose mobility pervades multiple scales of society. The political tales of waste can underline how mundane matters, when considered as more than just matters of fact, are not just matters of concern but also concerns embodied in matter. Garbage is relevant as a microcosm of politics scaled into the dynamics of everyday life. When we think of waste as assemblage, we are confronted with a fluid and multiscalar complex of materials and people that is precariously stabilized in relatively locked-in cosmopolitical arrangements. This means in practice that these arrangements ensure an everyday response to the problems of waste by maintaining some degree of routine and reliability in the connections between garbage, garbage producers, and garbage collectors. These assemblages are characterized by relatively stable "everyday geopolitics" that provide for some degree of coordination practices.[29] Yet new political projects and strategies such as Sydney 2030 destabilize these precarious structures, requiring more or less formalized adjustments in the assemblage of waste. The process of making international concerns like climate change into things of mundane proportion but assembled multiscalar reach, like garbage, also implies a partial reorganization of the geographies of responsibility of this assemblage. As Marres suggests, the issuefication of matters such as climate change and environmental sustainability into everyday realities is evidence of a "wider ontological process" in place to recast the connections between the geographies of the mundane and broader governance spheres. As stabilized cosmopolitical arrangements, the assemblages of waste are also underpinned by governance structures that ensure, to paraphrase Harold Laswell's famous formulation, who gets rid of what, how, and when—assemblages that are not just fixed in time but changing and being negotiated by all of us on a daily basis.

Notes

1. Bruno Latour, "From Realpolitik to Dingpolitik," in *Making Things Public: Atmospheres of Democracy,* ed. Bruno Latour and Peter Weibel (Cambridge, Mass.: MIT Press, 2005), 4–31; Noortje Marres, *Material Participation: Technology, the Environment, and Everyday Publics* (Basingstoke, Eng.: Palgrave Macmillan, 2012).

2. See Noortje Marres and Javier Lezaun, "Materials and Devices of the Public: An Introduction," *Economy and Society* 40, no. 4 (2011): 489–509. On cosmopolitics, see Isabelle Stengers, *Cosmopolitics,* vol. 1 (Minneapolis: University of Minnesota Press, 2010).

3. Jane Bennett, *Vibrant Matter: A Political Ecology of Things* (Durham, N.C.: Duke University Press, 2010).

4. Latour, "From Realpolitik to Dingpolitik," 31.

5. Nicky Gregson and Mike Crang, "Materiality and Waste," *Environment and Planning A* 42, no. 5 (2010): 1026.

6. Daniel Hoornweg and Perinaz Bhada-Tata, *What a Waste: A Global Review of Solid Waste Management* (Washington, D.C.: World Bank, 2012), 19.

7. Michel Callon and John Law, "Agency and the Hybrid Collectif," *South Atlantic Quarterly* 94, no. 4 (1995): 481–507.

8. Anders Blok, "Greening Cosmopolitan Urbanism?," *Environment and Planning A* 44, no. 10 (2012): 2327–43; Bruno Latour, "Whose Cosmos, Which Cosmopolitics?," *Common Knowledge* 10, no. 3 (2004): 450–62.

9. Nicky Gregson, "Recycling as Policy and Assemblage," *Geography* 94, no. 1 (2009): 64.

10. Hoornweg and Bhada-Tata, *What a Waste,* 24.

11. Sarah A. Moore, "Global Garbage," in *Global Political Ecology,* ed. Richard Peet, Paul Robbins, and Michael J. Watts (Abingdon, Eng.: Routledge, 2011), 133–44.

12. Kate O'Neill, "The Changing Nature of Global Waste Management for the 21st Century: A Mixed Blessing?," *Global Environmental Politics* 1, no. 1 (2001): 77–98.

13. Ibid., 78.

14. Edd de Coverly et al., "Hidden Mountain: The Social Avoidance of Waste," *Journal of Macromarketing* 28, no. 3 (2008): 289–303.

15. The CCAC founders are Canada, Denmark, the European Commission, Germany, the Netherlands, Norway, Sweden, and the United States.

16. Michele Acuto, "The New Climate Leaders?," *Review of International Studies* 39, no. 4 (2013): 835–57.

17. UN General Assembly, "The Future We Want," Resolution A/RES/66/288, July 27, 2012, pp. 41–42.

18. Noortje Marres, "The Environmental Teapot and Other Loaded Household Objects: Reconnecting the Politics of Technology, Issues, and Things," in *Objects and Materials: A Routledge Companion,* ed. Penny Harvey et al. (London: Routledge, 2013).

Marres's focus was on cases such as green teapots and smart meters, whereas with waste we are in the presence of a more complex entity that is continually reassembled and thus ever-changing.

19. City of Sydney Council, *Sustainable Sydney 2030: The Vision* (Sydney: City of Sydney Council, 2008), 29.

20. Ibid, 30.

21. I have illustrated this more at length in "Ain't about Politics? The wicked Power-Geometry of Sydney's Greening Governance," *International Journal of Urban and Regional Research* 36, no. 2 (2012): 381–99.

22. Recognized in a number of council statements since 2007, the call for a comprehensive waste strategy was echoed in several fieldwork interviews with the council I carried out leading to "Ain't about Politics?"

23. WSN Environmental Solutions, *Waste: It's a Climate Change Issue* (Sydney, Australia: WSN Environmental Solutions, 2007). WSN has since been privatized: SITA Australia, owned by French utility company Suez Environment and Singaporean multinational SembCorp, acquired WSN in February 2011 for AU$235 million, making SITA the country's largest waste-management network.

24. Ibid., 10.

25. As with Marres's smart meters, which stand in between global environmental challenges, local energy grids, and individual appreciation of energy consumption, waste has a variety of "interfaces" assembling multiple scales. See Marres, "Environmental Teapot."

26. Kersty Hobson, "Bins, Bulbs, and Shower Timers: On the Techno-ethics of Sustainable Living," *Ethics Place and Environment* 9, no. 3 (2006): 317–36.

27. Clover Moore, "Getting on the Garage Sale Trail," Environment, ABC, April 8, 2011, http://www.abc.net.au/environment/articles/2011/04/08/3185880.htm.

28. Anna R. Davies, "Geography and the Matter of Waste Mobilities," *Transactions of the Institute of British Geographers* 37, no. 2 (2012): 191.

29. Prompted by a growing interest in the geopolitics of home and of popular culture, human geographers are increasingly at the forefront of this focus. See Jason Dittmer and Nicholas Gray, "Popular Geopolitics 2.0: Towards New Methodologies of the Everyday," *Geography Compass* 4, no. 11 (2010): 1664–77; Katherine Brickell, "Geopolitics of Home," *Geography Compass* 6, no. 10 (2012): 575–88.

Carbon

Chris Methmann and Benjamin Stephan

Climate politics is obsessed with carbon. We disclose and offset carbon, we try to capture and store it, we strive to become carbon-neutral, we trade it and we tax it. We calculate the carbon contained in trees and peatlands, we label the carbon that goes into producing our food. Business elites gather in the Carbon War Room. Everyone has a carbon footprint. A huge and globalized scientific, political, economic, and social apparatus, a vast machine seeks to administer carbon on a global, regional, national, local, and individual scale.[1] Carbon has become the thing in climate politics. It is the currency as well as the concern, the object as well as the subject of climate change.[2] It is, so to speak, the material condensation of our concern with a warming world. Climate politics effectively is carbon politics.

How does carbon constitute, structure, legitimate, and affect international climate politics? Based on our research, we approach this question in four steps. First, we investigate how carbon became so central for all practices relating to global warming. Second, we investigate the core elements of today's carbon governmentality and how it shapes international politics. Third, we highlight that the carbontology of climate politics rests on shaky grounds; for carbon has a life of its own, it is uncooperative and so constantly undermines the foundations of carbon governmentality. Finally, we address the effects of carbon governmentality and its unruly object, namely, the dynamics of de- and repoliticization.

What We Mean by "Carbon"

Before we depart on our trip through carbon country, let us briefly spell out what exactly is meant by "carbon." Obviously, it has little to do with the composites used to produce lightweight airplane parts, bike frames, or smartphone casings. In the context of climate change, "carbon" has become the shorthand for a whole suit of molecules—carbon dioxide being the most abundant—that in their gaseous forms in the atmosphere contribute to the greenhouse-gas effect. The shorthand also refers to the element carbon itself where it exists in forms that can be converted into greenhouse gases: fossil fuels that are being burned, forests that are being destroyed, and so on. And the term *carbon* is used to refer to emission allowances and offset certificates that are being traded on the global carbon markets. As we will show in the subsequent sections, in neither of these instances is carbon tangible. Instead, carbon as the medium and object of climate governance is the result of multiple layers of imagination and abstraction.

Unearthing Carbon

Baron John Baptiste Joseph Fourier is usually recognized as the first person who, in 1827, established a connection between the composition of the atmosphere and its average temperature, terming it the "hothouse effect." The refinement and systematization of meteorological observations and knowledge throughout the nineteenth century, involving the First Meteorological Congress in 1873 and the foundation of the International Meteorological Organization in 1878, allowed for improvements in climate science. In 1863, John Tyndall described vapor as a greenhouse gas and was later the first one to explain past ice ages with a variation in atmospheric CO_2. In 1896, Swedish chemist Svante Arrhenius published a paper in which he argued that doubling the CO_2 concentration in the atmosphere would result in a mean temperature increase of 5°C–6°C. In 1908, he was the first to propose a relationship between industrialization and global warming, predicting that "the slight percentage of carbonic acid in the atmosphere may by the advances of industry be changed to a noticeable degree in the course of a few centuries."[3]

Global warming was not experienced as an existing problem yet. From its very beginning, it was "discovered" as a theoretical phenomenon imagined on the basis of the known physical attributes of carbon dioxide. Already here, climate change was made thinkable in terms of carbon. And carbon was understood as a planetary system, a carbon cycle between the atmosphere and the sea. The usual response of most scientists to Arrhenius's concerns was, however, that enough carbon would be absorbed by the oceans so as to balance the cycle.

It took six decades until the implications of Arrhenius's work were fully acknowledged. In 1957, Roger Revelle and Hans Suess argued that only a certain percentage of emitted carbon was bound by the oceans; the rest would remain in the atmosphere. Humanity, they feared, was in fact conducting "a large scale geophysical experiment."[4] As a result, Revelle persuaded one of his students to establish a permanent CO_2-measurement station at Mauna Loa, Hawaii.

Nonetheless, it took two more decades to establish a scientific consensus on climate change. In 1960, the Keeling Curve, based on the measurements at Mauna Loa, was able to prove that the increase in atmospheric carbon dioxide in one year roughly corresponded to the amount of fossil fuels burned—proving the sea-compensation hypothesis wrong. During the 1970s, efforts were made to establish reliable climate modeling (as opposed to mere weather forecasting, which had dominated international meteorological cooperation before). These efforts proved successful when, in 1979, both the U.S. National Academy of Sciences and the First World Climate Conference, organized by the World Meteorological Organization, concluded that these models appeared to be plausible in predicting further global warming through carbon emissions. Despite all these concerns, it was not until the 1980s that global warming came to be experienced as an actual phenomenon. Only then the historical temperature measurements were interpreted as displaying an anomalous warming since the middle of the twentieth century. In other words, global warming was discovered roughly a century before it was actually experienced, simply by extrapolating from the physical attributes of carbon and other greenhouse gases.

When climate science emerged, after World War II, it did so in two distinct and often-competing scientific discourses: carbon-cycle science and meteorological modeling. Only the growing concern about global warming toward the end of the twentieth century provided a window of opportunity for merging the two.[5] This combination, in turn, constituted the terrain of intelligibility for the emerging problem of climate change. Ever since Arrhenius's first publication, it had been rendered as a problem of stocks and flows of carbon: the earth's carbon cycle.[6] And today, "everything we know about the world's climate—past, present, and future—we know through models."[7] In other words, climate scientists try to capture the carbon history of the global atmosphere and correlate it to historical climate and other geophysical data in order to get a sense of how carbon dioxide and other greenhouse gases influence the atmosphere. These historical records then enable and refine the models that seek to predict the future climate. The aim of these measurements and calculations is crucial for understanding the carbon cycle of the earth.

As the research became more sophisticated, climate scientists increasingly considered greenhouse gases other than carbon dioxide and started to include them in their models.[8] However, it was necessary to compare the different greenhouse gases regarding their impact on the atmosphere. The Intergovernmental Panel on Climate Change (IPCC) thus introduced the Global Warming Potential (GWP) measure in its *First Assessment Report*.[9] For the calculation of the GWP, the IPCC chose CO_2 as the reference molecule, assigning it a GWP

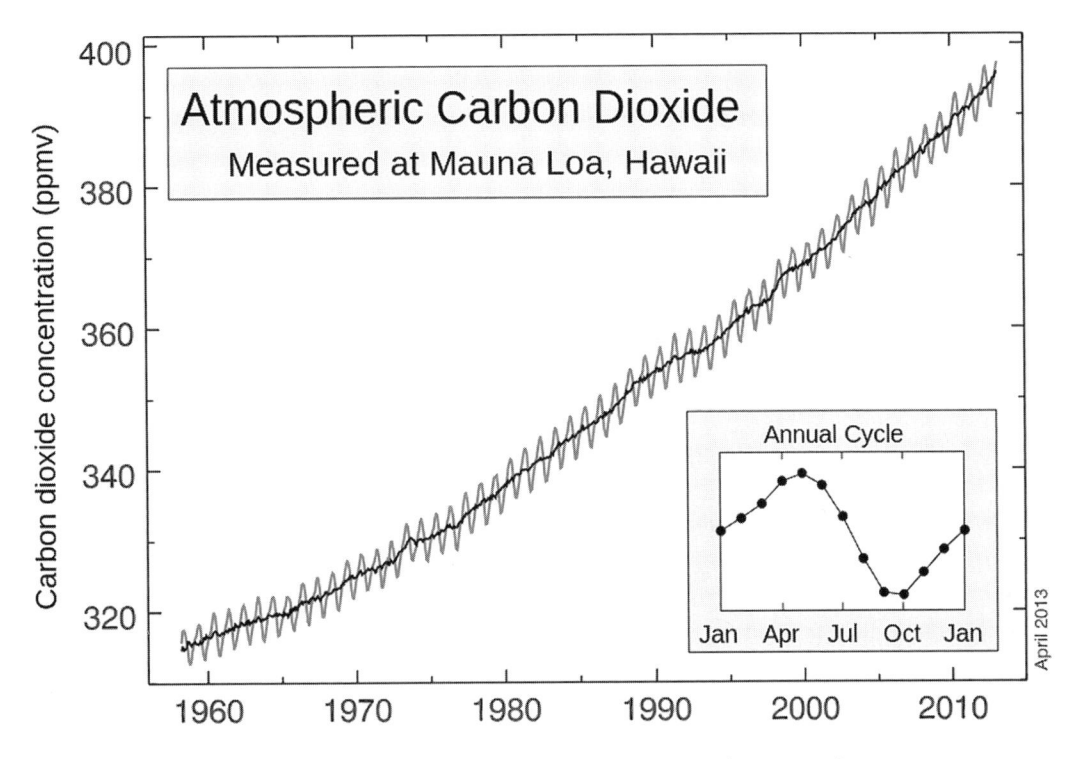

Atmospheric Carbon Dioxide
Measured at Mauna Loa, Hawaii

The history of atmospheric carbon-dioxide concentrations as directly measured at Mauna Loa, Hawaii. This curve is known as the Keeling Curve, and it is an essential piece of evidence of the human-made increases in greenhouse gases believed to be the cause of global warming. Wikimedia Commons. Narayanese, Sémhur, and NOAA.

of 1. Methane, for example, has a GWP of 25, meaning its global-warming effect is twenty-five times higher than that of CO_2. HFC-23, by contrast (a by-product in the production process of many refrigerants), displays a GWP as high as 14,800.[10] These numbers are the result of mathematical methods and spectroscopic measurements and can be disputed on different grounds. For example, comparing the warming effect only within the first one hundred years after releasing a gas into the atmosphere, as the Kyoto Protocol does, is "in a sense arbitrary."[11] The GWP values for the different gases would vary significantly if one chose a longer or shorter time period. Such a decision thus influences how damaging different gases appear to us and how much we value their mitigation, for example, in carbon markets. Furthermore, the calculation of the GWPs involves a significant level of uncertainty. The values for many gases had to be adjusted over the course of subsequent IPCC assessment reports. Despite these problematic aspects, the idea of linking different greenhouse gases through the concept of the GWP has been very influential. It has added an additional level of abstraction to our understanding of carbon, making it convertible for different greenhouse gases and so giving it an even more central place in today's governance of climate change.

Governing Carbon

Given the centrality of carbon for understanding global warming, it comes as no surprise that it also heavily informs the political approaches designed to mitigate global warming. The cornerstone of international climate politics is the United Nations Framework Convention on Climate Change (UNFCCC). It was adopted in 1992 at the Earth Summit in Rio de Janeiro and aims for a "stabilization of greenhouse gas concentrations in the atmosphere at a level that would prevent dangerous anthropogenic interference with the climate system."[12] Accordingly, climate politics, on most levels, clearly has a bias toward the output side of the problem, namely, carbon emissions. However, dissecting the politics of global warming in general, we discern different rationalities and dynamics that make climate change governable. They have evolved successively, both complementing and contradicting each other, and forming the more or less coherent whole of international climate governance as we know it today. In this section, we outline those rationalities and dynamics in detail.

The first rationale presents climate change as an inherently global problem. It directly emerges from the history of climate change as global carbon-cycle change and crystallizes in the UNFCCC. The whole treaty represents climate change as a problem of common, global concern for all countries alike.[13] Accordingly, climate change is subject to joint planetary surveillance, management, and control. Climate change is made governable on a global scale, backed up by large-scale scientific assessments, at the highest levels of international politics. This corresponds to the image of "spaceship earth," which "humankind is able to steer on the basis of data and models provided by the natural sciences."[14] In other words, the UNFCCC renders the earth's carbon cycle a complex organism comprising both human and nonhuman elements whose circulation has to be carefully managed from a god's-eye perspective, which has led a number of scholars to describe this process as "biopolitical management."[15]

This idea of global management soon came to be complemented by a second rationality for governing climate change. With the adoption of the Kyoto Protocol in 1997 as a concretization of the UNFCCC, a new approach was introduced. Whereas the aim of the UNFCCC could be described as a deterritorialization of the atmosphere, much of the Kyoto Protocol negotiations were concerned with reterritorializing the atmosphere into national carbon inventories:

Although this national greenhouse gas accounting made little sense to carbon cycle science at the time, it was the direct result of the interstate negotiations in the early 1990s. In order to allocate responsibility for climate mitigation efforts among the negotiating states, it was necessary to first know how much carbon is emitted and sequestered within respective state borders. Hence, the global cycling of carbon between the atmosphere, oceans and land—long a preoccupation for climate scientists—had to be broken down into the conventional geopolitical grammar of the nation-state.[16]

Next to scaling down the global carbon cycle into political entities such as nation states, the Kyoto Protocol also introduced a third rationale. With the creation of its flexible instruments for trading carbon-emission-reduction commitments—emissions trading, the clean-development mechanism (CDM), and joint implementation—the Kyoto Protocol introduced the market as a central logic for governing climate change. Having spoken about de- and reterritorialization before, we can best describe the impact of the introduction of the market logic as an atomization of climate politics. The idea of the carbon market spread beyond the formal international climate regime. Companies as well as cities or even individuals now track their carbon emissions and compete for reductions. If one looks beyond emission-trading systems such as the European Union Emission Trading System (EU ETS), this does not necessarily involve formal market competition. Instead, carbon is governed by liberal governmental technologies of agency and performance.[17] For example, cities voluntarily organize in global networks and agree on reduction targets.[18] Individuals compare their carbon footprints or form carbon-rationing action groups (CRAGs).[19] This activity instills competitive behavior in actors to reduce their carbon emissions, although it is not a formal market. The entirety of this mode of governing climate change, instilled through the Kyoto Protocol, has been characterized as advanced liberal government.

With the creation of the Kyoto mechanism came the "coinvention" of the ton of carbon-dioxide equivalents,[20] adding another level of abstraction to our current understanding of carbon. This coinvention took place when UNFCCC negotiators detailed the rules for Kyoto's flexible mechanisms in the Marrakech Accords (2001). In defining the units for the different trading and accounting mechanisms, the Marrakech Accords gave birth to the "metric ton of carbon dioxide equivalent" as an accounting metric by mentioning it sixteen times in the treaty text.[21] The concept builds on the IPCC's Global Warming Potential and enables the translation of greenhouse gases other than carbon dioxide into carbon-dioxide equivalents. The reduction of one ton of methane emissions, for example, equals that of twenty-five tons of carbon-dioxide equivalents. Known today by the abbreviation tCO_2e, it has been at least as important for climate governance since its creation as the definition of the different Kyoto units themselves: "It translated a scientific/nationalist way of accounting for carbon into one that could become fungible for market exchange."[22]

The different dynamics and rationalities that shape the climate regime—the simultaneous deterritorialization, reterritorialization, and atomization of climate politics as well as the biopolitical management and advanced liberal government of climate change—are obviously not without contradictions. For instance, one can either manage carbon globally or decentralize responsibility. All this is held together, though, through the matter of carbon. Carbon is the linchpin between the global and the local, between public and private, and between hierarchical, science-based management and the flexibility of markets. Carbon allows us to think, measure, and problematize the global carbon cycle. But it can also be disaggregated into national carbon inventories, companies' EU ETS allowances, project-based Certified Emission Reductions (CERs), or even an individual's carbon footprint and hence spur individually responsibility, competition, and market efficiency. And based on the promise of being able to exactly attribute carbon to these subunits, they can again be aggregated into global values, maintaining the fiction of precise top-down surveillance and management. In other words, global climate politics are constituted through a "carbon governmentality."[23]

The CDM, which was brought into existence by the Kyoto Protocol, demonstrates carbon governmentality and the combination of these different dynamics and rationalities. Project developers, which may be national authorities, business actors, international investors, or the like, set up an individual greenhouse-gas-emission project in a developing country, say, a single generator powered by landfill gas. After the project has accomplished a complex registration, verification, and validation procedure involving national authorities, external verifiers, and the international CDM executive board, the supposed amount of abated greenhouse-gas emissions is then issued as CERs. These are sold to the developed world, thereby generating an extra revenue, which makes the project profitable. Through the EU Linking Directive, however, CERs can be injected virtually directly into the European ETS, the world's biggest carbon market. In fact, as the CDM is based on an international treaty, it establishes a transnational market. For example, a project developer in Addis Ababa generates CERs with landfill gas, which she sells to the operator of a coal-fired power plant in North Rhine–Westphalia, Germany, who can then stick to burning coal. In the absence of an international post-2012 treaty, the EU has restricted the possibilities for introducing CERs into the EU ETS. Since the beginning of the EU ETS's third phase (2013–2020), only CERs generated through projects in least-developed countries are being accepted.

The CDM is thus based on the premise of climate change as a global problem, trying to link different locations on the earth in the struggle against it. At the same time, it draws on the market logic of advanced liberal government: emissions reductions should be made where they can be achieved in the most cost-effective manner. Although some of the advanced-liberal-government logics are epitomized in the trading aspect of the CDM, biopolitical management enables the methodologies that underwrite the projects by calculating the emissions reductions. The scientific methods used for this purpose are often derived from or closely linked to the attempts to measure carbon globally (see the following section for more details). The reterritorializing moment of the Kyoto Protocol is reflected in the CDM. It has been set up by governments, and its main body, the CDM Executive Board, consists of scientists and national bureaucrats appointed by governments. Nevertheless, it was companies that drove the CDM over the past years: companies that developed the majority of the CDM projects in the global South and companies in the global North—mainly European companies subject to the EU ETS—that bought the largest chunk of the CERs to fulfill emission-reduction requirements. However, once converted into EU allowances and surrendered by a company to its home government, the CERs are subject to a territorial logic. They show up in the national account and can be used by the government to cover the country's obligation under the Kyoto Protocol. Furthermore, CDM credits are also being bought by cities or individuals to offset their carbon footprint, indicating the atomization dynamic we mentioned earlier.

Unruly and Virtual Carbon

How does this amalgamation of biopolitical management and advanced liberal commodification and the simultaneous deterritoralization, reterritoralization, and atomization of climate politics actually work? To be honest, not very well. Governing climate change by focusing on carbon emissions presumes the ability to clearly disentangle it and measure it in sufficient detail. Turning to market mechanism, furthermore, assumes that we are able to equate the destruction of HFC-23 in a Chinese refrigerating plant with the distribution of efficient cookstoves in sub-Saharan Africa or the replacement of a lignite-fired power plant with a wind park in Germany. In this section, we demonstrate that measuring and accounting for carbon is not a simple and straightforward procedure. In many instances, carbon remains uncooperative, resisting our attempts to rule it. This "uncooperative carbon" provides a shaky ground for carbon governmentality.[24]

The central problem of the different regimes of practices that seek to govern carbon is a simple question: How do we actually know how much carbon is emitted and what amount of carbon emissions has been prevented by adopting a certain policy, practice, or behavior? Take, for example, the framework to reduce emissions from deforestation and degradation (REDD+) that is currently being negotiated under the UNFCCC. REDD+ seeks to reward tropical developing countries for avoiding deforestation and forest degradation. It will pay them a financial compensation based on the amount of avoided carbon emissions. It has not been decided yet whether or not REDD+ will be integrated into existing carbon markets. If this is the case, the financial donor would be a buyer that receives carbon credits in return, which could be used to achieve reduction commitments elsewhere. Even so, the basic problem remains the same: the avoided carbon emissions.[25] Two operations are crucial for any form of payment that is made on the basis of avoided emissions.

First, it is necessary to find out how much carbon (in this case it really is the element) is stored in the trees, which if not saved from deforestation would be released as CO_2 or methane into the atmosphere. Yet the only direct way of estimating the amount of carbon stored in a single tree—cutting down a tree, drying and weighing it, and then multiplying the weight of its dry matter by 0.5 (the default conversion rate from dry matter to carbon)—is too laborious for larger forested areas and would thwart the whole idea of saving forests. Therefore, scientists have developed a variety of different methodologies to measure the uncooperative carbon that otherwise resists its management and commodification. They draw on predominantly satellite-based remote-sensing techniques and combine them with biome averages or, to achieve more accurate results, with field inventories. Biome averages (numbers are provided, for example, by the IPCC) come in the form of average carbon content per hectare by ecosystems such as tropical rain forests, subtropical dry forests, or boreal coniferous forests.[26] Although this method is relatively inexpensive, it comes with large uncertainty in values. Most actors perceive it to be insufficient for a commodification of forest carbon on the carbon market. Remote-sensing experts and forest scientists are currently trying to develop an approach that is applicable on a large scale with the necessary accuracy by combining remote sensing and field inventories. Field inventories are conducted on sample spots across the targeted forest area to corroborate remote-sensing data. Scientists measure the breast-height diameter of every tree, determine its species, and map everything with the help of GPS devices. Through allometric equations, they can estimate its dry mass and hence its carbon content.[27] Depending on the method used, additional samples from deadwood, litter, and soil are taken. Satellite remote sensing and the field inventories are linked through mathematical and statistical models, for example, to determine the number and location of field-inventory sites. These approaches give scientists more accurate estimates of the amount of carbon stored in a forest, from which they can derive the amount of emissions that will be released to the atmosphere should the forest be destroyed or degraded.[28]

To know the amount of emissions that have been avoided, a second intellectual operation is needed: we have to find out how much deforestation would take place if certain measures are not taken. In other words, we have to compare the actual state of things to an alternative future—called the baseline or reference level. In the case of REDD+, this is done by extrapolating historical deforestation rates or by creating scenarios on the basis of predictions on aspects such as population growth or infrastructure development. Determining a baseline through counterfactual reasoning is not limited to REDD+ but forms the basis of any compensation or offset mechanism; one needs to determine the emission reductions that are additional to what will happen anyway in order to reward the party avoiding emissions.[29] To calculate the emissions that are being reduced through a CDM project installing wind turbines for the generation of electricity, for example, one has to know what form of electricity generation is being replaced (now and for the duration of the project); or, if the project is to be implemented in an area where electricity is not yet available, it is necessary to determine which form of electricity generation will be taken up otherwise. This counterfactual reasoning has been flagged as highly problematic: a single scenario out of an infinite number of alternatives is being selected as the baseline. In doing so, future societal decisions about infrastructure development or electricity generation, for example, are being foreclosed by transforming them into seemingly apolitical technical scenarios and calculations.

Imagining carbon (in the form of avoided emissions) through counterfactual reasoning adds another level of abstraction to our thinking about carbon. Considering the different levels of abstraction that are at play by the time carbon is being traded on carbon markets, it is fair to conclude that we are looking at a virtual object—virtual carbon.[30] Emissions allowances as the EU issues them are not actual greenhouse gases but simply the right to emit them. Similarly, offset credits like the CERs generated through the CDM represent carbon, which only hypothetically would be emitted in case particular measures are not taken.

This virtual carbon, however, has a few material characteristics that have resulted in problematic effects. Being virtual, a CER or any other offset credit claims to represent avoided greenhouse-gas emissions to the amount of one tCO_2e. As a buyer of this credit, one does not have any idea whether this claim actually holds up. From the serial number of the credit that shows up in one's account, it is not possible to judge the quality of the credit. Theoretically, the quality of the credit has already been verified by an external auditor. In the case of the CDM, the UNFCCC has licensed globally operating certification firms such as Det Norske Veritas (DNV), SGS, and TÜV NORD to do so. These firms, however, are competing for jobs and are hired and paid by the project developers. This mechanism leaves the project developer and verifying company without any fundamental interest in the quality of the credit. This is a particularly precarious situation considering that what the verifiers should scrutinize—the counterfactual assumptions in the baselines—are highly problematic constructions. In the end, nevertheless, most buyers of carbon credits do not mind; they are companies that need the credits to fulfill emission-reduction goals or to claim that they are carbon-neutral. As a result of this problematic structural setup, a high number of credits have been issued that do not fulfill everything they claim to do.[31]

In addition to undermining the environmental integrity of the very mechanisms that create it, the virtual character of carbon has invited bluntly fraudulent behavior. A company does not keep its carbon allowances (being virtual) in a safe but holds them in an online account with the respective carbon registry. In the case of the EU ETS, criminals have easily gained access to a number of these accounts by sending out simple phishing e-mails. Furthermore, as emissions allowances or reductions, credits do not need to be shipped in containers but are moved around via a few mouse clicks. This facilitates tax fraud. EU member states have lost billions of euros to VAT fraud in EU ETS allowances.

The different levels of abstraction underpinning the current understanding of carbon give the impression that we are dealing with clearly defined and neatly disentangled objects. As we have shown, however, these abstractions are unstable and subject to constant challenges, whether through the uncooperative carbon's resisting being measured and accounted for; disputes around issues concerning the larger scientific context, the relative climate impact of different greenhouse gases; disagreements on technical aspects such as appropriate methodologies for offset projects; or cases of carbon fraud and crime. The carefully boxed-in picture of carbon is at risk of breaking out through these challenges.[32] Substantial efforts have constantly to be undertaken by various actors—scientists, policy makers, standard bodies—to contain these overflows and to keep up the image of carbon as a fully disentangled object that can be governed independent of its immediate context.

Depoliticization through Carbonification

What happens if the materiality of carbon brings about a carbon governmentality that is located precisely at the intersection of the de- and reterritorialization of the atmosphere, biopolitical management, and advanced liberal commodification? What are the effects in terms of the political?

Underlying this question is a particular notion of the political. Usually, "politics" refers to what happens in parties, parliaments, and governments, or among nations. Our understanding of the political (as opposed to politics), however, is primarily interested not in the constellation of actors but in the scope and scale of what they argue about. It "has to do with the establishment of that very social order which sets out a particular, historically specific account of what counts as politics and defines other areas of social life as not politics."[33] Although we think that the materiality of carbon as such does not have much effect on the politics of climate change in the traditional meaning of the term, it has a lot of effect on how the boundaries of this very realm, politics, have been drawn. Who counts as in and out what is actually discussed? Where is the horizon of the possible and reasonable? In this sense, whether or not major actors such as the United States and China disagree fiercely on emissions-reductions commitments is not the factor that decides the level of politicization of climate politics. For us, politicization would highlight and question the fundamental causes of global warming: dependency on fossil fuels, industrial agriculture, increase in trade, and individual mobility. Furthermore, politicization would democratize climate politics and include actors such as social movements or the general public into the policy process. Depoliticization, by contrast, would limit the debate to the output side—carbon emissions—and leave it to the chosen few experts. In this sense, the struggle between China and the United States is only superficially politicized, as it basically is a numbers game only: which limits, what schedule, whose conditions. The structural causes of carbon emissions are rarely discussed.

For us, the "carbonification" of climate politics results in a double dynamic of politicization and depoliticization.[34] Acting as the linchpin of climate politics, carbon opens the arena of climate politics to a whole new range of actors. Carbon governmentality enables individuals, nongovernmental organizations (NGOs), and businesses—even cities—to work on their individual carbon footprints. Carbon links efforts on many different levels, in many different parts of the planet, and in various sectors of society with the global endeavor to counter climate change. Frustrated with your own government's climate policy? Working on your carbon footprint enables you to compensate the government's failure by optimizing your own carbon emissions. In this sense, the carbonification of global warming serves as a tool for those willing to show some action. It expands climate politics beyond the traditional climate-change regime and makes everyone a potential climate protector. This holds true even for individual carbon-accounting schemes that sidestep slow international diplomacy and empower individuals.[35] The efficiency narrative that comes with many of the carbon markets, moreover, draws actors into climate politics that usually take an antagonistic stance toward it, such as big transnational corporations.[36]

There is, however, a contradicting second effect. The multiple layers of abstraction we have outlined result in what Erik Swyngedouw has called the "fetishization of carbon."[37] Treating carbon as an object independent of its immediate context renders many aspects of the problem invisible. By governing climate change through carbon, we reduce climate politics to carbon management only; and this has a strong depoliticizing effect. On the one hand, profound questions of equity and justice are transformed into technical optimization questions. The history of the CDM is very illustrative in this regard. Originally, it was brought into the negotiations by some developing countries as a proposal for an equity fund compensating the global South for the historical climate debt of the North. These negotiations, however, soon came to revolve around technical questions of carbon accounting and measurement, and eventually resulted in the creation of the CDM, with the consequences sketched earlier.[38] On the other hand, governing climate change through carbon reduces the actual scope of climate politics to the output side, namely, carbon. Politicians and NGOs argue how much carbon cars may emit, for example, but rarely do they question structural mobility changes, such as car-free cities, which would be to question the use of cars in general. World Trade Organization director-general Pascal Lamy muses about the question whether it is better to import fresh flowers from Kenya instead of growing them in European greenhouses. Whether or not we need these flowers in winter at all is not at stake. We assume the worst (very dirty and very inefficient power plants) in order to justify the bad (dirty and inefficient power plants). Being obsessed with the measuring and accounting of the carbon stored in trees carbonifies forests.[39] It leaves us little room to see forest as a habitat for biological diversity, the home for the people who inhabit it, or anything other than merely carbon stocks. The ubiquitous fixation with carbon management effectively functions as blinders for climate policy. The quest for carbon efficiency obscures the need for effective climate policy. In effect, a growing carbon market with never-ending teething problems receives attention and at the same time distracts from the fact that emissions are still rising sharply. There are alternative options. However, demands such as leaving fossil fuels in the ground, striving for a sustainable life, and eating healthy and organic food are marginalized in the global debate—although their effect on halting global warming could be tremendous.

This is our contradictory conclusion: carbon politics broaden the arena of climate politics if and only if all the newly included actors are willing to subscribe to a particularly narrow form of climate politics that entirely revolves around managing carbon. This, however, forecloses complementary or alternative approaches to the problem. A very complex global carbon-management regime has evolved over the past twenty-five years. However, it has not enabled the stabilization, let alone reduction, of greenhouse-gas emissions. In our opinion, breaking with our obsession with carbon and developing alternative views will significantly increase our chances to tackle the fundamental problem: global warming.

Notes

1. Paul N. Edwards, *A Vast Machine: Computer Models, Climate Data, and the Politics of Global Warming* (Cambridge, Mass.: MIT Press, 2010), 1.
2. Philippe Descheneau, "The Currencies of Carbon: Carbon Money and Its Social Meaning," *Environmental Politics* 21, no. 4 (2012): 604.
3. Svante Arrhenius, quoted in Matthew Paterson, *Global Warming and Global Politics* (London: Routledge, 1996), 54.
4. Roger Revelle and Hans E. Suess, "Carbon Dioxide Exchange between Atmosphere and Ocean and the Question of an Increase of Atmospheric CO_2 during the Past Decades," *Tellus* 9, no. 1 (1957): 19.
5. David M. Hart and David G. Victor, "Scientific Elites and the Making of US Policy for Climate Change Research, 1957–74," *Social Studies of Science* 23, no. 4 (1993): 667.
6. Eva Lövbrand and Johannes Stripple, "The Climate as Political Space: On the Territorialisation of the Global Carbon Cycle," *Review of International Studies* 32, no. 2 (2006): 225–26; William Boyd, "Ways of Seeing in Environmental Law: How Deforestation became an Object of Climate Governance," *Environmental Law Quarterly* 37, no. 3 (2010): 880–84.
7. Edwards, *Vast Machine*, xiv.
8. Andrew Lacis et al., "Greenhouse Effect of Trace Gases, 1970–1980," *Geophysical Research Letters* 8, no. 10 (1981): 1035–38.
9. *IPCC First Assessment Report* (Geneva: Intergovernmental Panel on Climate Change, 1990), 58–61.

10. Susan Solomon et al., eds., *Climate Change 2007: The Physical Science Basis;* Contribution of Working Group I to the Fourth Assessment Report of the Intergovernmental Panel on Climate Change (Cambridge: Cambridge University Press, 2007).

11. Donald MacKenzie, "Making Things the Same: Gases, Emission Rights, and the Politics of Carbon Markets," *Accounting, Organizations, and Society* 34, nos. 3–4 (2009): 446.

12. United Nations Conference on Environment and Development (UNCED), 1992, *The Framework Convention on Climate Change* (Rio de Janeiro: UNCED, 1992), §2.

13. Lövbrand and Stripple, "Climate as Political Space," 25.

14. Angela Oels, "Rendering Climate Change Governable: From Biopower to Advanced Liberal Government," *Journal of Environmental Policy and Planning* 7, no. 3 (2005): 198.

15. Timothy W. Luke, "Environmentality as Green Governmentality: Geo-power, Eco-knowledge, and Enviro-discipline as Tactics of Normalisation," in *Discourses of the Environment*, ed. Éric Darier (Oxford: Blackwell, 1999), 121–51; Oels, ibid.

16. Eva Lövbrand and Johannes Stripple, "Making Climate Change Governable: Accounting for Sinks, Credits, and Personal Budgets," *Critical Policy Studies* 5, no. 2 (2011): 192.

17. Mitchell Dean, *Governmentality: Power and Rule in Modern Society* (London: Sage, 1999), 168.

18. Michele M. Betsill and Harriet Bulkeley, "Cities and the Multilevel Governance of Global Climate Change," *Global Governance* 12, no. 2 (2006): 141–59.

19. Matthew Paterson and Johannes Stripple, 2010. "My Space: Governing Individuals' Carbon Emissions," *Environment and Planning D: Society and Space* 28, no. 2 (2010): 341–62.

20. Matthew Paterson and Johannes Stripple, "Virtuous Carbon," *Environmental Politics* 21, no. 4 (2012): 573.

21. Ibid.

22. Ibid.

23. Chris Methmann, "The Sky Is the Limit: Global Warming as Global Governmentality," *European Journal of International Relations* 19, no. 1 (2013): 69–91.

24. Adam G. Bumpus, "The Matter of Carbon: Understanding the Materiality of tCO_2e in Carbon Offsets," *Antipode* 43, no. 3 (2011): 620.

25. Benjamin Stephan, "Governing the Forest Frontier" (unpublished manuscript).

26. Harald Aalde et al., 2006, "Forest Land," in *IPCC Guidelines for National Greenhouse Gas Inventories*, vol. 4, *Agriculture, Forestry, and Other Land Use*, ed. Simon Eggelston et al. (Geneva: Intergovernmental Panel on Climate Change, 2006), table 4.7.

27. Heather Lovell and Donald MacKenzie, "Allometric Equations and Timber Markets: An Important Forerunner of REDD+?," in *The Politics of Carbon Markets*, ed. Benjamin Stephan and Richard Lane (London: Routledge, 2014).

28. Benjamin Stephan, "Bringing Discourse to the Market: The Commodification of Avoided Deforestation," *Environmental Politics* 21, no. 4 (2012): 621–39.

29. Larry Lohmann, 2005. "Marketing and Making Carbon Dumps: Commodifica-

tion, Calculation, and Counterfactuals in Climate Change Mitigation," *Science as Culture* 14, no. 3 (2005): 203–35; Methmann, "Sky Is the Limit"; Stephan, ibid.

30. Paterson and Stripple, "Virtuous Carbon."

31. Lambert Schneider, *Is the CDM Fulfilling Its Environmental and Sustainable Development Objectives? An Evaluation of the CDM and Options for Improvement; Report Prepared for the WWF* (Berlin: Öko-Institut, 2007).

32. Lohmann, "Marketing and Making Carbon Dumps."

33. Jenny Edkins, 1999, *Poststructuralism and International Relations: Bringing the Political Back In* (Boulder, Colo.: Lynne Rienner, 1999), 2.

34. Ayşem Mert, "Partnerships for Sustainable Development as Discursive Practice: Shifts in Discourses of Environment and Democracy," *Forest Policy and Economics* 11, nos. 5–6 (2009): 326–39.

35. Paterson and Stripple, "Virtuous Carbon."

36. Chris Methmann and Benjamin Stephan, "Political Sellout! Carbon Markets between Depoliticising and Repoliticising Climate Politics," in Stephan and Lane, *Politics of Carbon Markets*.

37. Erik Swyngedouw, "Apocalypse Forever? Post-political Populism and the Spectre of Climate Change," *Theory, Culture, and Society* 27, nos. 2–3 (2010): 216.

38. Mathias Friman and Björn-Ola Linnér, "Technology Obscuring Equity: Historical Responsibility in UNFCCC Negotiations," *Climate Policy* 8, no. 4 (2008): 339–54.

39. Stephan, "Bringing Discourse to the Market"; Benjamin Stephan, "How to Trade Not Cutting Down Trees," in *Interpretative Approaches to Global Climate Governance: Deconstructing the Greenhouse*, ed. Chris Methmann, Delf Rothe, and Benjamin Stephan (London: Routledge, 2013), 57–71.

Currency

Emily Gilbert

Currency Protests

The Occupy George initiative in the United States, which is an offshoot of the broader Occupy movement, has been distributing banknotes stamped with bold red infographics that draw attention to economic disparities in U.S. society. The red overprinting on the one-dollar note in the following figure, for example, graphically illustrates that the four hundred wealthiest Americans hold as much wealth as the bottom half of the country. One of the most iconic symbols of the United States, and of its wealth, the dollar bill, is being used to undermine this very status. These overprinted notes are circulated at Occupy events to draw attention to wealth inequities. The Occupy George website also provides instructions on how to overprint your own currency with a photocopier or how to order custom-made stamps online. In another example, Seattle activists associated with MicCheckWallSt, an offshoot of Occupy Wall Street, dropped $5,000 in overprinted currency from the top of a Seattle building. The notes were printed with the words "Money as speech silences us all" to protest the influence of money in politics.[1] Still another related movement, the Move to Amend campaign, which is challenging corporate spending in U.S. elections, has encouraged its supporters to stamp U.S. currency with slogans such as "Corporations are not people," "Money is not speech," and "Not to be used for bribing politicians."[2]

Adorning currency with protest symbols is also taking place in Canada: the grassroots Idle No More movement for Aboriginal sovereignty has also generated compelling monetary embellishment. Images of modified five- and twenty-dollar bills surreptitiously appeared on Facebook. The notes, adorned with portraits of former prime minister Wilfred Laurier and the queen of England, respectively, depicted these figures in Native headdress, alongside the slogan "Idle No More."[3] Although the notes have been criticized, the Facebook posting encouraged the wide circulation of the currency. In Iran, activists used green ink to write protest slogans, such as "Death to the Dictator," on currency

The red overprinting on the one-dollar bill states that the four hundred wealthiest Americans hold as much wealth as the lower half of the country. From Occupy George, http://www.occupygeorge.com/.

during the 2009 Green Movement in Iran.[4] Other notes have depicted haunting images of those who have been persecuted by the state. In China, the Falun Gong has printed messages on currency to outline its harassment by the Chinese Communist Party.[5] In Zimbabwe, a newspaper critical of the Mugabe regime, the *Zimbabwean*, launched an award-winning advertising campaign to encourage new readership; posters made of the devalued Zimbabwean currency were hung on billboards with the following message: "Thanks to Mugabe this money is wallpaper."[6]

In these examples, currency is being used to convey a political message. At the same time, the inscriptions are drawing attention to the materiality of currency. Most of the time, currency and its materiality are taken for granted. Indeed, "[m]oney's reproduction as an institution depends on how unproblematically it is taken for granted."[7] This taken-for-grantedness is one of the things that makes money so interesting, and perhaps even possible. What, then, does it mean when it is interrupted? What happens when we take the "stuff" that is money seriously?[8] In line with Bruce Braun and Sarah Whatmore, I raise these questions in the hopes of articulating "a more fully materialist theory of politics" that is grounded in practice and attentive to social and political implications.[9] In the following section, I map out the contours of a materialist politics while also foregrounding some of the critical work that has sought to take the materiality of money seriously.[10] I then turn to examine how modern money is embedded in social, political, and economic institutions in ways that are grounded in time and space, in that they are nationally specific but also legible on the international scale. This focus on materiality helps to shed light on what is at stake in the contemporary protests on currency, which is the focus of the subsequent section. There, I consider the challenges to national currencies that have been unfolding over the last several decades, particularly since the collapse of Bretton Woods.

Material Politics of Money

Why study objects? As Braun and Whatmore argue, refocusing attention on the material helps us to understand how it "constitute[s] the common worlds that we share and the dense fabric of relations with others in and through which we live."[11] Human experience is constituted in and through the objects with which we coexist. Objects both enable and constrain meaning while they also they temporalize and spatialize social relations.[12] Jane Bennett has described this as the "enchanted materialism" of objects, and, in her later work, as "thing-power," which refers to "the curious ability of inanimate things to animate, to act, to produce effects dramatic and subtle."[13] For these authors, then, material objects are rooted in time and space and have their own agentic capacity: objects are not just static but becoming (or being rewritten), and their influence is iterative.

The agency that is ascribed to objects offers ways to rethink the political so that it encompasses more than human actors. This also requires rethinking how political space is constituted by and constitutive of humans and non-humans, including objects. Noortje Marres and Javier Lezaun appeal to the concept of the public to address instead "the role of materials and artefacts in the public organization of collectives" and to explore how the material becomes invested "with more or less explicitly political and moral capacities."[14] Bruno Latour meditates upon what an "object-oriented democracy" might look like. So doing, he argues, means taking account of how objects "bind all of us in ways that map out a public space profoundly different from what is usually recognized under the label of 'the political.'"[15] This current work on materialism is pushing toward a more-than-human politics that will reframe the political in light of both the agentic capacities of objects and the ways that objects structure the constitution of the public.

The centrality of money to contemporary life can hardly be overexaggerated. But we rarely stop to think about the money that passes through our hands. Karl Marx's characterization of money as originating out of the commodity form provides clues for critically analyzing its materiality.[16] Money, as with other commodities, is fetishized in that the social relations of production are masked. Drawing upon Marxist theory, we can examine money as a commodity and thus begin to demystify its role as a neutral representation of value and begin to pry open its affective dimensions.[17] Yet money is not just an über commodity but also a unit of account and a standard of value rooted in social relations of credit and debt, which are complicated, conflicted, and confusing.[18] Social theorists such as Georg Simmel and Max Weber have thus moved beyond Marx to examine money's multiple and diverse functions alongside its embeddedness in webs of social relations.[19] This tack has the advantage of moving away from a grand narrative that plots the rise of money as precipitating social decline, with monetary relations determined a priori to be individualizing and alienating.[20] These approaches also upend traditional, mainstream, and abstract theories of the economy that presume that money is simply a neutral tool wielded by rational actors.

Current thinking has also drawn from anthropological accounts of premodern money, such as those of Paul Einzig, Karl Polanyi, Jonathan Parry, and Maurice Bloch, which have examined cowrie shells, Yap stones, and other objects. Anthropological accounts breathe life into money objects, although they often insist on a radical separation between premodern and modern monies.[21] Modern monies such as coins and paper do not have the same liveliness in the eyes of these anthropologists, and the complexity of the social relations in which they circulated is kept obscure.[22] Nevertheless, these accounts have provided a useful frame for sociological and geographical work that attends to money as a social relation that circulates among people across different times and places and through different networks of trust.[23] This research has informed science-and-technology studies of the market that seek to recover the agency of both human and nonhuman actors in the economy.[24] It refocuses attention on money's circulation and, as such, on the performative aspects of money.[25] Money needs to circulate in order for it to work as currency, as a medium of exchange.[26] And as it circulates, it acts as an "intermediary" that knits together assemblages of nation-states, media, financial services, machine intelligence, and other factors.[27]

Modern life is structured around money and depends upon its smooth circulation as currency. It has become so naturalized that its mundane power is rendered invisible: we don't think twice about the coins and bills that we exchange daily or about the morass of social, political, and economic institutions upon which this smooth circulation relies. Disruptions such as the currency protests detailed at the beginning of this chapter are thus important to examine. Although the overprinted inscriptions do not derail the circulation of money, they do slow it down or interrupt its seamless flow. They are an occasion for interrogating the materiality of money that passes through our hands. And, in that they offer a public engagement with money as a form of circulating media, they solicit a public response. They thus both point to the institutions that make the circulation of money possible—banks, markets, states—and push back against these very institutions. In the following section, I examine some of the institutional structures that prop up the circulation of currency and which, in turn, are propped up by money.

Making Currency (Inter)National

The currency protests noted earlier all take aim at the state: its perpetuation of class inequities, its capitulation to moneyed interests, and its persecution of its peoples. In this section, I explore why state power is being targeted in this way and with what implications. I begin by setting out a brief chronology of the long imbrication of currencies and states that illustrates how, as Polanyi has argued, the rise of states, national currencies, and liberal markets has gone hand in hand.[28] This provides the context for my analysis in the following section, where I argue that currency protests are unfolding precisely because the state's relationship to currency is changing—largely because of internationalization and privatization—and because of people's frustration at how citizenship and politics are being reconstituted in its wake.

Modern money has a distinct temporality and spatiality that are rooted in the nation-state. As Eric Helleiner has documented, currencies have been progressively nationalized since the early nineteenth century, so much so that by the mid-twentieth century, national currencies had become axiomatic.[29] England established the first homogeneous national currency, in the early part of the nineteenth century. Other countries of Europe, notably France and Germany, were quick to follow. Although there was considerable internationalization in the late nineteenth century, WWI prompted a more nationalist focus, as "states turned inward to manage their economies in times of war and depression."[30] In this era, the United States and Japan consolidated their national currencies, followed by other independent countries in Europe, Asia, and the Americas in the interwar period. After WWII, national currencies were established in the newly decolonized countries of Latin America, the Middle East, Africa, and Asia.

National currencies were prompted both by domestic economic goals and by aspirations to articulate and strengthen national identities.[31] They were used to promote the "imagined community" of the nation through elaborate nationalist iconography.[32] In turn, national currency depended upon public trust in the state for it to circulate, which was especially true with the shift to paper currencies in the nineteenth century, as this money was not seen to be inherently valuable, as gold and silver were. The public had to trust in the state's capacity and promise to redeem its notes. Thus, just as national currencies spurred nationalist sentiment, they also helped to legitimize the role of the state. And more and more, as national currencies become the predominant form of the territorialization of money, it came to be accepted that a state would have its own national currency and, moreover, that a sovereign state would need to have its own currency to be internationally recognized. Thus, although national currencies may have originated out of domestic interests, by the mid-twentieth century they were integral to the structuring of the international in economic relations, state systems, and national identities.

The territorialization of currencies was not inevitable but the result of hard work, determination, and the imposition of state interests from above. Nowhere is this more evident than in colonial contexts where Western currencies were imposed. Never mind that these colonies would develop their own national currencies while they forged their independence; the colonial project of currency originated as an imposition to extend imperial frameworks that imparted Western ideologies of state and economy. As Jean Comaroff and John Comaroff describe, the introduction of coinage among the Tswana peoples in South Africa was wrapped up in the evangelical campaign to impose Christianity and to supplant African theories of value. In their words, "for 19th century colonial evangelists in South Africa, saving savages meant teaching savages to save."[33] In Papua New Guinea, as Robert J. Foster details, the Australian government went to great lengths to educate the population about Western forms of money, hard work, and saving by producing financial booklets, schoolbooks, and even films.[34] The impositions of Western money did not go uncontested. In Kenya, for example, the Mau Mau insurrection in the mid-twentieth century sought to claim more control over colonial symbolism; "Mau Mau Very Good" was found inscribed on a five-shilling note in circulation.[35] In the early twentieth century, in the Sudan area of West Africa, locals outright refused to use colonial money in protest against their own waning autonomy.[36]

These colonial examples shed light on the state power at work in the production of territorial currencies, but lurking behind the state were the interests of the merchant class. The rise of central banks elucidates this role. For the most part, central banks were expected to be independent of the state, but they really straddled the public and the private spheres and have played an important role as they have gained monopoly over the currency supply and have gradually taken control over other monetary levers, such as interest rates. The origins of the Bank of England help illustrate this hybridity. The Bank of England was formed in 1694 by a consortium of merchants who offered to lend money to King William III to help in his war against France in return for a monopoly over banknotes. This move dramatically changed debt: "money was no longer a debt owed *to* the king, but a debt owed *by* the king."[37] This relationship persists. When central banks print currency, they are financing government debt, in that the amount in circulation far surpasses state reserves. The trick that national currencies play "is thus to turn sovereign debt into public money."[38] In other words, national currencies are a public good, but they are designed in the interests of the government and the central bank, and the creditor class to which they are beholden.

Only in the twentieth century did concerns regarding the public good come to the attention of central banks. Innovations in consumer lending and credit installments had already started to transform private debt as early as the industrial era,[39] but the idea of a publicly oriented national debt (rather than a debt created for sovereign interests) really emerged only with WWII. Indeed, it was in this moment that the "national economy" arose, in that it became a geopolitically bounded "knowable, calculable and administrable object" in the purview of the nation-state.[40] This was made possible, in part, by the currency stability that was introduced with the 1944 Bretton Woods agreement, which pegged Allied currencies to the U.S. dollar, which was in turn pegged to gold. Western states planned their economies through maximizing national outputs and gross domestic product. This mandated a newfound interest in the population and in promoting its productivity, which became factored into national economic calculations.[41] "Social citizenship" emerged from this pivot, the provision of "a modicum of economic welfare and security" by the state for its citizens.[42] Capitalism was thus made palatable by smoothing out, to a certain degree, its intrinsic unevenness. Central banks and governments wielded the levers of both monetary and fiscal policy in the interests of the electorate.[43]

Benjamin Kunkel suggests that one way of understanding this post-WWII era is that Western governments were offsetting their sovereign debt by repaying a public debt through social programs such as pensions and health care, and through bonds.[44] National currencies thus not only came to exemplify national identity through their iconography but also articulated a sense of the national public. The design of modern money has been crucial to stitching together the assemblage of the nation-state. Turning our focus to money's materiality draws this agentic power to light and opens the way for thinking through the ways that we have constituted our understanding of the public in and through money. It also makes it possible to think about how and why there is interest in disrupting these very same structures.

Currency Crisis

For about twenty-five years, the precarious calibration of government debt, financier and creditor interests, and public productivity was mediated by international currency stability and various iterations of the welfare state in the West. In the 1970s, however, the bargain started to unravel. Histories of the era capture well the broad economic impact of the advent of neoliberalism and globalization. Changes to currency were the consequences of this transformation. In 1971, the convertibility of U.S. dollars into gold was suspended by President Richard Nixon, which signaled the end of the Bretton Woods agreement. Now, markets and international investors—not governments—would be responsible for determining exchange rates. The end of Bretton Woods also ushered in a wide range of monetary innovation. In the words of Phillip Coggan, "From that point on, the final link with gold was removed and the ability of governments to run deficits, on both the trade and budget accounts, was vastly increased. Money and debt exploded."[45] Some of this explosion was of state money: states were producing more money to ease repayments of their escalating debts, which led to the high inflation of the 1970s. The emergence of the euro in the 1990s also challenged national currencies, as national currency was reterritorialized at the transnational scale. But the demise of Bretton Woods also marked the beginning of the deregulation of the market and the rescinding of capital controls, which encouraged an explosion of new financial instruments such as Eurodollars, Eurobonds, and derivatives.[46] Moreover, there was a proliferation of alternative currencies, from private electronic currencies issued by leading companies such as Microsoft and Nintendo to local exchange-trading societies that issued their own units of accounting—either engaging with or exercising a backlash against globalization.[47] Finally, a revolution occurred in forms of payment, such as the rise of credit cards, which transformed individual debt and public accounting, and accentuated money's abstraction.

The decline of national currencies and the rise of international and private monetary innovations were seen to be a boon to global trade. Yet Susan Strange calls the financial innovations of the market "mad money": they are "erratic," "unpredictable," and "irrational," largely because they operate outside the purview of governments.[48] The monetary innovations also followed different scripts from national currencies, many of which were either "single-purpose" or did not fulfill all the roles of national currency.[49] Indeed, money and finance became so complicated that very few people, experts included, grasped the full extent of the implications of the emergent new forms of money. In this chaos, the new monetarist orthodoxies of Milton Friedman and his cohorts gained traction; they argued that the main role of central governments and banks was to determine money supply, and that fiscal management would not resuscitate the economy.[50] National currency was to be reined in and treated as a neutral tool of the state. The policies, however, had clear ideological underpinnings in that they ushered in the demise of social support and redistribution. Thus, just as money was going "mad,"—to use Strange's terminology—the social scaffolding that had been erected to prop up the economic system was dismantled. Economic actors were much more exposed to risk; some profited from risk taking, whereas others were more exposed to the vagaries of the market. Huge differentials between rich and poor have been the result.

The global financial crisis—itself in no small part a creation of monetary innovations such as subprime mortgages, collateralized debt obligations, and credit-default swaps—has exposed these fault lines. The remedies to the crisis have exacerbated the fault lines even further, sometimes characterized as the difference between Wall Street and Main Street or, in the language of the Occupy movement, the 1 percent versus the 99 percent. These discrepancies between rich and poor have been felt for years in other parts of the world; really, it is only now that Western economic policies are coming home to roost. The end of Bretton Woods had devastating effects on the so-called developing world. The devaluation of the dollar led to "a massive net transfer in wealth from poor countries, which lacked gold reserves, to rich countries, like the US and Great Britain, which retained them."[51] The rapid rise in the price of gold benefited those from countries that held gold reserves; in contrast, the plunging value of the U.S. dollar drained the more impoverished countries that held the dollar in reserve. Today, austere structural adjustment programs that were imposed by the World Bank and the International Monetary Fund to discipline indebted countries are being rolled out across the globe. Unfortunately, it continues to be the case that austerity politics bear down most heavily on those who are already suffering the most.

It is out of this fulcrum that political protest on currencies has erupted. Protests are not just about currency crisis and devaluation but also about attacking the political sway of moneyed interests, rising social inequities, and the abandonment of populations. Taking seriously the object that is money provides a vantage point for understanding the convergence of these concerns. It draws attention to money's role as an intermediary in networks that bind together states, finance, and citizens in fragile assemblages of power. What the inscriptions on currency suggest is that the assemblage is unraveling in the public circuits through which it travels: currency is no longer being taken for granted. Its institutions and its reproduction have come into question. Although state currencies have always been a mechanism for mediating the national and international, in recent years international pressures have become significantly more pronounced, alongside of and often informed by a greater shift toward privatization and market forces that are undermining the national. The contemporary fascination with the materiality of money can thus be understood, at least in part, as a response to the international pressures in which currency circulates, even as the primary target appears to be the national state. Yet the point is not to reify the state or state power over currency. Indeed, state control over national currency has never been as entirely coherent as has been thought. Moreover, "the infrastructure of money has already become decentralized and global, so a return to the national solutions of the 1930s or a Keynesian regime of managed exchange rates and capital flows is bound to fail."[52] Rather than looking backward to the state or retreading a common narrative of progress and decline, a materialist approach to money encourages renegotiating the international assemblage under different terms. At the very least, and what the currency protests that introduced this chapter demand, nation-states and the monetary structure need to be redesigned in the interests of the public.

Notes

1. Emily Heffler, "Dollar Bills Rain Down on Seattle Street," *Seattle Times*, July 4, 2012, http://seattletimes.com/html/localnews/2018604851_money05m.html.

2. "Ben & Jerry's Co-founder and Move to Amend to Stamp Dollar Bills with 'Corporations Are Not People' and 'Money Is Not Speech,'" Move to Amend, June 6, 2012, https://movetoamend.org/press-release/ben-jerry%E2%80%99s-co-founder-and-move-amend-stamp-dollar-bills-%E2%80%9Ccorporations-are-not-people%E2%80%9D.

3. Elyse Bruce, "Idle No More: Defacing Currency," *WC Native News*, January 15, 2013, http://westcoastnativenews.com/idle-no-more-defacing-currency/.

4. Theunis Bates, "Iran's Opposition Finds Imaginative Ways to Protest" *AOL News*, January 13, 2010, http://www.aolnews.com/2010/01/13/irans-opposition-finds-imaginative-ways-to-protest/.

5. "The Banknote Revolt Hits China," *No More Chinese Communist Party* (blog), accessed July 22, 2013, http://nomoreccp.wordpress.com/page/8/.

6. Mark Sweney, "Zimbabwean Newspaper Campaign Turns Worthless Banknotes into Gold," *Guardian* (Manchester), June 24, 2009, http://www.guardian.co.uk/media/2009/jun/23/zimbabwean-cannes-lions-award.

7. Bruce G. Carruthers and Sarah Babb, "The Color of Money and the Nature of

Value: Greenbacks and Gold in Postbellum America," *American Journal of Sociology* 101, no. 6 (1996): 1557.

8. Bruce Braun and Sarah J. Whatmore, eds., *Political Matter: Technoscience, Democracy, and Public Life* (Minneapolis: University of Minnesota Press, 2011), ix.

9. Ibid., x–xi.

10. Jane Bennett, *Vibrant Matter: A Political Ecology of Things* (Durham, N.C.: Duke University Press, 2010); Braun and Whatmore, ibid.; William E. Connolly, "The 'New Materialism' and the Fragility of Things," *Millennium: Journal of International Studies* 41, no. 3 (2013); Diana Coole and Samantha Frost, eds., *New Materialisms: Ontology, Agency, and Politics* (Durham, N.C.: Duke University Press, 2010); Emily Gilbert, "Common Cents: Situating Money in Time and Place," *Economy and Society* 34, no. 3 (2005); Bruno Latour, "From Realpolitik to Dingpolitik; or, How to Make Things Public," in *Making Things Public: Atmospheres of Democracy*, ed. Bruno Latour and Peter Weibel (Cambridge, Mass.: MIT Press, 2005); Noortje Marres and Javier Lezaun, "Materials and Devices of the Public: an Introduction," *Economy and Society* 40, no. 4 (2011).

11. Braun and Whatmore, *Political Matter*, ix.

12. Ibid., xxi.

13. Jane Bennett, *The Enchantment of Modern Life: Attachments, Crossings, and Ethics* (Princeton, N.J.: Princeton University Press, 2001); Bennett, *Vibrant Matter*, 6.

14. Marres and Lezaun, "Materials and Devices of the Public," 490, 495.

15. Latour, "From Realpolitik to Dingpolitik," 15.

16. Karl Marx, *Capital: A Critique of Political Economy*, vol. 1 (1867; repr., New York: International Publishers, 1974).

17. Brett Christophers, "Follow the Thing: Money," *Environment and Planning D: Society and Space* 29, no. 6 (2011).

18. Emily Gilbert, "Follow the Thing: Credit. Response to 'Follow the Thing: Money,'" *Environment and Planning D: Society and Space* 29, no. 6 (2011); Randall L. Wray, "Alternative Approaches to Money," *Theoretical Inquiries in Law* 11, no. 1 (2010).

19. Gilbert, "Common Cents."

20. Karin Knorr Cetina, "Sociality with Objects: Social Relations in Postsocial Knowledge Societies," *Theory, Culture, and Society* 14, no. 4 (1997).

21. Karl Polanyi, *The Great Transformation* (1944; repr., Boston: Beacon, 1971).

22. Jonathan Parry and Maurice Bloch, eds., *Money and the Morality of Exchange* (Cambridge: Cambridge University Press, 1989).

23. Nigel Dodd, *The Sociology of Money: Economics, Reason, and Contemporary Society* (New York: Continuum, 1994); Geoffrey Ingham, *The Nature of Money* (Cambridge, Eng.: Polity, 2004); Viviana Zelizer, *The Social Meaning of Money: Pin Money, Paychecks, Poor Relief, and Other Currencies* (New York: Basic Books, 1994).

24. See, for example, Michel Callon, Yuval Millo, and Fabian Muniesa, eds., *Market Devices* (Malden, Mass.: Blackwell, 2007); and Donald MacKenzie, *Material Markets: How Economic Agents Are Constructed* (Oxford: Oxford University Press, 2009).

25. Gilbert, "Common Cents."

26. Marx, *Capital*.

27. Andrew Leyshon and Nigel Thrift, *Money/Space: Geographies of Monetary Trans-*

formation (London: Routledge, 1997). See also Marieke de Goede, *Speculative Security: The Politics of Pursuing Terrorist Monies* (Minneapolis: University of Minnesota Press, 2011).

28. Polanyi, *Great Transformation*.

29. Eric Helleiner, *The Making of National Money: Territorial Currencies in Historical Perspective* (Ithaca, N.Y.: Cornell University Press, 2003). See also Emily Gilbert and Eric Helleiner, eds., *Nation-States and Money: The Past, Present, and Future of National Currencies* (London: Routledge, 1999). Some of these ideas are also discussed and developed in Emily Gilbert, "Currency in Crisis," *Scapegoat: Architecture/Landscape/Political Economy*, Winter–Spring 2013.

30. Keith Hart, "Why the Euro Crisis Matters to Us All," *Scapegoat: Architecture/Landscape/Political Economy*, Winter–Spring 2013, 40.

31. Helleiner, *Making of National Money*, 15.

32. Emily Gilbert, "Forging a National Currency: Money, State-Building, and Nation-Making in Canada," in Gilbert and Helleiner, *Nation-States and Money*. See also Jacques E. C. Hymans, "International Patterns in National Identity Content: The Case of Japanese Banknote Iconography," *Journal of East Asian Studies* 5, no. 2 (2005); Josh Lauer, "Money as Mass Communication: US Paper Currency and the Iconography of Nationalism," *Communication Review* 11, no. 2 (2005); Simon Hawkins, "National Symbols and National Identity: Currency and Constructing Cosmopolitans in Tunisia," *Identities: Global Studies in Culture and Power* 17, nos. 2–3 (2010).

33. Jean Comaroff and John L. Comaroff, "Beasts, Banknotes, and the Colour of Money in Colonial South Africa," *Archaeological Dialogues* 12, no. 2 (2005): 109.

34. Robert J. Foster, "In God We Trust? The Legitimacy of Melanesian Currencies," in *Money and Modernity: State and Local Currencies in Melanesia*, ed. David Akin and Joel Robbins (Pittsburgh: University of Pittsburgh Press, 1999).

35. Wambui Mwangi, "The Lion, the Native, and the Coffee Plant: Political Imagery and the Ambiguous Art of Currency Design in Colonial Kenya," *Geopolitics* 7, no. 1 (2002).

36. Mahir Saul, "Money in Colonial Transition: Cowries and Francs in West Africa," *American Anthropologist*, n.s., 106, no. 1 (2004).

37. David Graeber, *Debt: The First 5,000 Years* (Brooklyn: Melvin House, 2011), 339 (original emphasis).

38. Gilbert, "Currency in Crisis," 24.

39. Phillip Coggan, *Paper Promises: Money, Debt, and the New World Order* (Toronto: Penguin, 2012), 58.

40. Peter Miller and Nikolas Rose, "Governing Economic Life," *Economy and Society* 19, no. 1 (1990): 5. See also Timothy Mitchell, "Fixing the Economy," *Cultural Studies* 12, no. 1 (1998).

41. Miller and Rose, "Governing Economic Life," 13.

42. T. H. Marshall, *Citizenship and Social Class and Other Essays* (Cambridge: Cambridge University Press, 1950), 6.

43. Emily Gilbert, "Money, Citizenship, Territoriality, and the Proposals for North American Monetary Union," *Political Geography* 26, no. 2 (2007).

44. Benjamin Kunkel, "Forgive Us Our Debts," *London Review of Books*, May 10, 2012.

45. Coggan, *Paper Promises*, 236.
46. Ibid., 109.
47. Gill Seyfang and Ruth Pearson, "Time for Change: International Experience in Community Currencies," *Development* 43, no. 4 (2000).
48. Susan Strange, *Mad Money: When Markets Outgrow Governments* (Ann Arbor: University of Michigan Press, 2000), 1.
49. Hart, "Why the Euro Crisis Matters," 41.
50. Jamie Peck, *Constructions of Neoliberal Reason* (Toronto: Oxford University Press, 2010).
51. Graeber, *Debt*, 364–65.
52. Hart, "Why the Euro Crisis Matters," 42.

Biometric MasterCard

Elizabeth Cobbett

South Africa's African National Congress (ANC) government has called on MasterCard, a global payments network, to distribute the country's welfare benefits through the use of a biometric debit card. As a technology company in the payments industry, MasterCard facilitates the movement of money and transactions between millions of consumers and merchants.[1] Its cutting-edge technologies are now being applied in South Africa through a new biometric debit card, endorsed by MasterCard and the South African Social Security Agency (SASSA). SASSA is responsible for the management and distribution of the country's social grants. At present, nearly 16 million South Africans receive social grants. In 2012, 2.5 million SASSA MasterCard cards were issued to social-grant recipients; this number has now grown to 10 million grant recipients.[2]

Expenditure on social grants was around R105 billion (US$10 billion) for the 2012–13 fiscal year and was projected to grow to R122 billion (US$12 billion) in 2014–15.[3] The state has decided to outsource the administration of these grants to MasterCard. Through its cards and linked bank accounts, in this case to the South African–listed Grindrod Bank, MasterCard helps the state curtail fraud and simultaneously pulls poor people into the formal banking sector. MasterCard's card is operated by Net1, a South African international electronic-payments company trading on New York's Nasdaq (primary listing) and on Johannesburg's stock exchange (secondary listing). Net1's UEPS technology provides financial services in areas that have little to no financial infrastructure, as well to rural environments, even those which have little or no communications infrastructure.

What is taking place in South Africa is a departure from historic arrangements between the very poor and global finance. Capabilities developed in the spheres of technology, global finance, and governance enable the South African state to hand over the management of its monthly social-grant budget to MasterCard. Finance in South Africa is a mature and developed sector, and it has a dominant place in the country's political economy. However, the ANC government is going beyond its customary policy of aiding financial companies in

Officials mark ten million SASSA MasterCard cards issued to South African social-grant beneficiaries. Courtesy of MasterCard.

expanding their operations; the government is transferring the management of public funds to a global financial company. This is an unprecedented move and points toward new alignments between all layers and spheres of society and haute finance.

The under- or unbanked populations formerly held no interest for global finance, but technological innovations in biometric authentication and the growth of electronic money are changing this. Haute finance is looking beyond its habitual domains of lending to, investing in, and borrowing very large amounts of money from corporations and governments. It is making inroads into the bottom echelons of economic activity, that area of society which French historian Fernand Braudel called "material life."

But technology development is not enough incentive for financial firms to expand into new markets, let alone create them. State action is necessary to open up developing and emerging markets. This is taking place on a wide scale, and not only in South Africa. What is noteworthy in the South African case, however, is that MasterCard is using the route of managing public funds for the very poor as one way to create new financial markets and that this is an action being actively sponsored by the state.

MasterCard's new debit card for social-grant recipients therefore points to far-reaching changes taking place in relationships between populations, public authorities, and global finance. It is in this context that MasterCard's debit card for social grants makes the international. It brings together in an unprecedented manner people outside, or on the margins of, formal markets and global capitalism.

Thrusting Capital's Roots into Material Life

A way of thinking about these connections between the unbanked and global finance is through Braudel's rich ontology of civilizations and society. Two aspects of Braudel's work are of help here: world economies and a tiered typology of the economy, and world-economies. The tiered typology—capitalism, market economy, and material life—is particularly useful in understanding how two extremities of the broad economic sphere come to be connected through biometrics and digitalized money. Next, the analytical concept of world-economies demonstrates how the historical position of South Africa in the world political economy makes the relationship between MasterCard, the ANC, and South Africa's poorest section of the population possible.

Beginning with Braudel's tiered typology, typically the economy is taken as a homogeneous reality that can be measured apart from the messy societal context from which it emerges.[4] The market economy, which is usually studied and makes up most of our understanding of economic action, is analytically inserted by Braudel between two other zones of economic activity: material life and capitalism. These are more shadowy zones where economic actions and processes are less transparent, less calculable, and harder to quantify and qualify.

Material life is the rich zone of socioeconomic daily activity that, like a layer, covers the earth.[5] This ground floor of economic life is where we observe the everyday events of production, consumption, life, and death. This is the place of the busy multiplications of to-ing and fro-ing as people carry out their lives and find the means to secure a living. The peddler, the neighborhood shopkeeper, the local merchant, the family farmer, the miner, the migrant, and the domestic worker are all part of this thick movement of lives being lived. The outcome is a dense mass of interactions and transactions that meet with but are not always integrated into the transparent market economy.

Material life is often ignored at the expense of studying capitalism. We hold the impression "that capitalism equals growth; that capitalism is not one stimulus among many, but the stimulus, the tiger in the tank, the prime mover of progress."[6] Braudel, by contrast, sees everyday actions as the soil into which capitalism thrusts its roots:[7] "In reality, everything rested upon the very broad back of material life; when material life expanded, everything moved ahead, and the market economy also expanded rapidly and reached out at the expense of material life. Now capitalism always benefits from such expansion. . . . [T]he extensiveness of any capitalism is in direct proportion to the underlying economy."[8] Capitalism has been monopolistic and oligopolistic from its beginnings in the thirteenth century.[9] Sponsored by powerful economic and political decision makers, capitalism has always worn "seven-league boots."[10] Capitalism represents the high-profit zone; it is the domain of activity at the summit—that of long-distance trade, of public bonds and investment in megaprojects. Large-scale capitalism relies on the underlying double layer of material life and the coherent market economy.[11] Although capitalism expands, it does not aim to subsume all of economic activity; capitalism will adapt to more lucrative areas, leaving in its wake spheres or projects no longer considered worth exploiting.

It is with this analysis in mind that it is both interesting and relevant to think about MasterCard's ambition to bank the poor. It is also in the distinction between material life, the market economy, and capitalism that we can think about how ordinary, everyday-life actions in South Africa are drawn into large-scale capitalism. The biometric MasterCard / SASSA debit-payment card brings these two zones together as capitalism thrusts its roots into material life. It *makes* the international as the very local is connected directly with the global political economy.

This action is being enabled through the South African state. To understand this relationship between global finance, the state, and South Africans, it is necessary to examine the historic place South Africa has in world-economies.

World Economies

Braudel makes an important distinction between the world economy and a world-economy. "The *world economy* is an expression applied to the whole world."[12] World economy corresponds to "the market of the universe," "the human race or that part of the human race that is engaged in trade, and which today makes up a single market."[13] The world economy is analogous to the contemporary concept of globalization and the whole global political economy. The latter, a *world-economy*, is a zone of economic coherence that concerns a part of the world.[14] A world-economy can best be thought of as a political economic network that connects different areas of the globe to form an organic unity of trade, finance, and political contacts. Even in a globalized world economy, there continue to be historic networks—structures—built through long-distance trade, imperial dominance, alliances, and wars. Certain features define world-economies: they are centered on central zones, they have strong models of capitalism in place, and they are marked by hierarchies of regional economies. World-economies can exist simultaneously and overlap each other.

The history of banking and finance in South Africa was shaped through the country's role as world gold producer and its place in Britain's empire. It came to this place through its distinctive history as an important geopolitical station and strategic location at the southernmost extremity of Africa. The country lies on the ancient routes of long-distance trade between Europe and Asia. In the seventeenth century, the Vereenigde Oostindische Compagnie (VOC)—the Dutch East India Company—acquired the Dutch monopoly on all trade in Asian waters from the Cape of Good Hope onward. The VOC operated out of Amsterdam, the leading world financial center at the time. Dutch capitalists were closely associated with the state, and the state invested in these merchant capitalists, the perfect outcome of which was the VOC.[15] The Cape Colony, founded in 1657, was no more than a stopping point for the VOC, which remained fiercely watchful over this strategic position under Dutch control.[16]

The seventeenth- and eighteenth-century Anglo-Dutch Wars enabled England to control the major trade routes and to take over as the center of world trade and finance as London replaced Amsterdam as the beating heart of the European world-economy.[17] Credit markets were at the center of England's rise to financial power, underpinned by the sterling and the ability of England to obtain money in sufficient quantities and at the right time through the inflow of surplus capital from Dutch businessmen.[18]

By 1814, the British had taken over the Cape Colony from the Dutch East India Company and acquired full control of the strategic settlement. The discovery of coal in the 1840s, diamonds in 1867, and gold in 1885–1886 triggered a mineral revolution that thoroughly transformed South Africa's economy.[19] The discovery of gold coincided with the transition of the world, in 1870, to the British monetary system, based on the gold standard, and altered South Africa's place within the empire.[20] Gold brought the predominantly agricultural country into the core of the world's political economy. It became intimately linked to the world financial system operating out of London, the city at the heart of Britain's empire.[21]

By the end of the nineteenth century, South Africa was producing more than one quarter of the world's annual output of gold.[22] Gold transformed British colonialism into aggressive imperialism to secure its exploitation and its transfer to London for transformation and storage.[23] The Boer War (1899–1902) ensured that the mines in the Transvaal remained independent from Afrikaner nationalist interests and that South African gold would continue to underpin the Bank of England's authority of the world monetary system.[24] Gold made the international as it underpinned imperial Britain's financial system of the later nineteenth century.

Financial Making of Postapartheid South Africa

Globalization of finance, transnational structures of capital, and the historic balance of class power within South Africa enhanced the negotiating power of international financial agencies, foreign investors, and domestic business vis-à-vis the South African government during transition from apartheid to democracy.[25] The powerful group of capitalists who historically controlled the mines, energy, big industry, and the financial sector worked to ensure that the new government would create the kind of macroeconomic stability that would facilitate their activities.[26] The new ANC government adopted these policies as a means to regain credibility in the international arena of finance.

ANC political authority grew organically out of the struggle against apartheid, and it came to power in 1994 with the full and extensive backing of the majority. The fit between political legitimacy and the government was as tight as is possible in such circumstances. Two decades later, this fit is a whole lot looser, and waves of protest question the ANC's policies as poverty remains high while a new class of capitalists has joined established elites.

The state has applied a policy of development through financialization.[27] *Financialization* refers to the mounting role and influence of finance over domestic and international economies.[28] This policy is most visible in the rapacious commodification of public resources and services and in the deepening of capital markets through public and private debt. Tension is mounting between private financial interests and people living in escalating poverty and exclusion. This is expressed in nationwide protests as people fight to cover their basic survival needs. This extension of private capital gain over human security is being sponsored and endorsed by the state.[29] The protests in South Africa do more than point to a simple issue of malgovernance; they indicate a new locus of political confrontation between members of society and capital accumulation as promoted by the state.

History of South African Biometrics

South Africa is a country with a focus on measurement, classification, and control. Being a legacy of the apartheid regime, biometrics are not a thing of the past. South Africa is a leading example of a truly biometric state.[30] Although the country became politically independent through the Union State Act of 1910, it remained an imperial dependency directed by Britain's interests in global finance, gold, and mining.[31] Foreign investment and capital poured into South Africa, reorganizing mining as a highly capitalized and concentrated industry.[32]

Building on the racial apartheid already in place, South Africa's National Party brought the party to new levels of operation during 1948 and 1994. One way this was achieved was through the extension of the biometric identification system already in place. The first fingerprint sample was taken in 1900 by Sir Edward Henry, a British civil servant deployed in India who brought his system of fingerprinting criminals to South Africa.[33] It was subsequently used by the state to control and direct the movements of African, Chinese, and Indian populations and to buttress the national system of racialized capital accumulation.[34]

The postapartheid state used this biometrics lineage to create a single citizenship through the Home Affairs National Identification System (HANIS). The rhetoric for putting in place the biometric identity system was to create a new society of equal citizens. The Japanese company NEC, which developed the technology, put it this way: "In the past [South African] citizens were divided by race or belief. Today, everyone is integrated into a single digital archive, recognized through NEC technology by just one universal human feature, their fingerprint."[35] HANIS has become the world's largest citizen-identification database. Beyond enabling citizens to access public services, the HANIS identity card became an integral part of private-market transactions, from buying a car to trivial acts of renting a video on a Friday evening. A person does not exist in either the public or private sphere without this ID card.

Taking the HANIS card a step further, the South African Department of Home Affairs has just issued its first smart ID card. Replacing the current green, bar-coded paper ID, the new ID is a contactless card that has a microchip and will store fingerprint biometrics and biographic data. The new ID card will also be able to accommodate information from other public as well as private stakeholders. The ID card will effectively act as a platform that yokes the citizen to both public services and private companies.

President Jacob Zuma said that the new smart card was launched on former president Nelson Mandela's birthday, July 18, 2013, in honor of Madiba's legacy.[36] According to the Department of Home Affairs minister, Naledi Pandor, "President Mandela's lifelong struggle for freedom and human rights for South Africans and oppressed people all over the world was—at one level—dedicated to acknowledging the fundamental worth of each and every human being."[37] It is hard to understand how the issuance of this smart ID card would make Madiba proud and would honor his legacy of struggling for freedom. The state's argument is that this smart ID card represents a significant break with the past, as it buries the passbook legacy in South Africa. Passbooks were required by Blacks during apartheid to move around the country outside their designated residential area. Rather than a superficial break with the past, the new ID card seems very much to be continuing directly in step with South Africa's legacy of state surveillance and the use of technopolitical resources as governance tools. What is new is that these smart cards can seamlessly link the poor to high finance through digitalized biometrics.

Financial Inclusion in a "World beyond Cash"

The concept of what a bank is and how banking is being conducted is radically changing. Customer-branch banking is not only changing in industrialized economies, it is no longer an option for financial growth in Africa. The challenge for bankers and credit providers is to think how financial services can take place without the characteristic physical attributes of bricks-and-mortar banks. Banking is becoming more akin to process than to place. Challenges such as varying levels of literacy, great distances, the widespread absence of formal financial services, lack of official identification documentation, and the mobility of populations make the future of banking in Africa reliant on IT and digital technology. These features bring banking and credit services onto new terrain.

Financial inclusion, as set out by the Alliance for Financial Inclusion and adopted in its 2011 Maya Declaration, is an international strategy for development bringing together global financial institutions and national governments through private banking.[38] The rationale is that access to formal financial services is a key element in the process of socioeconomic empowerment and poverty alleviation. This policy would not be possible without biometrics and the digitalization of money. There are two principal ways of integrating biometrics into banking and finance: the point-of-sale (POS) system and smart cards. Both of these technologies offer financial actors a route to integrating the unbanked without relying on infrastructure or on particular attributes of the customer, for example, a high level of literacy. An example of POS is Rwanda's BioCash project, serviced by MobiCash, a mobile-payments company. Using existing national ID fingerprint records, the POS system connects people's fingerprints to a virtual bank account. Once a BioCash account has been set up, transactions are made without the ID card, and users use their fingerprints for authentication.

The second way is through smart cards, as in the case of South Africa's MasterCard system, which disburses government pension, disability, and public-assistance payments on a biometric debit card. MasterCard has an aggressive strategy for its operations in Africa and aims to make the continent part of its vision of a "world beyond cash."[39] The global company is mapping what it and the Center for Financial Inclusion call the invisible market of the under- or unbanked.[40] In line with this strategy, MasterCard is working with the Nigerian government to produce a MasterCard-branded national identity smart card with electronic-payment capability.[41] In a difference from South Africa's smart ID card, Nigeria's ID card will have immediate payment capability through prepaid-payment technology.

MasterCard is working with governments around the world to set up electronic-payment programs.[42] The South African government hopes that MasterCard's card will cut down on fraud,[43] and MasterCard expects that its card will bring millions of poor people with no bank accounts into its networks. It is wholly possible, however, to foresee South Africa following on developments in Nigeria and integrating MasterCard into its new smart ID card.

These arrangements allow MasterCard to work closely with leading African states and gain important footholds in Africa's economic powerhouses. The financial inclusion of millions of unbanked will be integrated into MasterCard's financial networks and not those of competing global financial operators, such as Visa.

From Armored Cars and Cash to Biometrics and Cashless

As adult recipients of state cash transfers, elderly pensioners were the first group of South Africans to be part of a large-scale study of fingerprint-based digital biometrics that began in 1990.[44] Before responsibility for delivering social grants was transferred to MasterCard, money was paid out at pay points. Only 46 percent of adults in South Africa had a bank account in 2004.[45] After eight years, about ten million more adults are in the formal banking system, bringing the percentage of the banked up to 56. The challenge for the state

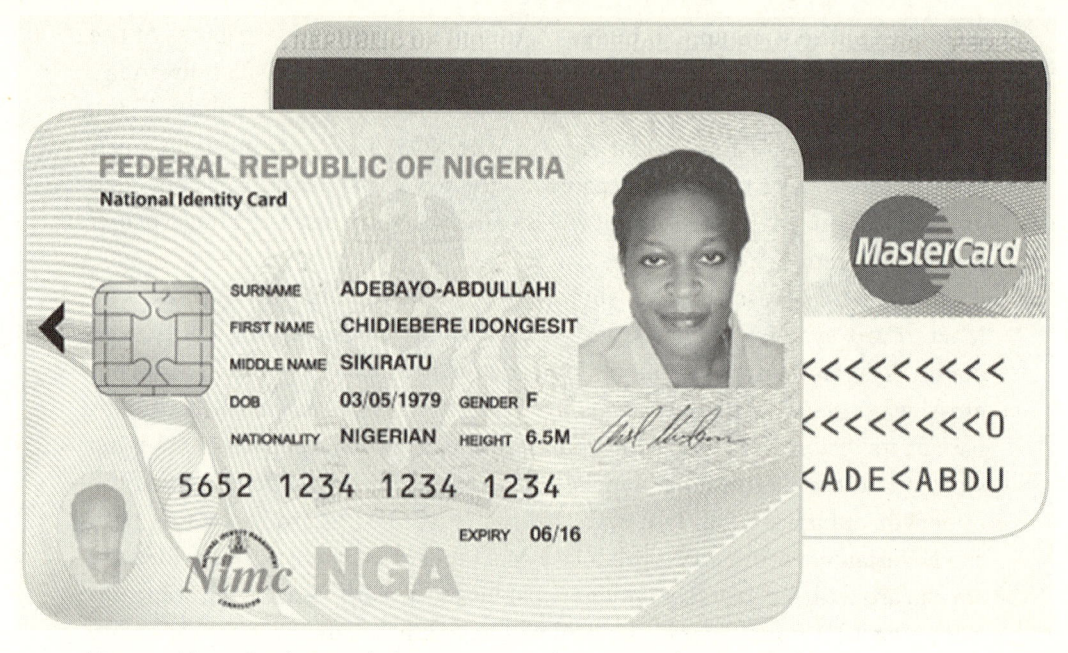

Nigerian MasterCard national identity smart card. Courtesy of MasterCard.

was to get grants to elderly persons who were living in communities with little financial-service infrastructure.

Social grants were delivered to pay points in truck-mounted dispensers and processed through authentication technology. Private security firms transported the money in onboard vaults inside industry-leading armored vehicles between processing centers and payout points, or ATMs. The majority of older persons interviewed by Black Sash, an advocacy group working in the domain of social protection and human rights, felt that not enough was done to keep the elderly safe at and around the pay-point sites and that the security provided was focused solely on protecting the cash and seeing it to its point of delivery.[46] Armed security did not prevent loan sharks on or close to the pay-point sites from offering loans or exacting payments. Black Sash suggested to SASSA that police-patrol vehicles should monitor activity immediately outside pay points and ensure the safety of beneficiaries when they left the premises.[47] This situation creates the strange, indeed surreal, image of state-of-the-art biometric and digital technology, fabricated in Japan, accompanied by heavily armed guards in military-like vehicles, dispensing money to the poorest members of the community in the middle of fields or townships where basic facilities such as shelter, seating, toilets, water, and fencing are not available for the frail pensioners.

MasterCard and SASSA's debit card sidesteps these difficulties and makes the need for armed security and mobile ATMs and the complications of delivering cash in rural locations with poor infrastructure a thing of the past. Now, enrollment in MasterCard's system is done by mobile payment facilities small enough to fit into a suitcase.[48] These mini enrollment facilities are installed by MasterCard's local partners, Grindrod Bank and Net1, to SASSA offices throughout South Africa. In order to have funds disbursed onto their cards, grant recipients have to verify their fingerprints and voices with a biometric reader. Biometric authentication and a personal identification number are required every month in order to access the grants.

Social-grant recipients are able to use their SASSA MasterCard debit cards to pay for goods and to check their account balances for free at till points. However, if they wish to withdraw cash at ATMs and till points, they are charged an initial free withdrawal. Normal banking charges of around R7 (US$0.80) are charged if beneficiaries use their SASSA cards outside the network of merchants supported by the payment system. Furthermore, and most importantly, social grants become the site of new financial relations as the poor establish relationships with MasterCard and with the national banks operating its payment system domestically.

Costs of Going Cashless

The way that people's everyday lives are entwined with global financial systems matters. This observation raises a question: How does technology permit a rearrangement of social order and foster new political-economic arrangements between different scales of the global economy?

Placing the pieces of the international network together, we have the following components in place in South Africa: new biometric technology, growth in digitalized forms of money, a new government policy requiring biometric citizen IDs,[49] desire by the ANC government to let global finance play a leading role in development, the internationalization of operations of South African banks and technologies, and the prevalence of welfare recipients who do not have bank accounts or do not live in areas where these services are available. Uppermost is the desire of international financial actors to expand their business into new markets, and these are primarily in developing and emerging economies.

SASSA's MasterCard debit card makes the international as it pulls excluded and poor members of South Africa's economy into a global financial arrangement through biometric technologies. Biometrics are replacing the customary practice of face-to-face encounters of formal banking by acting as a substitute for human verification. Concerns of fraud and risk are mitigated through new technologies that are able to guarantee recognition and to validate the authenticity of humans in financial transactions. These identification systems are also giving financial institutions a viable means to offer business services to the unbanked. Biometric technologies thus offer a portal between populations that were formerly outside of the formal banking system and global finance.

States participate in this process by empowering financial actors to obtain biometric readings of their citizens' bodies. The interest for governments is to transfer the risk associated with fraud and corruption of public funds to private financial markets. This is evident in South Africa, where the changing form of social grants, from cash delivered through mobile ATMs to cash withdrawn using MasterCard's biometric smart card, permits the ANC to transfer the risk attached to the distribution of social grants to private financial markets. This system has the additional advantage of deepening national capital markets and pushing citizens to find individual responses to socioeconomic problems.

These developments raise two important issues. First is the issue of public authority being transferred to the private sector and of citizens coming progressively under the direct control of global financial firms. States are opening up areas of their political economies to global capital governance. Among the sectors being pushed into this relationship are the most frail and vulnerable members of society. In Braudelian terms, the questions are these: Why is there a concerted move to bring everyday lives into contact with global finance? Why is it now that international financial institutions, central banks, private financial actors, and governments across the world find it imperative that the poor and very poor become banked when they have always been poor?

As well as depending on local contexts and national governments, financialization occupies a central place in understanding these changes. Digitalized biometric technology enables financialization that, in the case of the under- and unbanked, passes by the policy of financial inclusion. This would not be possible without state sanction and action. The state puts in place legal and social platforms that bring the under- and unbanked into contact with global financial systems. At the heart of this choice is the principle of reducing government's role in development and increasing that of financial instruments. World Bank Group president Robert Zoellick rationalizes this principle as "harnessing the power of financial services" to offer the poor and unbanked empowerment to improve their lives.[50] The World Bank also says that it allows "poor people to smooth their consumption and insure themselves against the many economic vulnerabilities they face—from illness and accidents to theft and unemployment."[51] The Maya Declaration describes the phenomenon as the means "to unlock the economic and social potential of the 2.5 billion 'unbanked.'"[52]

Keith Breckenridge, in an article on biometric banking in Ghana, offers an alternative analysis. For him, state administrative capacity, in the form of biometric identity surveillance, is being transferred to commercial actors. The state is able to withdraw "from the areas of society that have no prospect of paying their own way."[53] What does this process tell us about the transformation of transnational financial markets, their relation to governments and to the more vulnerable members of society?

This raises a second issue: whether the change in the form of money impacts and shapes international relations, politics, and government practices. The significance of this question can be understood by asking others: How does the provision of payments via a smart card equipped with biometric technology for authentication differ from cash? And what impact does this have on vulnerable members of society?

Over the last 150 years, three international financial systems have existed: the gold standard, the Bretton Woods agreement, and the current system of free-floating currencies and mobile capital. Each system has produced specific relations with domestic financial markets and therefore with societies. The British Empire operated through gold, directing its foreign policies to secure its provision in South Africa through coercive and often violent means. Twentieth-century American order was organized at first through the Bretton Woods system, which centered on the dollar. The capital controls of the Bretton Woods system allowed South Africa to manage capital flows, thus reducing the danger of capital flight and the risk of current-account imbalances. It also permitted the government to pursue its program of racial capitalism underpinned by foreign investment to exploit cheap labor and make profits. In the post–Bretton Woods system, the U.S. dollar retains its hegemony as the world's main reserve currency at the same time that we see the emergence of a global cashless system of digital banking and biometric technology. The increased fluidity and mobility of capital enable national public authorities to pass on the management of poor citizens' economic problems to global financial companies.

Critically, in this new world of cashless money, the *body* becomes the site of financial actions and authentication. What does this change? For the ordinary person, the move from cash, printed paper, and coins to e-money authenticated through the body raises critical questions about surveillance, control, and identity. E-money working through biometrics needs to be associated with a body, and people who use it henceforward become part of private databases operated by financial and security firms. There is a loss of anonymity with e-money that was historically possible with cash.

The fact that e-money does not allow people and their actions to go unobserved raises questions about social justice. Anonymity becomes the prerogative of the powerful and rich, as their money is circulated in great quantities through listed companies. It is harder to trace and link e-money back to a wealthy body than it is to a poor one. An elderly person whose small monthly allowance is dispensed on a biometric debit card will have her every transaction traced. She will also face severe cash constraints if she forgets her PIN to access her debit card. If a grant recipient enters the wrong PIN three times at an ATM, the card is retained by the ATM and the card will not be returned. There is a high likelihood of an elderly person forgetting her PIN, losing her card, and having no more funds for the rest of the month. Furthermore, SASSA will charge to replace the card.

Braudel's view that everything rests upon the very broad back of material life and that when material life expands, everything moves ahead, including capitalism, is an invaluable observation for understanding changes taking place throughout the global economy. Capitalism is thrusting its roots into material life that has so far been outside of its field of operations. This financial action is made possible by states and contributes to the growth of multiple layers of private and public governance. This complex system of regimes is pulling citizens out from the mantle of national public governance and into networks of private governance. Financial inclusion, sponsored by international financial institutions and national governments, spurs this movement as development, and social security is promoted through private capital markets. It is in this context that MasterCard's biometric debit card for social-grant recipients in South Africa makes the international.

Notes

1. Ajay Banga, "A World beyond Cash," Cornellcast video interview by Vrinda Kadiyali, April 16, 2013, Cornell University, http://www.cornell.edu/video/mastercard-ceo-ajay-banga-imagines-cashless-future.

2. "MasterCard to Power Nigerian Identity Card Program," May 8, 2013, http://Newsroom.mastercard.com/press-releases/mastercard-to-power-nigerian-identity-card-program.

3. Pravin Gordhan, "2012 Budget Speech," February 22, 2012, South African Treasury, http://www.treasury.gov.za/documents/national%20budget/2012/speech/speech.pdf.

4. Fernand Braudel, *The Structures of Everyday Life,* trans. Siân Reynolds, Civilization and Capitalism, 15th–18th Century 1 (New York: Harper & Row, 1982), 23.

5. Fernand Braudel, *The Wheels of Commerce,* trans. Siân Reynolds, Civilization and Capitalism, 15th–18th Century 2 (New York: Harper & Row, 1982), 23.

6. Ibid., 575.

7. Ibid., 21–22, 229.

8. Fernand Braudel, *Afterthoughts on Material Civilization and Capitalism,* trans. Patricia Ranum (Baltimore: John Hopkins University Press, 1977), 63.

9. Ibid., 57.

10. Braudel, *Wheels of Commerce,* 554. Seven-league boots permit the wearer to take great strides, a league being the equivalent to about one-half kilometer.

11. Braudel, *Afterthoughts on Material Civilization and Capitalism,* 112–13.

12. Ibid., 21 (original emphasis).

13. Ibid.

14. Ibid., 22.

15. Ibid., 175–76.

16. Ibid., 430.

17. Ibid., 352.

18. Ibid., 262.

19. Hobart D. Houghton, "Economic Development," in *The Oxford History of South Africa,* ed. Monica Wilson and Leonard Thompson (Oxford: Oxford University Press, 1971), 2:11.

20. Russell Ally, *Gold and Empire: The Bank of England and South Africa's Gold Producers, 1886–1926* (Johannesburg: Witwatersrand University Press, 1994), 12.

21. Ibid., 35.

22. Jean Jacques Van-Helten, "Empire and High Finance: South Africa and the International Gold Standard, 1890–1914," *Journal of African History* 23, no. 4 (1982): 529.

23. Sampie Terreblanche, *A History of Inequality in South Africa, 1652–2002,* 3rd ed. (Pietermaritzburg, S. Afr.: University of Natal Press, 2005).

24. Ally, *Gold and Empire,* 25.

25. Adam Habib and Vishnu Padayachee, "Economic Policy and Power Relations in South Africa's Transition to Democracy," *World Development* 28, no. 2 (2000): 245, 254.

26. Ibid., 260.

27. Seeraj Mohamed, "The State of the South African Economy," in *2010: Development or Decline?,* ed. John Daniel et al., New South African Review 1 (Johannesburg: Wits University Press, 2012), http://witspress.co.za/wp-content/uploads/2012/04/NSAR-1Watermark.pdf.

28. Seeraj Mohamed, "Financialisation of Economy Exacerbating Problems of Minerals Energy Complex in South Africa," South African Civil Society Information Service video, October 1, 2012, http://www.youtube.com/watch?v=Jll3HK29TGQ.

29. David A. McDonald, *World City Syndrome: Neoliberalism and Inequality in Cape Town* (New York: Routledge, 2008).

30. Keith Breckenridge, "The Biometric State: The Promise and Peril of Digital Gov-

ernment in the New South Africa," *Journal of Southern African Studies* 31, no. 2 (2005): 267–82.

31. South Africa's political economy is dominated by the minerals-energy complex (MEC). This concept was developed by Ben Fine and Zavareh Rustomjee, *The Political Economy of South Africa: From Minerals-Energy Complex to Industrialisation* (London: Westview, 1996), and refers to the interlocking relationship between the mining and energy sectors, the financial system, and the state. These three facets of the political economy have shaped the country's social and economic order over the last century.

32. Houghton, "Economic Development," 13.

33. Aurelia Wa Kabwe-Segatti, Nicolas Pejout, and Philippe Guillaume, "Ten Years of Democratic South Africa: Transition Accomplished?" (working paper, Institut français d'Afrique du Sud, Johannesburg, 2006), http://hal.archives-ouvertes.fr/docs/00/79/91/89/PDF/Cahiers-IFAS_8_Pejout.pdf, p. 45.

34. Keith Breckenridge, "The Elusive Panopticon: The HANIS Project and the Politics of Standards in South Africa," in *Playing the Identity Card: Surveillance, Security, and Identification in Global Perspective,* ed. Colin J. Bennett and David Lyon (London: Routledge, 2008), 39–56.

35. Nippon Electric Company, "NEC AFIS Created the World's Largest Civilian Fingerprint Identification Database for HANIS," NEC AFIS for Identity Management, 2008, accessed August 18, 2011, http://www.nec.com/global/cases/sa/pdf/catalogue.pdf.

36. "New IDs to Honour Madiba's Legacy," SANews.gov.za, July 4, 2013. http://www.sanews.gov.za/node/18188. Nelson Mandela is often affectionately referred to by his nickname, Madiba.

37. Ibid., para. 2.

38. Alliance for Financial Inclusion, "Maya Declaration: The AFI Member Commitment to Financial Inclusion," accessed March 22, 2013, http://www.afi-global.org/gpf/maya-declaration.

39. Nicole Ward, "MasterCard Morning Brew: Out of Africa Cashless Momentum," Cashless Pioneers, MasterCard, January 16, 2013, http://newsroom.mastercard.com/2013/01/16/mastercard-morning-brew-out-of-africa-cashless-momentum/.

40. Center for Financial Inclusion, "Financial Inclusion 2020: Mapping the Invisible Market," *Accion,* accessed March 23, 2013, http://www.centerforfinancialinclusion.org/fi2020/mapping-the-invisible-market.

41. Simon Echewofun Sunday, "Nigeria: Mastercard to Issue 13 Million National Identity Cards," Daily Trust, *AllAfrica,* May 15, 2013, http://allafrica.com/stories/201305151074.html.

42. Tim Murphy, "MasterCard Notes Growing Trend in Government Adoption of Electronic Payments as an Alternative to Cash and Checks" (press release), MasterCard, August 1, 2012, http://newsroom.mastercard.com/press-releases/mastercard-notes-growing-trend-in-government-adoption-of-electronic-payments-as-an-alternative-to-cash-and-checks/.

43. Stephen Timm, "Biometric Cards to Beat Social Grant Fraud," SouthAfrica.

info, June 14, 2012, http://www.southafrica.info/about/social/grants-140612.htm#.UVRi-VewXdE.

44. Breckenridge, "Biometric State," 272.

45. "South Africa's Banked Population Increases," *Ventures Africa*, October 31, 2012, http://www.ventures-africa.com/2012/10/south-africas-banked-population-increases/.

46. Black Sash. "Black Sash Submission to the South African Social Security Agency," accessed August 18, 2011, http://www.blacksash.org.za/files/sassasubmission011010.pdf, p. 4.

47. Ibid., 7.

48. Neal Ungerleider, "Social Security and Welfare Payments Go Biometric," *Fast Company*, September 26, 2012, http://www.fastcompany.com/3001575/social-security-and-welfare-payments-go-biometric.

49. Keith Breckenridge, "The World's First Biometric Money: Ghana's e-Zwich and the Contemporary Influence of South African Biometrics," *Africa* 80, no. 4 (2010): 642–62.

50. Robert B. Zoellick, "Three Quarters of the World's Poor Are Unbanked," Data and Research, World Bank, April 19, 2012, http://econ.worldbank.org/WBSITE/EXTERNAL/EXTDEC/0,,contentMDK:23173842~pagePK:64165401~piPK:64165026~theSitePK:469372,00.html.

51. World Bank, "About Global Findex," Global Financial Inclusion (Global Findex) Database, Data and Research, World Bank, accessed March 12, 2013, http://econ.worldbank.org/WBSITE/EXTERNAL/EXTDEC/EXTRESEARCH/EXTPROGRAMS/EXTFINRES/EXTGLOBALFIN/0,,contentMDK:23172730~pagePK:64168182~piPK:64168060~theSitePK:8519639,00.html.

52. Alliance for Financial Inclusion, "Maya Declaration."

53. Breckenridge, "World's First Biometric Money," 644.

Cocaine

Mike Bourne

As a psychoactive substance, cocaine is already recognized as active rather than passive material. But this action is entangled in a constantly forming and re-forming, dissolving and escaping assemblage of relations in which agency is attributable not to chemicals or people (traffickers, addicts, smugglers, policy makers) but to their intra-active coconstitution. In this way, cocaine is active and acted upon beyond its psychoactive properties in ways neglected by anthropocentric politics and theorizing. Cocaine is many things as it moves, and it is never only cocaine. Cocaine is a potential ally in the development of new materialism in international relations (IR).

This chapter addresses one question posed in two seemingly different ways: What is cocaine? What or where is the cocaine international, that is, the internationalization of cocaine control and prohibition, and the globalization of narcotics trafficking? Cocaine and the cocaine international are not the same as each other, and they are not always and everywhere the same in themselves, but both are fragile emergent assemblages devoid of innate ordering and a priori spatiality. Both cocaine and the cocaine international are sociochemical processes. Chemistry and society from a new materialist perspective are both processes of the constitution of relations. For Alfred North Whitehead, "a molecule is a historic route of actual occasions; and such a route is an event."[1] Viewed in this way, cocaine and the cocaine international are not mere molecules of chemicals or elements or levels of society but assemblages of physical, chemical, legal, informational, and other things (also assemblages). This chapter explores these routes through what Michel Serres calls "topology" that contrasts with the geometric fixed political and economic geography of state territories, borders, and transnational flows that is familiar to IR:

> If you take a handkerchief and spread it out in order to iron it, you can see in it certain fixed distances and proximities. If you sketch a circle in one area, you can mark out nearby points and measure far-off distances. Then take the same handkerchief and crumple it, by putting

it in your pocket. Two distant points suddenly are close, even superimposed. If, further, you tear it in certain places, two points that were close can become very distant. This science of nearness and rifts is called topology, while the science of stable and well-defined distances is called metrical geometry.[2]

Karen Barad's notions of intra-action (as opposed to the interaction of preformed, separate entities) and particularly the enacting of "cuts" echo this topological method. In contrast to inherent distinctions that exist (a "Cartesian cut"), the enacting of "agential cuts" is the process through which "the boundaries and properties of the components of phenomena become determinate and . . . particular concepts (that is, particular material articulations of the world) become meaningful."[3]

Again resisting a priori distinctions and proximities, Bruno Latour's concept of purification is also useful. Common dichotomies such as subject and object, real and constructed, structure and agent; and realms such as the social, political, economic, natural, and so forth are not reflections of natural differences but acts of purification that create "two entirely distinct ontological zones: that of human beings on the one hand and nonhumans on the other."[4] Cocaine and the cocaine international, then, are not present in separate, stable ontological zones of purely chemical or social relations. Indeed, chemistry and social science do not derive properties from abstract general laws but from processes of association and disassociation, distribution, and only ever provisional ordering. Importantly, from a new materialist perspective, there are no fundamental elements; "chemical atoms and human individuals" are "only fundamental from the point of view of specific scientific disciplines."[5] Rather, these entities are "quasi-objects" that are composed by movement and that circulate and resist easy ontological divisions as they move from nature to chemical to bodies, from scientific laboratories to the use of technologies "in the wild."[6] Both chemistry and IR, then, are not sciences of the combination of invariant and immutable entities but routes of becoming, of associations with uncertainties, overflows, and path-dependent trajectories irreducible to the a priori properties (including spatiality and politics) of components now seen as lacking a fixed essence. However, purification is no mere error of thought but a productive action (a cut). This chapter draws attention to how purification often entails other movements: how enacting a cut draws things closer together. Purification and the production and stabilization of hybrid cocaine objects require work by human-nonhuman associations. Drugs, drug trafficking, and drug policies assemble expertise, practices, technologies, models, techniques, people, political support, and other things through processes such as mobilization, enrollment, translation, channeling, borrowing, and learning.[7] Cocaine's relational, specific spatiality creates an international that is not found in the traditional scales and assumptions of IR but rather is emergent as multiple (hence reference to specific cocaine internationals) and "small and non-coherent," with varying reach and distribution.[8]

Through four sections, this chapter explores the ontological status of cocaine and the cocaine international and moves beyond assumptions of the singular natural (the coca leaf, the bodies of smugglers and "addicts"), the multiple social (meanings, identities, cultural practices), and the a priori Cartesian spatialities of the national and international control apparatus. In doing so, it does not simply assert nonhuman agency but reflects on how that agency distributes the political, stabilizes divisions, shapes regulations, and resists and enables categorizations. The chapter moves beyond asking what new materialism might mean for IR, to question how the practice of international politics already encounters and seeks to channel the active ambiguities of material objects.

Stuff

Cocaine was the "first modern drug," as Paul Gootenberg claims; "although plant-based, its discovery, profile and applications all derived from evolving laboratory science."[9] Cocaine resists and mobilizes divisions between nature and technology or nature and culture. It is one of fourteen alkaloids found in the leaf of the coca plant. As with other drug-alkaloid-containing plants, such as tobacco and opium, coca has a long history of human cultivation, in which purely natural specimens have been rare or nonexistent. Rather, as Pierre Chouvy argues with respect to the opium poppy, the human history of settlement, cultivation, and consumption produced a "symbiosis . . . between the plant and humans" that occurred within "a very old selection process of drug plants by humans over many thousands of years."[10] Already, then, there are too many verbs (*settle, cultivate, select*) and actants to refer to a single nature utilized but unchanged by human action.

Coca-leaf-bearing shrubs belong to the *Erythroxylum* genus, which contains hundreds of species. Only four major varieties of coca are cultivated. Some are indistinguishable from wild species, whereas others are highly isolated and genetically and morphologically distinct, with specific cultivation histories. Most of the world's cocaine in the 1980s came from Huánuco coca, believed to be the first domesticated coca variety, which grows in wetter areas of the Andes and has a limited geographical area of cultivation because of its ecological range. This variety was unknown in Colombia; early Colombian cocaine was of an Amazonian species that is genetically similar to Huánuco but is found only in cultivation and does not survive or interbreed with wild populations. Genetics and climate, then, can enact cuts or circulation between nature and cultivation. They do not do so alone, however, and although Amazonian coca has less than half the cocaine content of other cultivated varieties, its importance rose in the 1970s because it is easier to extract cocaine from—and its geographic areas of production were better suited to—clandestine refining. The other two cultivated varieties are harder to extract cocaine from but have a strong human–plant history, having been cultivated for medicinal and later pharmaceutical and Coca-Cola production.[11] The concentration and purity of cocaine in coca leaf and its products are a function of genetics, climate, altitude, horticultural methods, and drying, packing, and shipping methods. Cocaine is not, therefore, easily attributable to natural, technological, or social practices but dissolves the distinctions between them. Likewise, the contemporary geographical concentration of coca and cocaine production in the Andes is not a function of climate and nature alone but is inseparable from them.

The biochemical characteristics of leaves and alkaloids have had extended material effects in the geographical distribution of illicit cocaine production. Nineteenth-century scientific debate on whether coca leaf contained an active substance was generated by the decay of cocaine alkaloids in harvested leaf such that when leaves were examined in Europe, the alkaloid had diminished in potency.[12] This discovery concentrated previously international divisions of labor in legal coca processing into areas near zones of cultivation. In these places, processes of global trade and chemical decay combined with the Peruvian "scientific nationalism" of Alfredo Bignon, the French Peruvian scientist who created a simple method of cocaine extraction. Blending modernist themes of universalist science and medicine with nature and tradition, cocaine became a "modernizing good" in Peru.[13] Separated from Bignon by more than a century, contemporary cocaine laboratories owe much to his method; it has been assimilated and modified into tacit knowledge in plastic-lined pits filled with kerosene, cement lime, soda ash, and "other household supplies."[14]

The extraction and purification of cocaine from coca leaf also entangle humans and nonhumans and short and long histories, and bring distant places into relational proximity. The current status and history of cocaine are inseparable from transformations in the pharmaceutical industry and medical professions, themselves translations of wider modernist discourses of science, nature, and control. In cocaine's late nineteenth-century history, trajectories and differences in scientific and medical practices and preferences shaped the division of coca and cocaine into separate supply chains, with German consumers preferring the more "scientific" and "therapeutically precise" cocaine alkaloid, and consumers in other European as well as U.S. markets served more by coca-leaf-based tonics such as the famous Vin Mariani and, eventually, the nonalcoholic Coca-Cola.[15] Cocaine's contemporary technological history began with the advent of alkaloid science in German laboratories in the nineteenth century. Since then, the development of modern medicine has shaped the form and function of cocaine and the international politics of its prohibition. From a new materialist perspective, the use of chemical molecules is a process of not merely application but also adaptation and invention, of translation from distinct closed environments of the laboratory to the open systems of distribution and use in the wild.[16] Moving from the experimental laboratory to medical usage, cocaine became entangled with processes of the professionalization of medicine and emergent preferences for synthetic rather than natural drugs that enabled and reflected separation and prohibition. As cocaine became more popular in the nineteenth century, these processes enacted and embedded a distinction between cocaine hydrochloride as the stronger but legitimated substance of professional medical administration and elite quasi-therapeutic consumption, and the weaker coca-based tonics that were consumed more widely in an emergent culture of self-medication in the United States but were portrayed as a danger to moral and physical public health.[17] This cut between forms of cocaine was partly enacted by a professionalizing medical and pharmaceutical sector and reflected evolving wider preferences for synthetic rather than natural drugs, itself shaped by commercial pressures for isolatable intellectual property rights that have more widely discouraged pharmaceutical research into naturally occurring drug compounds.[18]

Beyond coca leaf, a number of other substances are required to manufacture cocaine. Making 1 kilogram of cocaine requires approximately 590 liters of gasoline or kerosene and 16–25 liters of the other chemicals.[19] The global geopolitical map of chemical precursors presents a different cocaine international than the usual focus on Andean coca cultivation and concentrated refining. It extends the political geography of cocaine chemistry to more powerful states and commercial interests, as essential chemicals are mainly produced in China and Europe, especially Germany and the Netherlands, and their relative neglect arguably stems from wider economic and political relations.[20]

Precursor chemicals shape their international regulation. Because they have many other uses, unlike coca crops subject to prohibition and eradication practices, precursors are distributed further along the supply chain and are encountered as a problem of diversion from the legal to illicit spheres or as a problem of illicit manufacture rather than of the substances themselves. The connections between global legal trade and cocaine production vary more by the specific chemicals than by the states and industries involved. For instance, potassium permanganate, a common oxidizing agent used to turn coca paste into cocaine hydrochloride, is exported legally from thirty-one major producing countries, especially China, but little legal trade occurs with customers in the Andean region, and 80 percent of seized illicit stocks derive from illegal manufacturing in zones of cocaine production. Other precursors are more widely diverted from legal trade; in 2011, Panama seized acetone, hydrochloric acid, toluene, and acetic acid totaling 3.6 tons hidden in a single shipment originating in Italy.[21] As well as having wide legal uses, many precursors are highly substitutable in cocaine purification, and many nonscheduled chemicals are also used to refine cocaine. This ambiguity and the changing patterns of chemical use compel regulatory bodies to cooperate with the chemical industry, thus shaping the range of actors enrolled in their control.

Enacting a cut between legal and illicit precursors is a relatively recent dimension of the cocaine international, introduced at the global level in the UN Convention against Illicit Traffic in Narcotic Drugs and Psychotropic Substances of 1988.[22] The contemporary chemical-control regime assembles diverse sites, processes, and materials in complex relations, including widely varying national export- and import-licensing systems, information exchange and reporting, and continual adaptation prompted by the substitutability and ambiguity of chemicals. This expansive assemblage is sustained and evolves through extensive reporting of production, exports and seizures by member states, and growing homogenization and strengthening of national export controls and technical and informational harmonization and capacity building. Thus, geographically dispersed processes enroll and distribute the regulatory burden in the construction of a precursor international that enacts a cut of a different kind on this chemical stuff. Whereas coca leaf is rhetorically and practically isolated in regulatory regimes that distribute the control apparatus around the plant itself in zones of production, precursor chemicals are characterized by substitutability and embedded in wider assemblages of legitimated production, trade, and use that shift the regulatory frame to one of adaptation, monitoring, and facilitation rather than prohibition and eradication.

Bodies

New materialism does not assert the importance of the physical and separate it from ideas, but rather shows their inseparability and heterogeneity as objects.

Such objects may be differentiated, enacting cuts, but they may also create topological proximities. They are "affiliative objects" that gather, unify, and stabilize action.[23] The national, global, international, and local trafficking and policy assemblages (the cocaine international) meet, act, and are produced with two such affiliative objects: the archetypal bodies of the addict and the smuggler.

Cocaine in coca leaf is not the same thing as cocaine in powder or crack, which is not the same thing as cocaine in the body. Further, the body is "a living labyrinth whose topology varies in time, where partial and circumstantial causalities are so intertwined that they escape any *a priori* intelligibility."[24] In the body of the cocaine user, the alkaloid passes through and opens up the blood-brain barrier, attaches to proteins, and blocks the flow of dopamine, creating an accumulation that induces a high. But this process is not universally the same in its effects and implications. Cocaine effects are often stronger in men than in women because of metabolic differences and gonadal hormone levels.[25] Women experience stronger cravings, suffer more trauma, and show greater willingness to engage in treatment.[26] Sex, ethnicity, age, and education all affect the cognitive impairment resulting from cocaine abuse.[27] Cocaine use before successful suicide is strongly associated with men rather than women, with more African Americans than white victims, and with younger rather than older suicide victims.[28]

Addiction is rare among cocaine users. Unlike some other drugs, cocaine was initially thought not to produce physical dependency and withdrawal. Addiction, then, is not inherent or reducible to cocaine's chemical composition or biological effects. The other purified option—that cocaine addiction is entirely psychological and not physical—however, is also debated. Neurophysical alterations in particular parts of the central nervous system (of a different type than those experienced for other drugs) generate different patterns of abuse and withdrawal.[29] Physical changes in brain activity both result from and contribute to cocaine use. However, cocaine dependency is not reducible to this intra-action but relates also to the method of use. Injection and smoking show similar dependency patterns that are more severe than intranasal use.[30] Cocaine enters the circulatory system more slowly through the nose than through the lungs or veins, and this differing rate affects dependency and the intensity and duration of the high.[31] Dose, duration of use, and the user's previous attendance at drug-treatment programs also affect dependence.[32]

Cocaine has become meaningful in its two major consumed forms, powder and crack, such that they are somewhat distinct assemblages. Cocaine powder is usually snorted or injected, but crack is smoked, as it is more volatile, vaporizing at a lower temperature.[33] Although crack is no longer thought to be instantaneously addictive, its use does correlate with addiction, though this may be due merely to addicts preferring the more efficient modes of ingestion it offers.[34] Crack cocaine is one of the "most criminalized of all illicit substances."[35] Not only does this derive from the appearance of enhanced addictive potential, but it is also an emergent effect of how race and class intra-act with physical form. Crack enabled cocaine to become available to "lower levels of the American socioeconomic racial order," and it became associated more with "desperation than recreation," less a status symbol of the consumptive habits of the rich than one of the degraded dependencies of the poor and dangerous.[36] For Dawn Moore, the short, intense high that crack produces "feeds its chaotic, dangerous status, making it a substance deemed worthy of more extreme interventions."[37] The intensity of high extends beyond users' bodies into the relocation of the "preferred purifying solution to the cocaine problem . . . from the therapeutic branch of the medical-industrial complex to its armed disciplinary forces."[38]

The categorical chemical separation of powder and rock is intra-active with law and with race and class via its materialization in differential criminal punishments that have been a source of recent controversy and reconstruction. In 2010, the U.S. government introduced the Fair Sentencing Act, which reduced the disparity in sentencing for the amount of crack cocaine required for the imposition of minimum mandatory sentences versus the amount of powder cocaine required, from a ratio of 100:1 to 18:1. This disparity tended to affect African American offenders adversely; they often served the same sentence for nonviolent cocaine offenses as white offenders did for violent offenses.[39] Further, laws contribute to harm by participating in the construction of an "affective climate" in which regulatory practices contribute to stigmatization and place a burden on care providers that can reduce access to clinical services and undermine trust between patients and treatment providers.[40] Treatment itself combines emerging research into pharmaceutical tools and psychological techniques. Cocaine was initially used and sold as a cure for other addictions, but there is currently no approved pharmaceutical treatment for cocaine addiction. Emerging pharmacological dimensions of treatment involve the enrollment and adaptation of molecules composed for other purposes. These forms of treatment combine with behavioral processes such as cognitive behavioral therapy to intervene in the learning processes developed in becoming an addict, or motivational incentives such as movie tickets and gym memberships given as prizes for clean drug tests, each materializing particular assumptions about universal human rationality.[41]

Looking at the "drug-using body" as an assemblage gives a different view of addiction. Addiction as an embodied assemblage resists the psychological, biological, or chemical reductionism that produces most controversies of what addiction is.[42] The drug-using body is not a stable, unified, bounded biological entity but "an ephemeral entity . . . that exists only in the event, in its moment of connection with the drug and the specific affects it enables."[43] Addiction is a process, not a condition. Initially, voluntary action becomes a process of physiological, psychological, and behavioral change and perhaps domination. The boundaries and composition of addicted entities are processes of temporary enrollment across the blurred, multiple, and fragmented drug-using bodies. This problematizes the view of the existence of a fixed, separate, a priori "standard human" found in both biochemical and sociopolitical models.[44] An assemblage view points instead to the process and relational spatial and temporal flows and sedimentations of addiction and asks what connections they make or disallow. As such, the conceptual diagram of levels of analysis familiar to IR (individual-society-state-international/regional/global) has no single, a priori, stable referent position for either cocaine or its international but instead has processes of flow, connection and disconnection, nearness and rifts, the production of provisional order, and stabilization never permanent or totalized. However, processes of purification and separation are not merely conceptual errors but active material processes that constitute the cocaine international. This can be seen in the bodies of smugglers as well as addicts.

Body Stuffers

Most frameworks for understanding illicit flows produce explanation through purification, including fixed points such as the supposedly inherent tendency of markets toward equilibrium, which explains the rise of illicit markets and the displacement of trafficking routes in response to enforcement efforts, and including its foundation in assumptions of the rational, calculating *Homo economicus*. Increasingly, this combines with a deterritorialized geopolitical imagination in which the spatialities of flows derive from the capacities of networked criminals able to choose their location at will. However, illicit flows of cocaine combine the production of the good itself, the transportation networks, and the movement of information and finance, which constitute a single flow but have multiple trajectories, anchorages, and arrangements of humans and nonhumans.[45] Here, movement and meaning are intra-actively constituted as processes of the commodification of cocaine; and its flows, for Gootenberg, "are accompanied and facilitated by flows of culture, science, politics, prestige and other immaterial exchanges."[46] Such flows bridge the gap between the national and the transnational and the "categorical domains of material culture, commerce, and consumption."[47]

These flows often, but not always, converge in the bodies of body stuffers, who ingest packages of cocaine to smuggle them. Considerable distributed action is necessary to integrate or separate bodies and stuff in ways that show how movement is central to an object-oriented IR. Body stuffing involves the ingestion of 100–200 small pellets of 5–12 grams of cocaine (several times the lethal dose of between 1 and 3 grams) covered by condoms or the fingers of surgical gloves.[48] Packaging techniques once varied considerably in size and construction, but they have become more similar, suggesting an informal standardization.[49] Body stuffing requires some training of the body, including swallowing relatively large objects (grapes, banana pieces) and following a diet to regularize the digestive cycle. Two days before the trip, only light meals are eaten, and a painful process of "loading" occurs over a period of hours, assisted by massages, olive oil, and petroleum jelly. Most body stuffers are from poor backgrounds. The fragmentation of the cocaine industry into small trafficking groups after the dismantling of the (always somewhat mythical) Colombian cartels also contributed to a rise in the use of such "boleros" (body stuffers).[50] Thus, the body stuffer is a human embodiment of the trained body, cocaine, condoms or the fingers of surgical gloves, poverty, and the fragmentation of illicit trade resulting from the U.S. war on drugs and criminal competition.

Customs and border officials may identify suspected smugglers primarily through observation (being attentive to shaking hands, excessive perspiration, clothing to disguise body contours, and inconsistencies in statements), intelligence gathered from other law enforcement agencies or from airline crews (because body packers will not eat on a flight), and the use of trained dogs. When identified by this assemblage, body stuffers are inserted into a different complex assemblage of sites and processes that enrolls hospitals, ultrasound and X-ray scans (conducted with consent), and routes of action determined in part by risks of cocaine toxicity from broken packets (50 percent of which cases result in death) or bowel obstruction, resulting in surgical action; otherwise, they are asymptomatic (in contained or stabilized body stuffing) and are managed through monitoring while awaiting "spontaneous drug passage."[51]

Whether cocaine is smuggled in bodies or elsewhere, the initial governmental encounter with it assembles other actants in other relational geographies of drug testing and identification. In this regard, cocaine may be active, but it also needs to be made actionable. After identification and extraction, processes of purification are required for legal action. Methods for analyzing cocaine enroll chemical processes and standardization practices to enact a cut between cocaine and other substances. Enacting this cut occurs through drawing other things into relational proximity, such as laboratories, law enforcement agencies, scientific knowledge, and equipment. Numerous testing methods exist, so identifying cocaine involves a series of selections. Internationally agreed-upon guidelines shape the selection of methods and the process of sample selection, which themselves invoke and enroll principles of analytical chemistry codified elsewhere.[52] The UN Office on Drugs and Crime (UNODC) has standards or guidelines on processes for identifying cocaine, analyzing impurities, testing for drugs in hair, sweat, and saliva, and impurity profiling, to name a few. Guidelines also shape geographic relations, as they require that presumptive tests conducted at sites of seizure and identification must be confirmed by an additional laboratory. Such guidelines are continually enacted as the UNODC facilitates harmonization through documents (multilingual dictionaries of drugs, precursors, and other terminology), extensive monitoring based on information exchange between states, and formalized methods for crop-yield assessment. Vienna, the base of the UNODC, is thus connected to cocaine as documents, training, numbers, guidelines, and wider scientific and political practices are brought together, distributed, and translated into action. Such action requires further linking and translation in relations among scientists and law enforcement organizations. As the editors of the *Bulletin on Narcotic Drugs* recently argued, "Laboratories produce results, and they need to have access to, and apply, quality standards to produce them, but those results also have to be turned into information and knowledge. Despite all technological advances, and contrary to popular belief, this does not happen automatically."[53] Bolstering drug-control assemblages through science entails a process of translation that is delegated to scientists, their results, and their expertise. Wider government knowledge also enrolls scientists and cocaine in more dispersed assemblages, as monitoring of cocaine consumption in European countries has begun sampling unprocessed wastewater to identify levels of cocaine and its metabolites after it passes through human bodies.[54]

Although the embodied enactments of body-stuffer assemblages constitute aspects of cocaine flows, and the embodied tacit knowledge of scientists stabilizes and strengthens regulatory assemblages intra-actively with politically and scientifically defined standards, neither cocaine smuggling nor its control is geographically fixed; it is merely somewhat stabilized. Here, processes of disembodiment and different processes of purification are enrolled as assumptions that simplify, stabilize, circulate, and affiliate international action on cocaine. In the contemporary cocaine international, illicit flows are simplified as they are brought into categorical and logical proximity with other transnational flows (crime and terror) and are claimed to correlate with the geopolitical map of state weakness and failure.[55] This draws in IR's recent measures and reconfigurations of anarchy and disorder in failed states discourse to stabilize the geographical focus of counternarcotics practice by claiming, as Antonio Maria Costa did as executive director of the UNODC, that "illicit commodities usually originate in trouble spots . . . [and] are then trafficked through vulnerable regions, to affluent markets. . . . [I]f you take a map of global conflicts, and then superimpose a map of global trafficking routes they overlap almost perfectly."[56] This is a picture of trafficking free from territorial links, except for those made by choice, and thus it neglects the rootedness of illicit flows in climatic, social, plant-genetics, or global chemical trades. However, it is a picture active in enabling expansive rearticulation and reinforcement of the focus on production rather than consumption, and the related political-chemical purification of essentialized coca/cocaine. It thus reconfigures an older purification in the drug-control regime that stemmed from the externalization of the portrayal of the drug problem and the associated global distribution of the adjustment costs.[57] Such externalization has often been materialized in militarization and coercion, most notably with President Richard Nixon's declaration of a war on drugs in 1971 and the later certification systems in which failure to comply with counternarcotics efforts resulted in mandatory sanctions against and a loss of aid for countries including Colombia, and controversial mass aerial herbicide attacks on coca fields.[58] The purification of cocaine trafficking from much of its chemical, political, and embodied constitution is itself an actant that circulates in academic theories and official reports to stabilize an assemblage of hypocrisy and violence such that the dematerialized figure of the criminal materializes and renders resilient the specific forms and trajectories of the cocaine international.

The International of Cocaine Regulation

International action on cocaine is always intra-active with cocaine. If cocaine is complex, heterogeneous, overflowing, and ambiguous in itself, then what does it mean to prohibit cocaine? Prohibition appears to be an action upon a substance without its relations: the location of harm in the chemical characteristics or at least the identification of chemical composition as the most easily manipulable dimension of wider social problems of addiction, crime, and death. Prohibitions rely on regulatory distinctions that travel and also on the enactment of smooth spaces and the distribution of ambiguities, both highly political acts even if not accompanied by formal political controversy.

The international of cocaine regulation seeks to make a cut by drawing other things together. The 1961 UN Single Convention on Narcotic Drugs separates two affiliated figures that it seeks to distinguish and stabilize but that resist as well as materialize their purification.[59] It recognizes, first, that "medical use of narcotic drugs is indispensable" and that "addiction to narcotic drugs constitutes a serious evil for the individual and is fraught with social and economic danger to mankind."[60] The founding cut of the contemporary drug regime, then, is the enactment of a distinction between medicine and addiction.

This cut is complicated by a further distinction between natural and synthetic drugs. The 1961 convention applies primarily to plant-based drugs, such as opium/heroin, coca/cocaine, and cannabis. The later Convention on Psychotropic Substances (1971) was a response to the growing use of synthetic drugs such as amphetamines, barbiturates, and psychedelics, but these are subject to weaker controls because of the influence of pharmaceutical companies and economic interests of Western states. This 1971 convention enacts a further distinction between "street drugs" and other categories of pharmaceutical drugs that are subject to less stringent controls. The growing illicit trade and consumption of drugs in the 1970s and 1980s in Western countries led to the 1988 UN Convention against Illicit Traffic in Narcotic Drugs and Psychotropic Substances, which unified the two earlier conventions and articulated a greater distinction between legal and illicit flows and a stronger focus on criminalizing illicit production, trafficking, and possession of drugs.[61] All of these cuts effect a progressive integration of cocaine with other drugs and serve to make cocaine a single substance.

Whereas early twentieth-century drug controls were targeted at particular substances, notably opium and cocaine in separate systems, by the second half of the century, the term *drugs* became a generalized category in which the social and biological effects of numerous substances were grouped together, disaggregated into categorical schedules, and brought into regulatory proximity.[62] The 1961 convention established a strong zero-tolerance approach to all such drugs. Moore argues that this generalization of the category of "drugs" was "an attempt to move anti-substance rhetorics away from targeted claims on specific substances and therefore specific populations."[63] In becoming a drug, cocaine was brought into proximity with heroin and other substances, because assumptions and practices of control bleed across boundaries and structure each other. This generalization away from specificity is part of an ongoing process of the creation of regulatory space in which attempts at standardization and harmonization seek to maximize the speed of the circulation of controls and render cocaine actionable.[64] Classifications and standards are "objects for cooperation" that "inhabit several communities of practice and satisfy the informational requirements of each. . . . [T]hey are objects that are able both to travel across borders and maintain some sort of constant identity."[65] Smoothing the flow of immutable regulatory objects, then, is integral to the contemporary assemblage of the cocaine international.[66] Processes of measurement, standardization, and harmonization enable practices of protection (of people and property rights) and mobility. It is perhaps not surprising that enhanced methodologies for estimation of cocaine crops, production, trafficking, and consumption are produced in the regulatory cocaine international. After all, a generalized drive for precision has been argued to be integral to "extend[ing] uniform order and control over large territories" and to bringing "wide domains of experience under systematic order."[67]

Although cocaine may be coca leaf, cocaine powder, or crack, the cocaine international seeks to erase these chemical/physical cuts with an assertion of a cocaine essence in all three. It does so by enrolling both the harms of drug addiction and specific "expert" opinion and research. Thus, the 1961 convention authorizes an essentialist account of coca leaf that was built upon a 1950 report of "experts" (though they had no expertise in coca, cocaine, or the Andean region) led by a prominent member of the U.S. pharmaceutical industry, who claimed that coca chewing was the cause of "racial degeneration" and "decadence" in "Indian" populations.[68] On this basis, the traditional use of coca, such as the chewing of coca leaf, including for medicinal purposes, was declared "quasi-medical" in the 1961 convention and was called to be terminated within twenty-five years.

Coca leaf is part of other assemblages that resist this singularity of cocaine. As a result, controversies centering on coca leaf remain a periodic feature of the cocaine international. In 2009, the government of Bolivia proposed the removal of the obligation to abolish coca-leaf chewing. This claimed coca leaf as a part of Bolivia's cultural heritage and biodiversity, declared it a harmless and nonaddictive substance, and enrolled assertions of cultural practices, indigenous peoples' rights, and medicinal practices. However, the proposed amendment was defeated by formal objections from eighteen UN member states, led by the United States. This group included no Latin American states, some of which submitted approvals for the amendment; Colombia initially objected and then withdrew its objection. Formal international political process, then, is sometimes needed to maintain essentialism and the relations it enables and stabilizes. Similarly, in 1995, a World Health Organization (WHO) and UN Interregional Crime and Justice Research Institute study on cocaine use found positive health and economic associations from coca leaf. The release of this report was prevented by the U.S. government, which argued that the study and the WHO's support for harm-reduction programs were "heading in the wrong direction" and that "if WHO activities relating to drugs failed to reinforce proven drug control approaches, funds for the relevant programmes should be curtailed."[69]

Although standardization and harmonization are central to the cocaine international, they rely upon a number of exceptions or cuts. Such exceptions take a lot of work (of both humans and nonhumans) and do a lot of work in stabilizing the cocaine international in the presence of resistances of coca and cocaine to universal and essentializing controls. Some exceptions stabilize prohibition by offsetting commercial and medicinal costs. Traditional coca chewing remains a target of eradication, yet coca leaf is produced and traded legally in exceptions codified in the 1961 convention. Hundreds of tons of coca leaf are imported to the United States by the Illinois-based firm Stepan, which removes the cocaine alkaloid and supplies Coca-Cola with "decocainized flavor essence" for its products.[70] Likewise, cocaine remains in legal use as a topical anesthetic and vasoconstrictor by some ear, nose, and throat surgeons, continuing a long history since its first use as a local anesthetic in 1884, though the search for alternatives began in 1905.

Further exceptions are integral to the operation of prohibition. The dispersed assemblages of identification discussed earlier enroll cocaine in a new form, as a comparator, that creates its own flows of cocaine. Laboratories require reference samples of cocaine in order to engage in processes of identification. This is achieved by requests to national drug laboratories that are regulated by global procedures established at the UNODC, which authorizes and provides samples.[71] The UNODC also coordinates a quality-assurance program in which laboratories send their results for review by an independent panel of forensic experts in order to harmonize drug testing: a process that involves further transfers of sample materials.[72] In this regard, the need for legal transfers of cocaine requires that states have competent national authorization bodies and makes the UNODC a cocaine supplier in order for the drug regime to operate. Such exceptions are not contradictions in regulatory frameworks but are essential to the materialization of control in its encounter with the overflowing ambiguities of cocaine.

What Is Cocaine? What or Where Is the Cocaine International?

Cocaine's history and contemporary global life are a story of incomplete purification and overflows. Cocaine mediates, translates, enables, and resists political controversies and trajectories in which the material and cultural, the natural and technological, the traditional and the modern are produced, reinforced, and resisted; in which divisions of geopolitics, class, race, and nationality are materialized; and in which the contemporary political encounter with human values and bodies, nonhuman substances and processes, and their entanglements articulate specific relational spatialities through which we think the international governing of material objects.

Cocaine is what it does. It is encountered and acts as and with many things: active substance, addict, addiction treatment, alkaloid, anesthetic, assemblage, authorization, behavior, body, body-brain crosser and alterer, bolero, border, chemistry, class, Colombia(n), commodity, comparator, crack, crime, dopamine accumulation, error, estimate, event, expertise, false positive, flavor, flow, graph, harm-reduction strategy, herbicide, high, hospital, kerosene, intellectual-property rights, laboratory practice, leaf, metabolite, modernity, Peru, plant, politics, potassium permanganate, potential addiction, powder, preferences, prison, process, punishment, race, Richard Nixon, sample, screen, sex, shit, smell, standard human, tea, threats, tradition, treatment, treaty, United States, UNODC, wad-mouth-chew-lime, waste, WHO, vasoconstrictor, x-ray. But although cocaine is multiple, it is never everything all at once. It moves from one thing to another—becoming and being translated, represented, and thus transformed.

That these processes are only periodically subject to formal political debate is testament to the stability of inchoate assemblages, but the distribution of agency and controversies is a political act made by assemblages. Where are the inherent ambiguities of cocaine distributed, resolved, and contested? Sometimes in the bodies of addicts and traffickers, sometimes in zones of production (produced by climate, altitude, war, genetics, distant governments, and local practices), and sometimes in the "international," whether embedded in treaties, in the UNODC, or in the threats and incentives issued by governments to one another and to international organizations. This distribution is sometimes shaped by chemical processes (alkaloid decay rates, the substitutability of precursors and their entanglement in legitimated modes of consumption and trade); other times it is enabled and shaped by the mobilization and materialization of other nonhumans (standards, methodologies, preferences, numbers, guidelines, treaties). Here it should be noted that both human and nonhuman emerge as blurred categories themselves, given that each is generative and affiliative of the other and that many entities are neither wholly human nor wholly nonhuman (think of numbers, documents and guidelines, and procedures). The cocaine international continually vacillates between the pure and the impure, the essence and the totality, the natural and the cultural, the human and the nonhuman; and never attains a permanent home in any one realm. Cocaine and the cocaine international are, then, social; economic; political; natural; cultural; large; small; moving and still; position and trajectory; action and reaction; linear and nonlinear; legal, illegal, and neither; and none of the above.

Notes

1. Alfred North Whitehead, quoted in Andrew Barry, "Pharmaceutical Matters: The Invention of Informed Materials," *Theory, Culture, and Society* 22, no. 1 (2005): 56.

2. Michel Serres and Bruno Latour, *Conversations on Science, Culture, and Time* (Ann Arbor: University of Michigan Press, 1995), 60.

3. Karen Barad, *Meeting the Universe Halfway: Quantum Physics and the Entanglement of Matter and Meaning* (Durham, N.C.: Duke University Press, 2007), 139.

4. Bruno Latour, *We Have Never Been Modern* (Cambridge, Mass.: Harvard University Press, 1993), 10–11.

5. Barry, "Pharmaceutical Matters," 54.

6. Michel Callon, Pierre Lascoumes, and Yannick Barthe, *Acting in an Uncertain World: An Essay on Technical Democracy* (Cambridge, Mass.: MIT Press, 2011), 10.

7. Eugene McCann, "Veritable Inventions: Cities, Policies, and Assemblage," *Area* 43, no. 2 (2011): 143–47.

8. John Law, "And If the Global Were Small and Noncoherent? Method, Complexity, and the Baroque," *Environment and Planning D* 22, no. 1 (2004): 13–26.

9. Paul Gootenberg, *Andean Cocaine: The Making of a Global Drug* (Chapel Hill: University of North Carolina Press, 2008), 23.

10. Pierre A. Chouvy, *Opium: Uncovering the Politics of the Poppy* (London: Tauris, 2009), 1.

11. Timothy Plowman, "Coca Chewing and the Botanical Origins of Coca (Erythroxylum SPP.) in South America," in *Coca and Cocaine: Effects on People and Policy in Latin America,* ed. Deborah Pacini and Christine Franquemont (Cambridge, Mass.: Cultural Survival, 1985), 5–33.

12. Gootenberg, *Andean Cocaine*.

13. Ibid., 101.

14. Paul Gootenberg, "A Forgotten Case of 'Scientific Excellence on the Periphery': The Nationalist Cocaine Science of Alfredo Bignon, 1884–1887," *Comparative Studies in Society and History* 49, no. 1 (2007): 202–32.

15. Gootenberg, *Andean Cocaine*, 60–62.

16. Barry, "Pharmaceutical Matters"; Callon, Lascoumes, and Barthe, *Acting in an Uncertain World*.

17. Joseph Spillane, "Making a Modern Drug: The Manufacture, Sale, and Control of Cocaine in the United States, 1880–1920," in *Cocaine: Global Histories,* ed. Paul Gootenberg (New York: Routledge, 1999), 21–45.

18. Alain Pottage, "The Inscription of Life in Law: Genes, Patents, and Bio-politics," *Modern Law Review* 61, no. 5 (1998): 740–65; Brett Neilson and Mohammed Bamyeh,

"Drugs in Motion: Toward a Materialist Tracking of Global Mobilities," *Cultural Critique* 71 (Winter 2009): 1–12.

19. Damian Zaitch, *Trafficking Cocaine: Colombian Drug Entrepreneurs in the Netherlands* (The Hague: Kluwer Law International, 2002), 42.

20. Sayaka Fukumi, *Cocaine Trafficking in Latin America: EU and US Policy Responses* (Aldershot, Eng.: Ashgate, 2008), 116. For more in-depth discussion, see Mike Bourne, "Netwar Geopolitics: Security, Failed States, and Illicit Flows," *British Journal of Politics and International Relations* 13, no. 4 (2011): 490–513.

21. International Narcotics Control Board, *Precursors and Chemicals Frequently Used in the Illicit Manufacture of Narcotic Drugs and Psychotropic Substances* (Vienna: INCB, 2012), 20–22.

22. Francisco Thoumi, *Political Economy and Illegal Drugs in Colombia* (Boulder, Colo.: Lynne Rienner, 1995), 132.

23. Lucy Suchman, "Affiliative Objects," *Organization* 12, no. 3 (2005): 379–99.

24. Bernadette Bensaude-Vincent and Isabelle Stengers, *A History of Chemistry* (Cambridge, Mass.: Harvard University Press, 1996), 263.

25. Scott E. Lukas et al., "Sex Differences in Plasma Cocaine Levels and Subjective Effects after Acute Cocaine Administration in Human Volunteers," *Psychopharmacology* 125, no. 4 (1996): 346–54.

26. David Walker et al., "Sex Differences in Cocaine-Stimulated Motor Behavior: Disparate Effects of Gonadectomy," *Neuropsychopharmacology* 25, no. 1 (2001): 118–30.

27. Arthur Masneill Horton and Charles Roberts, "Sex, Ethnicity, Age, and Education Effects on the Trail Making Test in a Sample of Cocaine Abusers," *International Journal of Neuroscience* 108, nos. 3–4 (2001): 281–90.

28. Steven Garlow, "Age, Gender, and Ethnicity Differences in Patterns of Cocaine and Ethanol Use Preceding Suicide," *American Journal of Psychiatry* 159, no. 4 (2002): 615–19.

29. Frank H. Gawin and Herbert D. Kleber, "Evolving Conceptualizations of Cocaine Dependence," *Yale Journal of Biology and Medicine* 61, no. 2 (1988): 123–36.

30. Michael Gossop et al., "Severity of Dependence and Route of Administration of Heroin, Cocaine, and Amphetamines," *British Journal of Addiction* 87, no. 11 (1992): 1527–36.

31. Fabrizio Schifano and John Corkery, "Cocaine/Crack Cocaine Consumption, Treatment Demand, Seizures, Related Offences, Prices, Average Purity Levels, and Deaths in the UK (1990–2004)," *Journal of Psychopharmacology* 22, no. 1 (2008): 71–79.

32. Gossop et al., "Severity of Dependence."

33. Schifano and Corkery, "Cocaine/Crack Cocaine Consumption," 71.

34. John Morgan and Lynn Zimmer, "The Social Pharmacology of Smokeable Cocaine: Not All It's Cracked Up to Be," in *Crack in America: Demon Drugs and Social Justice*, ed. Craig Reinarman and Harry Gene Levine (Berkeley: University of California Press, 1997), 144.

35. Dawn Moore, "Drugalities: The Generative Capabilities of Criminalized "Drugs," *International Journal of Drug Policy* 15, nos. 5–6 (2004): 424.

36. Jimmie Reeves and Richard Campbell, *Cracked Coverage: Television News, the*

Anti-cocaine Crusade, and the Reagan Legacy (Durham, N.C.: Duke University Press, 1994), 130.

37. Moore, "Drugalities," 425.

38. Reeves and Campbell, Cracked Coverage, 130.

39. American Civil Liberties Union, "Fair Sentencing Act," http://www.aclu.org/fair-sentencing-act.

40. Kane Race, "Drug Effects, Performativity, and the Law," International Journal of Drug Policy 22, no. 6 (2011): 410–12.

41. National Institute on Drug Abuse, Cocaine: Abuse and Addiction (Bethesda, Md.: NIDA, 2010), 4–5.

42. Darin Weinberg, "On the Embodiment of Addiction," Body and Society 8, no. 4 (2002): 1–19.

43. Peta Malins, "Machinic Assemblages: Deleuze, Guattari, and an Ethico-aesthetics of Drug Use," Janus Head 7, no. 1 (2004): 88.

44. Steven Epstein, "Beyond the Standard Human?," in Standards and Their Stories: How Quantifying, Classifying, and Formalizing Practices Shape Everyday Life, ed. Martha Lampland and Susan Leigh Star (Ithaca, N.Y.: Cornell University Press, 2009), 35–53.

45. Bourne, "Netwar Geopolitics."

46. Gootenberg, Andean Cocaine, 105.

47. Ibid., 105–6.

48. Shamir O. Cawich et al., "Occupational Hazard: Treating Cocaine Body Packers in Caribbean Countries," International Journal of Drug Policy 20, no. 4 (2009): 377–80; and Zaitch, Trafficking Cocaine.

49. Stephen Traub, Robert Hoffman, and Lewis Nelson, "Body Packing—The Internal Concealment of Illicit Drugs," New England Journal of Medicine 349 (2003): 2519–26.

50. Zaitch, Trafficking Cocaine, 152.

51. Cawich et al., "Occupational Hazard," 378; Traub, Hoffman, and Nelson, "Body Packing."

52. UN Office on Drugs and Crime, Recommended Methods for the Identification and Analysis of Cocaine in Seized Materials (Vienna: UNODC, 2012), 17.

53. B. Remberg, and A. H. Stead, "Science in Drug Control" (editorial), Bulletin on Narcotics 57, nos. 1–2 (2005): 2.

54. Ibid.

55. Bourne, "Netwar Geopolitics."

56. UN Office on Drugs and Crime, Crime and Instability: Case Studies of Transnational Threats (Vienna: UNODC, 2010), iii.

57. H. Richard Friman, "Externalizing the Costs of Prohibition," in Crime and the Global Political Economy, ed. H. Richard Friman (Boulder, Colo.: Lynne Rienner, 2007), 49–66.

58. Martin Jelsma, "The Development of International Drug Control: Lessons Learned and Strategic Challenges for the Future" (working paper prepared for the First Meeting of the Global Commission on Drug Policies, 2011), http://www.globalcommissionondrugs.org/wp-content/themes/gcdp_v1/pdf/Global_Com_Martin_Jelsma.pdf.

59. Suchman, "Affiliative Objects."

60. United Nations, Single Convention on Narcotic Drugs, 1961, https://www.unodc.org/pdf/convention_1961_en.pdf, p. 1.

61. Jelsma, "Development of International Drug Control."

62. Moore, "Drugalities."

63. Ibid., 423.

64. Andrew Barry, *Political Machines: Governing and Technological Society* (London: Athalone, 2001).

65. Geoffrey Bowker and Susan Leigh Star, *Sorting Things Out: Classification and Its Consequences* (Cambridge, Mass.: MIT Press, 2000), 16.

66. Ibid.; Vaughan Higgins and Wendy Larner, eds., *Calculating the Social: Standards and the Reconfiguration of Governing* (Basingstoke, Eng.: Palgrave Macmillan, 2010).

67. M. Norton Wise, introduction to *The Values of Precision*, ed. M. Norton Wise (Princeton, N.J.: Princeton University Press, 1995), 5.

68. Martin Jelsma, "Lifting the Ban on Coca Chewing: Bolivia Proposal to Amend the 1961 Single Convention," Transnational Institute Series on Legislative Reform of Drug Policies, no. 11 (2011), http://www.tni.org/sites/www.tni.org/files/download/dlr11.pdf.

69. World Health Organization, *Forty-Eighth World Health Assembly, Geneva, 1–12 May 1995: Summary Records and Reports of Committees,* document WHA48/19995/REC/3, p. 229.

70. Clifford May, "How Coca-Cola Obtains Its Coca," *New York Times*, July 1, 1988, http://www.nytimes.com/1988/07/01/business/how-coca-cola-obtains-its-coca.html; Lynn Sikkink, *New Cures, Old Medicines: Women and the Commercialization of Traditional Medicine in Bolivia* (Belmont, Calif.: Wadsworth/Cengage, 2010), 122; Gootenberg, *Andean Cocaine*, 198–205.

71. International Narcotics Control Board and the UN Office on Drugs and Crime, *Guidelines for the Import and Export of Drug and Precursor Reference Standards for Use by National Drug Testing Laboratories and Competent National Authorities* (New York: UNODC, 2007).

72. UN Office on Drugs and Crime, "International Quality Assurance Programme (IQAP)," http://www.unodc.org/unodc/en/scientists/quality-assurance-support.html.

Clocks

Yvgeny Yanovsky

On November 11, 1918, at 4:50 a.m., the warring sides of the Great War signed the armistice agreement that was supposed to bring the fighting to an end. The terms stipulated that an armistice should commence exactly at 11:00 a.m. that day.[1] The news about the agreement was transmitted from the Eiffel Tower to all combatants on the field, and orders of this matter were delivered to all the links in the command chain. In fact, most of the soldiers on both sides were aware that they were supposed to stop fighting at 11:00 exactly. But what was unclear is what they should do until 11:00, and many of them wondered if they needed to keep on fighting or to stop at once. According to the most conservative estimates, in those last hours of the Great War, both sides suffered 10,994 casualties, of which 2,738 were deaths—a fatality rate that was higher than in an average day of that war.[2] The most fascinating—and disturbing—fact is that that day was supposed to be the last day of the war, and most of those who kept fighting knew that the war was de facto over. If so, why did soldiers from both sides continue to kill each other with such devotion? To some, it might look completely absurd that people willingly put their lives at risk when they already knew that the war had ended. What caused people to fight, kill, and risk being killed up to the last literal moment?

Timekeeping devices, clocks, and watches shape the notion of war. How do the material changes in a measurement and the representation of the dimension of time shape the way people run a war in our era? The ability to precisely measure and represent time also changes the way in which we fight, and the notion of security itself is now largely interconnected with precise timekeeping. In this chapter, I will focus on wars, because they are strictly connected to the practice of security. I highlight one of the dominant patterns of the political rule today: our lives become more dependent on the ability to be precise.

Accordingly, this study investigates the specific changes that have occurred with regard to war as a result of the intrusion of precision into politics. Toward the end of the nineteenth century, a number of entrepreneurs, along with technological improvements, led to an essential change in time measurement. As a result, the presence and understanding of time have been completely changed, turning time into a thing that has the ability to deeply influence political processes. This happened especially after time measurement became accurate and handy to growing sections of society. Nowadays, time is no longer an amorphous dimension but a tangible, valid image that exists on its own, and the clock has become one of the main political tools to run order and obedience. Yet only a few hundred years ago, time was less present in everyday life. It was usually not measured accurately, because there was no practical ability to do so and no political will in this regard. The accessibility to a precise timekeeping device constructed new modes, practices, and conducts of war.

In order to make my intervention more specific and materialistic, I will focus primarily on the micro level of the conduct of war. This study examines the impact of time on human behavior during wartime and directly in combat. I offer a possible exploration of people's motivation to behave in a mode that at first glance makes no sense, when they are prepared to go on fighting and to endanger their lives even though it is obvious to them not only that the outcome of the confrontation has been already determined but also that they belong with the winning side. We may understand the motivation of the German side at the last hours of the Great War as an attempt to defend itself or maintain its honor, whereas the behavior of the Allies' soldiers can hardly be understood only in these terms. What caused the officers to give orders to fight to the last minute, and why did the soldiers obey these orders? Why sacrifice themselves in a battle that would not change the political end result, which was already decided and known by the time they engaged the enemy?

The soldiers were surely motivated by mechanisms of profound self-discipline to engage in behavior that seems logical when considered this way. Here, direct access to precise timekeeping devices plays an important role: the ability to measure time in an accurate way creates the possibility of referring to the war as a cognitive framework defined by terms of time. Accurate timekeeping devices materialize war into this frame by defining its time boundaries in a precise way. This possibility changes the mode in which the war is conducted, because the precise differentiation between wartime and peacetime enables a change in the expectations and demands of the state's population. Time measurement and representation in means such as clocks and personal watches enable a governmentality that also applies at times of war, when people are prepared to act extremely, as is expected from them, and even to put their lives in danger. By doing so, they transmit governmentality to the battlefield. In principle, it is possible to argue that a soldier who fought in the Hundred Years War could not fight until the last minute even if he was requested and willing to do so, not only because the concept of minutes was dim for him but also because measuring minutes within the field conditions was beyond his abilities. The idea of engaging in war until an exact point in time was completely absent in the world he lived in.

Time and Politics

Despite some efforts to investigate the social origins of time in the nineteenth century, the main work in this regard began on the eve of the twentieth, in the work of Émile Durkheim.[3] Based on anthropological assumptions, Durkheim tried to uncover the origins of a social time. He claimed that in different cultures we can find compatibility between calendars and means of time measurement (based on cycles of nature) and social events that are supposed to be predicted by them. In *Elementary Forms of Religious Life*, he argues that there is a clear separation between time as an objective dimension and people's time, that is, time based on social ground. The earliest time units derive from the rhythm of social life and the division of time into those units (minute, week, year), a reaction to a circle of human life and the events that surround it. Societies tend to organize their lives according to that constant rate, which they choose, and each time frame reflects that rate in accuracy.[4] If so, every society develops some sort of time-measuring device that reflects the tempo of social activities and assures their repetition. Today's society uses clocks that are able to calculate time to a precision of a second, reflecting the increasing tempo of our life and splitting time into tiny units, based on a social construction.

The increasing availability of a means of time measurement is one of the most important signs of progress in the field of measurement of time. Time has never been so close and accessible. We can trace its development along a line from Stonehenge, through various timekeeping machines of the ancient world, through the fire clock of the medieval church, through mechanical clocks of the thirteenth century, and up to the scientific discoveries that led clock makers to use a constantly vibrating quartz atom as a heart for cheap analogical watches.[5] As a result of this process, time has become more accessible to us, and one can know without any effort precisely what time it is. Additionally, the distribution of timekeeping devices, wristwatches especially, teaches us about another process. If in the ancient world people were mostly disconnected from timekeeping devices, and in the Middle Ages they were literally submitted to it (because of the position of the device itself in a central place and its connection to the inescapable toll of bells), now we keep time by our own will. The knowledge of time follows us on a personal level in everyday life practices.

The clock has become a practical and common device that embodies the power of time in modern society, and by which people are constantly examined. The clock, an ultimate representation of precision, receives greater importance than merely being a tool for showing time.[6] Modern society demands its members be loyal to precision and hence to the clock.[7] Today, we can talk about a wide variety of precise clocks that surround us: on our wrists, on our screens, and on our walls. They have become so common that their existence simply cannot be ignored. Today no one can run his or her life without being exposed to the measurement of the dimension of time, and to the means designed for this purpose. By buying a time-measuring device, we signal to society that we are willing to be examined by that device.

As the number of time-measuring devices rises, precision acquires a profound social significance.[8] In 1909, George Beard claimed that the precision that is dictated by the clock makes people nervous, because it creates an atmosphere in which expectations can be broken constantly.[9] From a social standpoint, precision has become even more important. Two types of precision can be distinguished. The first type can be called "natural," because it is provided by timekeeping devices that are based on repeated natural cycles and that allow much flexibility in their measurement. The appearance of the second type occurred shortly after the invention of the mechanical mechanism. This type is fully different from the previous one, because it can provide frames that are based on permanent lengths, and it can continue measuring constantly, without any change. Mechanisms of our clocks allow numberless repeating, as long as certain conditions stay the same.

Mechanical clocks provided the best solution to a single problem that all those who worked on measuring time faced: assuring that equal periods of time would have a uniform length. This was unattainable without undergoing a process of fundamental change in timekeeping, which replaced natural precision with instruments of mechanical precision. This mechanical precision creates a conceptual basis for the creation of steady cognitive units defined in terms of time. People get used to the comfort these units afford and learn about their rigidity in the process of social interaction. These cognitive units have become more and more rigid, with increasing precision in the devices capable of measuring them, until we see that today there is no real way to escape them. The dichotomy of precision expressed by these means, capable of unambiguously defining where the border passes between units, creates a human understanding that all cognitive frames have a clear beginning and end, predictable and defined, a situation that was not taken for granted in a world where it was impossible to be precise.

This dichotomous division was transferred to the battlefield. People who are used to an order and security created by precision find it increasingly difficult to accept vague definitions in war and warfare. They want to know for sure when war begins and ends. The state, in turn, provides answers to these demands, as it creates a situation in which precision is also applied in regard to security action. Think, in this regard, of the various criticisms and negative emotions with regard to the U.S. "war on terror" as an open-ended conflict, that is, one that is not bounded in time. Here, one can notice the connection between politics and precision: technological capabilities create political incentives to use precision and accuracy in order to achieve political goals, and politicians realize the benefits that might arise.

Today we live in a world of global and standardized time in two ways. First, the length of the main units of time are the same everywhere in the world; and second, a system of global time—based on Greenwich mean time—exists. This system defines in a precise and homogeneous way the division of time in every place on earth. There is a global certainty about the dimension of time that constitutes a new and unprecedented social order.

The process of the global unification of time was the outcome of the initiatives of several private entrepreneurs. In this context, the work of Stanford Fleming, the chief engineer of the Canadian Railway Company, is of great importance. In 1876, he promoted the idea of calculating time based on the division of the world into twenty-four units, each fifteen degrees wide. By his initiative, the 1884 International Meridian Conference was convened, in Washington, D.C. The new standard for measuring time became based on equal intervals of the Greenwich Meridian.[10] Fleming's enthusiasm took his ideas still further, and he encouraged their ratification in different countries by continuing contact with political leaders all over the world. His contacts with Count Helmuth von Moltke, who once was a chief of the German General Staff and later became the best-known promoter of the idea of standardization of time in the German parliament, is very significant. Moltke's point of view was unique, mostly because of his position as a soldier-politician, as Fleming put it. Moltke understood that the ability to take advantage of precise timekeeping would have great importance in the following wars. In his last speech in the parliament, he argued that it was a matter of military importance to embrace the system of standardized time in Germany, because victory in the next war would be largely dependent on fast mobilization.[11] From his point of view, the ability to schedule railway operation in a precise manner and on a large scale would be decisive. In his opinion, a country that would not implement the system of standardized time would be at a disadvantage in modern warfare, and maybe would not be able to fight at all.[12]

Only by the process of standardization does timekeeping become global. Before standardization, each and every place located on the same longitude had its own twenty-four-hour cycle, measured by solar time. To figure out the impact of the change, it will be very useful to use a counterfactual method by which we can try to understand what would be the result of the attempt to declare armistice at 11:00 in a world before the standardization of time.[13] Considering the Great War, there was no practical ability to declare an armistice at 11:00 exactly without a standard time; only because of standardization can 11:00 be measured simultaneously along the entire western front. The front line as it was in November 1918 starts in Ostend, a Belgian city on the approximate longitude 3, and continues to the Swiss border, east of Nancy, at longitude 7. According to solar time, there is a twenty-minute difference between those two points. This means that soldiers in Nancy would fight for twenty minutes longer than those in Ostend in order to finish the war exactly at 11:00. Hence, any attempt to end the war at an exact point in time, before standardization, was doomed to failure.

Standardization provides the clock with global power, because it is known for sure what the exact time is anyplace on the globe. Today, looking at your clock, you can know what time it is not only at your location but at every other location as well. Standardization of the dimension of time opened a new era; from then on, time reckoning was international, accessible, and precise. In this sense, modern clocks are global or international devices.

By using the work of French philosopher Michel Foucault, I will try to connect the concepts of precise timekeeping and security. Foucault claims that rulers, because of political and social changes that have happened throughout history (especially the rise of the Westphalian order), have been forced to change the way in which they implement power. The old way—the material and highly visible power of sovereignty—was replaced by "governmentality." The purpose of its operation is to ensure the proper management of the ruled thing. It can be done by improving people's lives and inducing an overwhelming commitment to making their lives better, especially by providing them a sense of security. This governmental power becomes such a usual thing that most of us feel it as something natural. Rulers understood not only that usage of power in a suppressing manner is expensive but also that it will never ensure stability, unlike the disciplining form of power.[14] Governmental power operates at the level of the individual, causing one to change her or his ways, apparently voluntarily. It does not limit one's actions but creates an expected pattern for them.[15] Because everyone enjoys the benefits of security, we tacitly agree to become a part of this mechanism and to allow constant observation in order that deviants that can harm the social body might be recognized.[16]

Observation occurs constantly, by various methods. People are so used to being watched that they do not imagine that observation takes place at all, when the observer and the observed subject are actually both parts of governmental power fields. The use of exact time indexes in the process of observation and examination provides the necessary anchor to ensure a continuity of time's cycle, while it fixes its predetermined points. In his *Discipline and Punish*, Foucault shows how young criminals are disciplined by time measurement in prison, where they are compelled to do some actions limited by fixed and hardened time limits. We can see that it is not just the action but the time frames themselves that make people behave properly; the disciplining power is hidden not in the outcome but in the accurate execution of the process itself.[17] Even though Foucault does not mention it directly, it is clear that this process cannot be maintained without timekeeping devices of some kind, as it is impossible to expect the criminals to calculate time in any other way. The idea of using a timekeeping device and the ability to be precise in the context of education and discipline is revolutionary, reflecting a profound change with respect to time in general and its measurement in particular. After a while, educational methods based on accurate timekeeping left the boundaries of the prison and were applied to society as a whole.

Following Foucault, I claim that exact timekeeping puts a sense of security to the test in a growing numbers of focal points as clocks and other modern timekeeping devices are embedded with governmental power. Time is a source of power that is used to enforce discipline on the population. In a world of standard time, where clocks are accessible to everyone and the time units become smaller, there is a growing number of possibilities for examination. Modern time systems allow division of a human action into small units, precisely numbered, that are easy to test. In those time frames, people learn how to behave themselves properly, namely, to fit their actions to what is expected from them by the state and society. The ability of exact timekeeping adds a concept of security with a great deal of immediacy, embodied in a human demand to get security at any given time. Security and precision are strongly connected to each other. We like the feel of security and want it to be instantly provided. We have become used to the benefits it provides and behave ourselves in expected ways in order to acquire it.[18] Security has been transformed from a vague concept to something that has constantly and instantly been supplied to every individual in society. By the rules dictated by precision, one can easily feel whether he or she is secure, and can act accordingly. Security, in its turn, provides order. As the WWI case clearly shows, in order to sustain temporal order, dictated by their clocks, people may even put their lives in danger.

In the modern era, the clock is a tool to create a clear distinction between peace and war. Using the concept of precise timekeeping, one can distinguish when peace begins and ends. Time measurement enables a dichotomous distinction between order and chaos, between normalcy and anomaly. Given that we live in a world of global timekeeping in which time is accessible to all, the importance of the ability to count it in a precise way grows, as society becomes more and more dependent on it. It is expected of the people to act precisely, and they expect the state to act in a similar manner with respect to them. A person who owns a watch is expected to behave himself according to the well-known rules of precision taught to him during his life, in times of peace as well as in times of war. Embedded into modern clocks, time becomes a physical entity. Today we do not think of time as a metaphysical dimension but as an object that gets meaning every time one looks at a clock; in this sense, I can claim that clocks (and other modern timekeeping devices) implement governmental power and serve as political tools of self-disciplining, by providing the ability to be precise.

Time at War

As the ability to legitimately use force in the international system shrinks because of the rise of various rules and laws of international society, the idea of clearly demarcating wartime is ever more internalized by states and peoples.[19] In the past, when violence between actors on an international level was frequent, in a world of overlapping loyalties when there were fewer restraining factors on the use of international force, there was no political importance in limiting war to clearly defined time borders, and there was no known effort to end war by an exact time. So it is possible not only that the Hundred Years War was a reality of actual conflict but also that it could last for so much time without clearly identifiable points of beginning and end.[20] Today, most states are forced to make war faster, at least outwardly, because continuous insecurity undermines the legitimacy of the state itself, by proving it ineffective.

The accessibility to precise timekeeping devices makes it impossible to fight an endless war. The state is almost all about security, because it is this institution's main purpose to provide it. Therefore, there is a clear political desire to reduce the length of wars, and precise timekeeping devices provide tools to do so. Now the sovereign state cannot drive violence constantly for a long period of time, so it needs precise time to clarify exactly when war is not only allowed but demanded. War itself becomes a conceptual framework defined in terms of accurate time, and for its participants, for the troops on the field, clocks provide critical information in this matter. When a time frame is defined as a "war" and one is defined as a "soldier," there is a set of clearly defined rules that dictates the behavior required of her. In some way, war becomes similar to a game. Both wars and games are placed in well-known time frames and are conducted according to shared and agreed-upon rules.[21] Like people in a game, who act according to the various rules that differentiate the situation from normal life, people in a state of war are governed by other rules that command to kill and destroy in order to achieve victory. In the end, the ability to define frames by the exact time allows a clear border to be created between these two different situations. Today, we hardly can imagine a war that lasts one hundred years. In the world of precise timekeeping, there is a demand to define the limits of war in terms of accurate time. Even when we examine cases of a long conflict, we can see that states use this tactic and clearly define peace and wartime. For example, in the Middle East, there has been a tension between Israel and Arab countries since 1948, but there is no constant war, at least according to the way the situation is presented in Israel.[22] By dividing its history into precise periods of wartime and peacetime, Israel continues to ensure its citizens' sense of security and can sustain order, in the reality of a continuous war that has lasted some sixty-six years.[23]

So far, I have presented the effects of changes in the measurement of time regarding war at the macro level, but to complete the whole picture I have to present the changes at the micro level as well, that is, to look at how changes in the way we measure time have influenced the conduct of war and the people who are actually fighting it. Since the measurement of time became precise, accessible, and standardized, the way we fight has changed, too, because from the logic of winning war we have now moved to the logic of attempting to meet objectives, defined by precise time. Here, the practices of self-discipline and self-governing that are embedded in the world of governmentality play a major part, because they promise that the population will be well trained. I shall illustrate my argument with the example of World War I because of, first, its place on the timeline—close to the acceptance of standard time (1884)—and second, the anomalous and puzzling way it ended—exactly on the eleventh hour of the eleventh day of the eleventh month of 1918.

World War I (1914–1918) was different from other conflicts, in both the scope of human losses and the intensity of fighting. It was the first precise war, at both macro and micro levels. Because it was a war planned by people who openly sided for the unity of time, it became precise before it even started.[24] A critical part of the German preparation for this war was the Schlieffen Plan, which was based on the extremely precise arrival of trains and troops.[25] Once executed, the timetable, precise to the minute, dictated its own logic.[26] This was also the first war in which soldiers carried watches with them, and since then they have become an essential piece of personal equipment.[27] The war's end, at 11:00, demonstrates that because of the shift in the way people calculate time, the way they practiced war also changed. The fighting continued exactly until 11:00, when at that time, fire finally ceased. Places in which minutes before there was bitter fighting were taken by the Entente forces without fighting after 11:00. Precise time measurement for the soldiers had a disciplining effect that made them act as expected, even in the last minutes of war.

History shows that the way wars ended before the standardization of time was different from the way WWI ended. For example, the American Civil War ended in several stages; the main Confederate forces surrendered on April 9, 1865, but other armies, spread over the South, surrendered only by the end of the month.[28] The Franco-Prussian War faded away as defeated French forces surrendered in stages over a period of several months.[29] In other words, those wars ended when the main obstacle was removed, meaning when a victory was achieved. This is different compared to the way in which WWI ended, when soldiers continued to fight even when it was clear to everyone not only who was the winner but also when the war was supposed to end.

Of course, one can argue that there were other reasons why the wars before the standardization of time ended differently from WWI. Yet I believe that this case symbolizes the rise of the importance of exact timing regarding the management of war, on the micro level, and this is reflected in the willingness of soldiers to discipline themselves and fight until the last moment. Not all wars after WWI have ended in the way that the Great War did (even though some did, such as the Winter War); indeed, there are many cases that did not end at a precise time. Even in these cases, however, there are clear political incentives to present those wars as having a clear end. War becomes a normal matter as its temporal boundaries are clearly defined. In this way, people can live at peace with a war.

With increasing accessibility to precise time-measuring devices, war objectives became to be defined in terms of precision, and the ability to meet the target they defined turned into a measure of success and operational-planning guidance. Thus, when Stalin urged his generals to finish the Battle of Berlin on Lenin's birthday, April 22, 1945, what guided him was not clear strategic considerations but the importance of the day in the Soviet calendar. In the eyes of those who planned the operations over Berlin, that order became almost sacred, and they did not spare human lives in order to conquer the city by that date. Although the predetermined target was not achieved, the very willingness to plan such extensive military operations, considering the target defined in terms of accurate time regardless of military interests and capabilities, is proof of the rise of the dimension of time's influence in matters of war.[30] The change indicates, above all, the political importance of precise timekeeping devices.

When it was decided that the Great War's armistice would begin at 11:00, Marshal Ferdinand Foch, commander in chief of the Allied armies, ordered that a message be conveyed to all forces fighting on the front by broadcasting news about it from the Eiffel Tower.[31] At this stage, German forces received the same orders.[32] So basically everyone, from the highest echelons to the common soldiers in the front-line trenches, knew that the war would end at 11:00. Everyone also knew exactly what time it was. In his command, Foch stated that the armistice was to take effect at 11:00.[33] Marshal Douglas Haig, the commander of the British Expeditionary Force, in his own command set, demanded that "troops will stand fast on the line reached at that hour," meaning at the point where they might be at the last minute before 11:00.[34] This kind of command created uncertainty about the actions soldiers were expected to take up until the armistice went into effect. When asked for clarification, many commanders at various levels acted as they saw fit. Self-discipline mechanisms played a key role in the decision of most of the division commanders to keep on fighting until the last moment.[35] For example, the commander of the 28th U.S. Infantry Division, Major General William Hay, had been aware of the armistice agreement since 8:30, yet he chose to continue fighting. The commander of the 55th Brigade, which was a part of the 28th Division, Brigadier General Frederic Evans, remembered that the soldiers kept on fighting because he had received an order to fight exactly until 10:59. In a similar situation was Colonel Cassius M. Dowell, commander of the 103rd Brigade of the 23rd U.S. Infantry Division, who said that he was ordered as late as 9:45 to keep on fighting until 11:00. Dowell wondered why he was required to put his men in severe danger in a battle the outcome of which had already been determined. He recalled, "I stood there a few seconds, debating as to whether I should send my men forward.... I expected my casualties to be very heavy"; and still he warned his artillery unit that if any shells were left unfired at 11:00, he would court-martial the responsible battery commander. Lieutenant Harry Rennagel, who served under Dowell, recalled that "when the orders came to go on top, we thought it was a joke," and still they did as commanded.[36] General Joseph Kuhn was much more decisive when he commanded the 79th U.S. Infantry Division: "Hostilities will cease on whole front at 11 hours today. . . . Until that hour the operations previously ordered will be pressed in vigor."[37]

When General Charles Pelot Summerall tried to spur his troops before the attempt to cross the Meuse River that morning, he claimed that "[o]nly by increasing the pressure can we bring about enemy's defeat."[38] General Hunter Liggett's later attempt to justify the Meuse crossing sounds as if it were taken from that world of military thought in which the new way to see time makes it impossible to break a frame constructed by it. He justified pressing on by citing the future assault on Metz (expected on November 14): "This offensive would cut the German army in two . . . and in my judgment would have brought about a capitulation in the field of everything north of that break"; the same capitulation was achieved on the eleventh.[39]

In many cases, the decision whether or not to act was given to the lower command levels (lieutenants or squad leaders) or even to the soldier himself, and still most of them chose to follow orders to the very end. They continued to fight, kill, and be killed even though most of them had watches and knew what the exact time was—or perhaps because of that. By acting this way, they continued to behave properly, as expected in a time of war, and stopped exactly when "peacetime" came into play. There is no historical evidence of any organized attempt to refuse to obey those orders or to avoid fighting, even though during the war there were at least two well-documented events of this type: The Christmas Truce mutiny (1914) and the French Army Mutinies (1917). Also there is no evidence that soldiers were not aware of the exact time. It seems that they simply felt that there was only one right way to act, according to the rules of the temporal unit they were a part of.

During the 1919 House of Representatives Committee on Military Affairs investigation of the efficiency of the war's conduct on that last day, Republican committee member Alvan T. Fuller observed "that those troops who were not killed or wounded marched peacefully into Germany at 11 o'clock. Is that true?"[40] General John Pershing at first argued that he wasn't sure that the armistice would have held, but in the end he agreed that he ordered American soldiers to fight until 11:00 because he simply followed the orders of his superior, Marshal Foch, issued on November 9, to keep up the pressure against the retreating enemy until the cease-fire went into effect. Other officers who were responsible for the carnage in the last hours of WWI gave even more bizarre justifications. For example, 89th Division commander Major General William Wright claimed that he ordered his troops to take Stenay in order to provide them "probable bathing facilities there."

Clocks | 359

What unites all these justifications is that all those army officers had a desire to stay in the service, so it is very interesting to see the difference between their testimony and that of the Brigadier General John Sherburne, former artillery commander of the black 92nd Division, who returned to civilian life. He gave the committee members what they most wanted: the views of a decorated noncareer officer who felt no obligation to absolve the army. He argued that soldiers knew about the armistice even before 6:00 a.m. and were stunned when the attack planned for that morning wasn't canceled by the division command; as they saw it, the fighting was "an absolutely needless waste of life." As he said, most of the officers' perspectives on war were those of "a child who had been given a toy . . . that he knows within a day or two is going to be taken away." Moreover he believed that "these men were looking upon this whole thing as, perhaps one looks upon a game of chess, or a game of football."[41] This testimony provides confirmation that because of the changes in the way we calculate time, in our day war becomes a framework defined by precise time such that people's actions within its borders are driven by a different logic, produced by those limits.

One can argue that the reason for continuing the fighting lies in a common desire to finish the war in an honorable way. I think this claim may be right, at least regarding the German side, which at that point could only defend itself and lacked an ability to create an offensive initiative. It seems logical that soldiers did not want to be captured or killed, so they defended themselves when being attacked. But it is more difficult to establish this claim in relation to the Entente forces, which had a strategic advantage on the eve of signing the armistice agreement, and had a good chance to end the war with a "real" victory. Some of the Entente commanders strongly supported this "decisive victory" idea.[42] Something similar can be understood from the report published by General Sir Frederick Maurice after the war, in which he claimed that in the last months of the war, it was clear to both sides that Germany was on the brink of collapse and that the Allies should take advantage of this.[43] Despite all this, the armistice agreement was signed without taking that opinion into account. Hence, I think that the Allied soldiers' behavior in the last moments of WWI cannot be explained in terms of honor, prestige, or "brothers in arms" behavior alone. There is no other way but to see this behavior as a massive act of self-discipline caused by direct access to precise timekeeping devices.

This chapter has uncovered the manner in which the measurement of time reshaped the concept of war. Toward the end of the eighteenth century, a fundamental shift occurred in the way time was measured and represented. War can be treated as a conceptual framework defined in terms of precise time, a move that changed the logic that rules it. The expectation from the people to act as the Allied soldiers did lies in the ability to define a specific period of time as "war," with its own rules and demands. This definition cannot be made without international access to precise time and timekeeping devices. My research has two main conclusions. First, I argue that there are political incentives to limit war within a framework defined by time. Second, there is a strong self-disciplinary mechanism embedded in everyday practices and devices that makes people obedient to the rules of precision even in moments of critical endangerment of their own lives. Today, the state's power relies on the ability to create an expected behavior among the population of its citizens, and precise timekeeping devices are one of the primary tools to do so.

Notes

1. Martin Gilbert, *The First World War: A Complete History* (New York: Holt, 1994), 500–501.
2. Joseph Persico, *Eleventh Month, Eleventh Day, Eleventh Hour: Armistice Day, 1918, World War One, and Its Violent Climax* (New York: Random House, 2004), 378.
3. Stephen Kern, *The Culture of Time and Space, 1880–1918* (Cambridge, Mass.: Harvard University Press, 2003), 17.
4. Émile Durkheim, *The Elementary Forms of the Religious Life* (1912; repr., London: Biddles, 1976), 22–32.
5. Anthony Aveni, *Empires of Time: Calendars, Clocks, and Cultures* (New York: HarperCollins, 1989), 74–78; Robert Hannah, *Time in Antiquity* (New York: Routledge, 2009), 72–77; Daniel Boorstin, *The Discoverers* (New York: Vintage, 1985), 37; David Saul Landes, *Revolution in Time: Clocks and the Making of the Modern World* (Cambridge, Mass.: Harvard / Belknap, 1983), 118–20.
6. Eviatar Zerubavel, "Private Time and Public Time: The Temporal Structure of Social Accessibility and Professional Commitments," *Social Forces* 58, no. 1 (1979): 40.
7. David Lewis and Andrew Weigert, "The Structure and Meaning of Social Time," *Social Forces* 60, no. 2 (1981): 439–42.
8. Kurt Wolff, ed., *The Sociology of Georg Simmel* (New York: Free Press, 1950), 412–13.
9. Kern, *Culture of Time and Space*, 13.
10. Stanford Fleming, "According to the Protocols of the Congress," *Documents Relating to the Fixing of a Standard of Time and the Legalization Thereof*, Canada Parliament Session, 1891, n. 8, pp.16–19.
11. Richard Hamilton and Holger Herving, *The Origins of World War One* (Cambridge: Cambridge University Press, 2003), 154.
12. Fleming, "According to the Protocols," 24–25.
13. Amir Lupovici, "Constructivist Methods: A Plea and Manifesto for Pluralism," *Review of International Studies* 35, no. 1 (2009): 203.
14. Michel Foucault, *Power/Knowledge: Selected Interviews and Other Writings,*

1972–1979, ed. Colin Gordon, trans. Colin Gordon et al. (Brighton, Eng.: Harvester, 1980), 156.

15. Timothy Mitchell, "The Limits of the State: Beyond Statist Approaches and Their Critics," *American Political Science Review* 85, no. 1 (1991): 93.

16. Foucault, *Power/Knowledge*, 56.

17. Michel Foucault, *Discipline and Punish: The Birth of the Prison*, trans. Alan Sheridan (New York: Pantheon, 1977), 6–7.

18. Michel Foucault, "Governmentality," in *The Foucault Effect: Studies in Governmentality*, ed. Graham Burchell, Colin Gordon, and Peter Miller (Chicago: University of Chicago Press, 1991), 102.

19. Janice Thomson, *Mercenaries, Pirates, and Sovereigns: State-Building and Extraterritorial Violence in Early Modern Europe* (Princeton, N.J.: Princeton University Press, 1994), 21–26.

20. The Hundred Years War officially started in 1337. In the year 1419, after some years of continuous fighting, the two sides signed the Treaty of Troyes. However, the war continued to rage. After many vicissitudes and a series of military disasters, French forces conquered the city of Bordeaux, capital of Aquitaine, and by this action fulfilled the mission of getting the English out of the area. Historians consider that this act ended the war. Yet the peace treaty was not signed until 1492, and in between, in the years 1474, 1488, and 1492, English forces invaded France once again. Furthermore, Calais remained English until 1558, and the English gave up demands for the French throne only in 1802. Kenneth Fowler, "War and Change in Late Medieval France and England," introduction to *The Hundred Years War*, ed. Kenneth Fowler (London: St. Martin's, 1971), 1–3.

21. Johan Huizinga, Homo ludens: *A Study of the Play Elements in Culture* (Boston: Beacon, 1955), 8–13, 89–91.

22. The history of Israel begins with the War of Independence (1948), followed by the Sinai War (1956), the Six-Day War (1967), the War of Attrition (1967–1970), and the First (1982) and Second (2006) Lebanon Wars. Meanwhile, Israel suffered two civil wars, the First (1987–1993) and Second (2000–2006) Intifadas, and its soldiers were stationed on Lebanese soil for twenty years.

23. We can see a very similar attitude in regard to the long conflict in Thucydides: "It would be certainly of judgment to consider the intervals of the agreement as anything else except a period of war." Thucydides, *History of the Peloponnesian War* (New York: Penguin, 1972), 363–64.

24. Gerhard Dohrn-van Rossum, *History of the Hour: Clocks and Modern Temporal Orders*, trans. Thomas Dunlap (Chicago: University of Chicago Press, 1996), 349–50.

25. Terence Zuber, *German War Planning, 1891–1914: Sources and Interpretations* (New York: Boydell, 2004), 210, 246.

26. Barbara Tuchman, *The Guns of August* (New York: Macmillan, 1962), 79–80.

27. Kern, *Culture of Time and Space*, 288.

28. Frederic Paxson, *The American Civil War* (London: Williams & Norgate, 1911), 242–43.

29. Geoffrey Wawro, *The Franco-Prussian War: The German Conquest of France in 1870–1871* (Cambridge: Cambridge University Press, 2003), 298.

30. Antony Beevor, *Berlin: The Downfall, 1945* (New York: Penguin, 2007), 46–147, 229. During the Battle of Moscow (1941), Marshal Georgy Zhukov's order to General K. D. Golubev, the 10th Army commander, stands as another example: "Tomorrow will be the birthday of Stalin. Try to mark this day by the capture of Balabanovo. To include this message in our report to Stalin, inform us of its fulfillment not later than 7 p.m. 21 December." Simon S. Montefiore, *Stalin: The Court of the Red Tsar* (London: Weidenfeld & Nicolson, 2003), 414.

31. Persico, *Eleventh Month*, 202.

32. Ibid., 328.

33. Ibid., 324.

34. Ibid., 329.

35. Ibid., 347.

36. Ibid., 7–8.

37. Ibid., 133–36.

38. "World War I: Wasted Lives on Armistice Day," HistoryNet.com, June 12, 2006, http://www.historynet.com/world-war-i-wasted-lives-on-armistice-day.htm.

39. During the Meuse crossing, November 11, 1918, on the last day of WWI, the 2nd, 89th, and 90th Divisions suffered 1,130 casualties, including 792 severely wounded and 127 dead. Persico, *Eleventh Month*, 335–36.

40. "World War I."

41. Ibid.

42. As John Pershing put it in an Allied Supreme War Council meeting, "Germany's morale is undoubtedly low. . . . [W]e should take full advantage of the situation and continue the offensive until we compel her unconditional surrender." Persico, *Eleventh Month*, 132–33.

43. "Sir Frederick Maurice on the Allies' Decision to Accept an Armistice," FirstWorldWar.com, http://www.firstworldwar.com/source/armistice_maurice.htm.

Acknowledgments

The Social Sciences and Humanities Research Council of Canada provided support for this research and for the presentation of this research at the International Studies Association, the Canadian Political Science Association, and the Millennium Conference. Earlier versions of the Introduction were presented at the invitation of Goethe University, Queen's University, Queen's University Belfast, and the Université de Quebec in Montréal. I am grateful to my friends and colleagues who gave essential feedback and support, particularly Can Mutlu, Jairus Grove, Miguel de Larrinaga, Benjamin Muller, Pete Adey, Emily Gilbert, and Debbie Lisle.

I thank Pieter Martin at the University of Minnesota Press for his sincere, enthusiastic support for this project at every stage.

Contributors

Michele Acuto is research director and senior lecturer in global networks and diplomacy at University College London's Department of Science, Technology, Engineering, and Public Policy (STEaPP) and coordinator of the City Leadership Initiative at University College London. He is a fellow of the Institute for Science, Innovation, and Society (InSIS) at the University of Oxford.

Peter Adey is professor of geography at Royal Holloway, University of London, and chair of the Social and Cultural Geography Research Group with the Royal Geographical Society.

Rune Saugmann Andersen is a guest lecturer at the University of Helsinki. His interdisciplinary work on the role of video in relation to security has appeared in *Security Dialogue* and *Journalism Practice*. He explores visual methodology and leads workshops on video as a research method.

Jessica Auchter is assistant professor of political science at the University of Tennessee at Chattanooga. Her research focuses on visual politics. She is author of *The Politics of Haunting and Memory in International Relations,* and she has published in the *Review of International Studies* and the *International Feminist Journal of Politics.*

Mike Bourne is senior lecturer in international security studies in the School of Politics, International Studies, and Philosophy at Queen's University Belfast. His research interests concern new materialism and posthumanism in security studies, security theory and critical approaches to security, and security technologies. His publications include *Understanding Security* and *Arming Conflict: The Proliferation of Small Arms.*

Kathleen P. J. Brennan is a doctoral candidate in political science at the University of Hawai'i–Mānoa. Her dissertation addresses the intersection of political theory, popular culture, and international relations, with a focus on overlapping online and offline spaces in which the world becomes more complex, equally filled with humor and despair, and promotes unexpected connections.

Elizabeth Cobbett is lecturer at the University of East Anglia. Her research explores how global finance seeks profitable opportunities within social structures and focuses on global financial flows within sub-Saharan Africa, especially South Africa, Kenya, and Nigeria.

Stefanie Fishel is visiting assistant professor at Hobart and William Smith Colleges. She approaches international relations with techniques and inspirations from Continental theory; science, technology, and society (STS) studies; the philosophy of science; and new approaches to biology.

Emily Gilbert is director of the Canadian studies program at the University of Toronto, with a cross-appointment in the Department of Geography and Planning. Her research concerns money, citizenship, mobility, borders, security, and militaries. She is coeditor, with Deborah Cowen, of *War, Citizenship, Territory*.

Jairus Grove is assistant professor of political science and director of the Center for Futures Studies at the University of Hawai'i. His research focuses on the ecologies and futures of global warfare, in particular the ways war continues to expand, bringing more participants and technologies into the gravitational pull of violent conflict.

Charlie Hailey teaches design, history, theory, and design/build at the University of Florida. A licensed architect, he has worked with the designer/builders Jersey Devil. His books examine camping as place making *(Campsite)*, camps as contemporary spaces *(Camps)*, and islands as manufactured cultural landscapes *(Spoil Island)*.

John Law is professor of sociology at the Open University and codirector of the ESRC Centre for Research on Socio-Cultural Change (CRESC) and the director of the Social Life of Method theme within CRESC. He is the author of *After Method: Mess in Social Science Research* and *Aircraft Stories: Centering the Object in Technoscience*.

Wen-yuan Lin is associate professor at the National Tsing Hua University, Taiwan. He uses STS material-semiotic approaches to explore the emerging alternative-knowledge places and politics of empirical ontology in technological and medical practices. His website is wylin.gec.nthu.edu.tw.

Oded Löwenheim is senior lecturer in international relations at the Hebrew University of Jerusalem. He is the author of *The Politics of the Trail: Reflexive Mountain Biking along the Frontier of Jerusalem* and *Predators and Parasites: Persistent Agents of Transnational Harm and Great Power Authority*. He explores emotions in politics and studies how conflict permeates the lives of ordinary individuals. He writes autoethnographies about the Israeli–Palestinian conflict to elucidate the daily workings of conflict, and he studies the personal, social, and political price for trying to disengage from a culture of violence and militancy.

Chris Methmann holds a PhD in political science from the University of Hamburg, Germany, and works as a campaigner for the German advocacy group Campact. His research centers on poststructuralist approaches to climate politics, climate security, and climate-induced migration. He has published in *European Journal for International Relations*, *International Political Sociology*, *Millennium*, and *Security Dialogue*.

Benjamin J. Muller is associate professor in political science at King's University College and a faculty member in the Centre for American Studies at Western University in Ontario, Canada. He has written *Security, Risk, and the Biometric State: Governing Borders and Bodies* and, with Samer Abboud, *Rethinking Hizballah: Legitimacy, Authority, Violence*. He held the first visiting research fellowship at the Border Policy Research Institute at Western Washington University and was the first visiting fellow at Confluence: Center for Creative Inquiry at the University of Arizona in Tucson.

Can E. Mutlu is assistant professor of international relations at Bilkent University. His research interests concern technology, security, and political sociology of global mobility regimes, especially the practices, technologies, and materialities of border security and mobility. He has published in *Comparative European Politics*, *European Journal of Social Theory*, *Environment and Planning D*, and *Review of International Studies,* and he is coeditor of *Research Methods in Critical Security Studies: An Introduction*.

Geneviève Piché is a PhD candidate in political science at Carleton University. Her research applies governmentality to examine the role of vehicles in maritime-migration incidents.

Joseph Pugliese is research director of the Department of Media, Music, Communication, and Cultural Studies, Macquarie University, Sydney, Australia. He has published on bodies and technologies; state violence; cultural studies of law, race, and whiteness; and migration and refugee studies. He was awarded the Faculty of Arts Excellence in Teaching Award in 2010 and was nominated for the Macquarie University Excellence in Research Award in 2014. His publications include *State Violence and the Execution of Law: Biopolitical Caesurae of Torture, Black Sites, Drones*; *Biometrics: Bodies, Technologies, Biopolitics*; and the edited collection *Transmediterranean: Diasporas, Histories, Geopolitical Spaces*.

Katherine Reese is a doctoral student at American University's School of International Service in Washington, D.C. Her interests integrate mobility, environmental politics, and social concepts of time. She has published on automobility in *Global Environmental Politics* and has taught courses on sustainability and technology. She was managing editor for *Journal of International Relations and Development*.

Mark B. Salter is full professor in political studies at the University of Ottawa. He is editor of *Politics at the Airport* (Minnesota, 2008) and coeditor of *Research Methods in Critical Security Studies: An Introduction*, with Can Mutlu. He is associate editor of *Security Dialogue* and *International Political Sociology*. In 2014, he was awarded the Canadian Political Science Association Prize for Teaching Excellence.

Michael J. Shapiro is professor of political science at the University of Hawai'i–Mānoa. Among his recent publications are *The Time of the City: Politics, Philosophy, and Genre* and *Studies in Trans-disciplinary Method: After the Aesthetic Turn*.

Benjamin Stephan holds a PhD in political science from the University of Hamburg, Germany. His research is concerned with transnational climate governance, particularly carbon markets and reducing emissions from deforestation and degradation (REDD+), as well as energy and transportation policy. His research has been published in *Environmental Politics,* and his most recent book is *The Politics of Carbon Markets,* coedited with Richard Lane.

Daniel Vanderlip recently completed his master of arts degree in political science at Carleton University. He is a senior policy adviser with the government of Canada.

William Walters teaches political sociology at Carleton University. He has published on migration and citizenship studies, the political sociology of states and international government, and the history of the present. His most recent book is *Governmentality: Critical Encounters.* He is now researching the traffic of migration.

Melissa Autumn White is faculty lecturer at the Institute for Gender, Sexuality, and Feminist Studies at McGill University. Her work has been published in *WSQ: Women's Studies Quarterly*, *Radical History Review,* and *Interventions: International Journal of Postcolonial Studies.*

Lauren Wilcox is university lecturer in gender studies at the University of Cambridge. She was a Charles and Amy Scharf Postdoctoral Fellow in political science at Johns Hopkins University. She is the author of *Bodies of Violence: Theorizing Embodied Subjects in International Relations,* and she has published in *Security Studies, Politics and Gender,* and *International Feminist Journal of Politics.*

Yvgeny Yanovsky is a PhD student in social studies at the Hebrew University of Jerusalem. His doctorate investigates the constitutive power of precision on the politics of war. His interests include time, length, precision, power relationships, politics, and wars.